T0344404

Lipid Biochemistry

Lipid Biochemistry

An introduction

Fourth edition

M.I. Gurr
Nutrition Consultant and Visiting Professor
University of Reading and
Oxford Polytechnic

J.L. Harwood
Professor of Biochemistry
University of Wales
Cardiff

 SPRINGER-SCIENCE+BUSINESS MEDIA, B.V.

First edition 1971
Second edition 1975
Third edition 1980
Fourth edition 1991

© 1971, 1975, 1980 M.I. Gurr and A.T. James
   1991 M.I. Gurr and J.L. Harwood
Originally published by Chapman & Hall 1991

Typeset in 10/12pt Times by EJS Chemical Composition,
Midsomer Norton, Bath, Avon

ISBN 978-0-412-26610-2

British Library Cataloguing in Publication Data

Gurr, M. I. (Michael Ian)
  Lipid biochemistry.–4th ed.
  1. Organisms. Lipids
  I. Title II. Harwood, J.L.
  574.19247
  ISBN 978-0-412-26610-2   ISBN 978-94-011-3062-2 (eBook)
  DOI 10.1007/978-94-011-3062-2

Library of Congress Cataloging-in-Publication Data

Available

We are honoured to dedicate this new edition to Dr Tony James, whose influence on lipid biochemistry has been profound and is still stamped on this book.

Contents

1 The nature of lipids and their place in living things

1.1 DEFINITIONS

The word 'lipid' (in older literature spelled also as lipide or lipoid) is used by chemists to denote a chemically heterogeneous group of substances having in common the property of insolubility in water, but solubility in non-polar solvents such as chloroform, hydrocarbons or alcohols. It is necessary to use this definition based on physical properties since there may be little or no chemical relationship between the numerous different compounds now classified as lipids, many of which are described in this book. It is not always possible to discern a clear distinction between the terms fat and lipid. The term fat is more familiar to the layman and brings to mind substances that are clearly fatty in nature, greasy in texture and immiscible with water. Familiar examples are butter and the fatty parts of meats. Fats are thought of as solid in texture as distinct from oils which are liquid at ambient temperatures. Chemically, however, there is little distinction between a fat and an oil, since the substances that the layman thinks of as edible fats and oils are composed predominantly of esters of glycerol with fatty acids. These are called triacylglycerols and are chemically quite distinct from the oils used in the petroleum industry, which are generally hydrocarbons. Lipid to the chemist embraces the wider range of fatty substances that are described in this book.

1.2 STRUCTURAL CHEMISTRY AND NOMENCLATURE

Lipids occur throughout the living world in microorganisms, higher plants and animals. In this book, they will be described mainly in terms of their functions although from time to time it will be convenient, even necessary, to deal with lipid classes based on their chemical structures and properties.

The naming of lipids often poses problems. When the subject was in its infancy, research workers would give names to substances that they had newly discovered. Often, these substances would turn out to be impure mixtures and as the chemical structures of individual lipids became established, rather more systematic naming systems came into being. Later, these were formalized further under naming conventions laid down by the International Union of Pure and Applied Chemistry (IUPAC) and the International Union of Biochemistry (IUB). Thus, triacylglycerol is now preferred to triglyceride but the latter is still frequently used especially by clinical workers and you will need to learn both (Chapter 4). Likewise, outdated names for phospholipids: Lecithin (for phosphatidylcholine (Chapter 6) and Cephalin (for phosphatidylethanolamine and phosphatidylserine) will be avoided in this book but you should be aware of their existence. Further reference to lipid naming and structures will be given in Chapter 6.

The very complex naming of the fatty acids is discussed in detail in Chapter 3. Their main structural features are their chain lengths, the presence of unsaturation (double bonds) and of substituent groups. In regard to chain length, it is cumbersome to have to say every time: 'a chain length of ten carbon atoms' and we shall, therefore, refer to a '10C fatty acid'. If we wish to refer to a specific carbon atom in a chain, we shall write, for example: 'the substituent at C10'. The numbering of fatty acid carbon atoms is done from the carboxyl end of the chain with the carboxyl carbon as C1. An important aspect of unsaturated fatty acids is the opportunity for isomerism, which may be either positional or geometric. Positional isomers occur when double bonds are located at different positions in the carbon chain. Thus, for example, a 16C monounsaturated fatty acid may have positional isomeric forms with double bonds at C7 and C9, sometimes written $\Delta 7$ and $\Delta 9$. (The position of unsaturation is numbered with reference to the first of the pair of carbon atoms between which the double bond occurs.) Geometric isomerism refers to the possibility that the configuration at the double bond can be *cis* or *trans*. (Although the convention *Z/E* is now preferred by chemists instead of *cis/trans*, we shall use the more traditional and more common *cis/trans* nomenclature throughout this book.) In the *cis* form, the two hydrogen substituents are on the same side of the molecule, while in the *trans* form they are on opposite sides (Figure 1.1).

Another important feature of biological molecules is their stereochemistry. In lipids based on glycerol, for example, there is an inherent asymmetry at the central carbon atom of glycerol. Thus, chemical synthesis of phosphoglycerides yields an equal mixture of two stereoisomeric forms

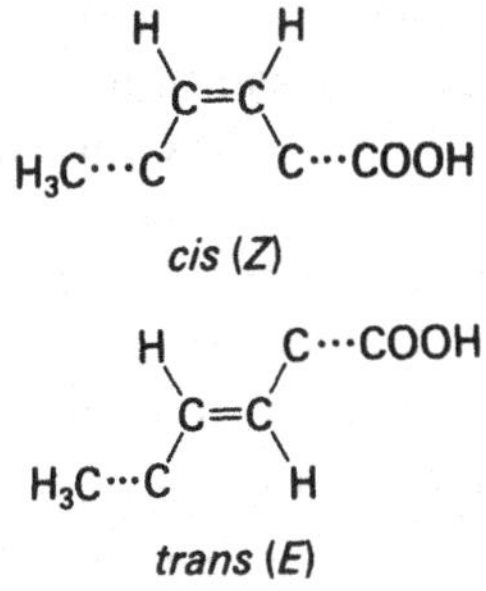

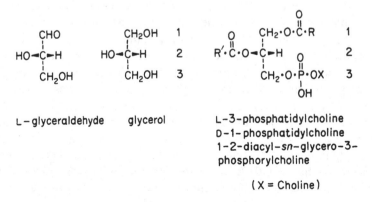

Figure 1.1 Geometrical isomerism in unsaturated fatty acids.

whereas almost all naturally occurring phosphoglycerides have a single stereochemical configuration, much in the same way as most natural amino acids are of the L (or *S*) series. In the past, naturally occurring compounds were designated L-*α*- and represented by the Fisher projection (Figure 1.2). The glycerol derivative was put into the same category as that glyceraldehyde into which it would be transformed by oxidation, without any alteration or removal of substituents. Phosphatidylcholine was therefore named: L-*α*-phosphatidylcholine. The IUPAC-IUB convention has now abolished the DL (or even the more recent *RS*) terminology and has provided rules for the unambiguous numbering of the glycerol carbon atoms. Under this system, phosphatidylcholine becomes, 1,2-diacyl-*sn*-glycero-3-phosphorylcholine or more shortly, 3-*sn*-phosphatidylcholine. The letters *sn* stand for stereochemical numbering and indicate that this system is being used. The stereochemical numbering system is too cumbersome to use routinely in a book of this type and, therefore, we shall normally use the terms 'phosphatidylcholine' etc. but introduce the more precise name when necessary.

Figure 1.2 The stereochemical numbering of acylglycerols.

Another field in which nomenclature has grown up haphazardly is that of the enzymes of lipid metabolism. This has now been formalized to some extent under the Enzyme Commission (EC) nomenclature. The system is incomplete and not all lipid enzymes have EC names and numbers. Moreover, the system is very cumbersome for routine use and we have decided not to use it here. You will find a reference to this nomenclature in the reading list should you wish to learn about it. Enzymes that catalyse the biosynthesis of certain molecules are sometimes called synthetases or alternatively synthases. We shall standardize on the term **synthetase** in this book.

1.3 FUNCTIONS OF LIPIDS

The major roles of lipids can be described conveniently as structural, storage and metabolic, although individual lipids may have several different roles at different times or even at one and the same time.

1.3.1 Structural lipids: these are important at surfaces and in membranes – barriers between one environment and another

Lipids play an important part in biological structures whose purpose is to provide barriers that protect organisms against their environment. The simplest type of barrier is simply a layer of lipids on the surface of the skin or fur of animals, the surface of leaves in plants, or associated with the walls of microorganisms.

The importance of lipids in such barriers lies in their ability to exclude water and other molecules. The characteristic physical feature of lipids, namely their water insolubility, derives from the chemical structure of part of the lipid molecule which is described as hydrophobic (Greek: water-hating). In lipids that are esters of fatty acids, the hydrophobic moiety is the hydrocarbon chain of the fatty acid. Some idea of the varied chemistry of the fatty acids can be obtained by reading Chapter 3. The nature of the fatty acid chain plays a major role in determining the physical properties of those lipids of which they are part. Thus the larger the number of double bonds (higher degree of unsaturation), the lower the melting point of the acyl chains. Within the groups of saturated fatty acids, the melting point is also lowered as the chain length decreases, or if the chain is branched. Similar hydrocarbon chains with hydrophobic properties are also seen in fatty alcohols which are normally present as components of wax esters (Chapter 4). The aliphatic hydrocarbon chain is not the only hydrophobic structure found in nature. The sterol ring system is widespread and the most abundant sterol in the animal kingdom is cholesterol. Other sterols, such as β-sitosterol are major constituents of plants. Because they contain an alcohol function, they may also form esters with fatty acids (sterol esters) which are amongst the most hydrophobic of all body lipids (Chapter 6).

Lipids also form an integral part of biological membranes. All living cells are surrounded by a membrane that provides a barrier between the cell and its environment. They also occur within the cell, providing a structure in which many metabolic reactions take place. In mammals, the lipids involved in membrane structures are mainly the glycerophospholipids and unesterified (free) cholesterol, while in plants, the glycosylglycerides are predominant, especially in the chloroplasts and β-sitosterol is the most abundant sterol rather than cholesterol. The chemistry of these structural lipids and their role in membrane architecture are described in Chapter 6, while their biochemistry forms the basis for Chapter 7.

The importance of these compounds lies in their possession of chemical groupings that associate with water (hydrophilic groups) in juxtaposition with hydrophobic moieties. These sorts of lipids are often called polar lipids, or more technically amphiphilic (Greek: liking both) and this amphiphilic nature is of immense importance in respect of their properties in membranes and in foods. (In contrast, hydrophobic fats, without polar groups, such as triacylglycerols, wax esters and sterols are often called neutral, apolar or non-polar lipids, but these are imprecise terms and best avoided.)

Phosphoglycerides are amphiphilic lipids in which the polar moiety is the phosphate group plus an organic base, such as choline. In glycosylglycerides the polar group is a sugar. Brain and nervous tissue are particularly rich in glycolipids based on the alcohol, sphingosine, as distinct from glycerol.

Current theories of biological membrane structure envisage that most of the lipid is present as a bimolecular sheet with the fatty acid chains in the interior of the bilayer. Membrane proteins are located at intervals at the internal or external face of the membrane or projecting through from one side to the other (Chapter 6). There may be polar interactions between the phospholipid headgroups and ionic groups on the proteins as well as hydrophobic interactions between fatty acid chains and hydrophobic amino acid sequences. Lipid molecules are quite mobile along the plane of the membrane but there is limited movement across the membrane. Indeed the composition of lipid molecules on each side of some membranes is quite different, a phenomenon called **membrane asymmetry**.

The physical properties of the membrane, which are strongly influenced by the lipid composition, seem to be important in so far as they are regulated in the face of environmental changes (diet, temperature, etc.) by subtle changes in the proportions of amphiphilic lipids, sterols and fatty acids. Such changes in physical properties may modulate the activities of membrane proteins, such as enzymes, transporters of small molecules across the membrane or **receptors** for substances such as hormones, antigens or nutrients.

1.3.2 Storage lipids: the high energy density of triacylglycerols makes them ideal as long-term fuel stores

Fatty acids in the form of simple glycerides constitute an important source of fuel in mammals and in many plants. The triacylglycerols are by far the most important storage form (Chapter 4).

Whereas structural lipids have a fairly conservative fatty acid composition, with a high proportion of unsaturated acids, the fatty acid composition of storage lipids is more variable, reflecting to a large extent the composition of the diet in simple stomached animals, such as man. In general, storage lipids in mammals tend to have a preponderance of saturated and monounsaturated fatty acids. In the plant kingdom, the storage lipids of seed oils have a wide variety of fatty acids. In a particular plant family, one fatty acid tends to predominate in the seed oil and this is frequently of unusual structure. The biggest reservoir of fatty acids to supply the long term needs of human beings for energy is the adipose tissue. Fatty acids are mobilized from this tissue to meet the demands for energy at times when dietary energy is limiting, for example in starvation or in strenuous exercise. The release of stored energy is regulated by the amounts and types of different dietary components and by hormones, whose secretion may also be regulated in part by diet. Other tissues, such as the liver of mammals, can accommodate fat in the form of small globules but only in the short term. The excessive accumulation of fat in mammalian liver is a pathological condition. However, many species of fish normally store fat in the liver or the flesh rather than in the adipose tissue.

Milk fat can also be regarded as a kind of energy store, for the benefit of the new-born, and like adipose tissue fat, is composed mainly of triacylglycerols. Egg yolk lipids likewise provide a store of fuel and nutrients for the developing embryo.

Higher plants also make use of various kinds of lipids as storage fuels. Many seeds store triacylglycerols to provide energy for the germination process. These normally occupy the embryo of the seed, although in some cases like the avocado, the lipid is contained in the mesocarp of the fruit. Some fruits, like the jojoba, use wax esters as the fuel storage form.

In animals, storage fat may be derived directly from the fat in the diet or it may be synthesized in the adipose tissue, mammary gland or liver from simple sugars. The capacity of these tissues to synthesize fatty acids is geared to the animal's needs and is under strict dietary and hormonal control. When there is little fat in the diet, the fatty acid pattern of a storage tissue is dependent on its own biosynthetic activity or the fatty acids supplied to it by the liver. The introduction of fat into the diet suppresses to varying extents the synthetic activity of the tissues and mechanisms operate to transport dietary fat into the storage fat so that the fat composition is more characteristic of the diet.

Plants of course must synthesize all their storage fat from simple precursors. This occurs at a certain stage after flowering and the resulting seed oil is frequently characterized by a very specific pattern of fatty acids in which one, often of unusual structure, predominates.

An important aspect of lipids in a living organism is their dynamic state. In all parts of the organism, lipids are continually being broken down or removed from the tissue and replaced. This continual breakdown and resynthesis is called turnover. Lipid replacement may occur by complete synthesis of the lipid from its simplest precursors, by replacement of parts of the molecules or by replacement of whole lipid molecules that have been transported from another site. Turnover allows a finer degree of metabolic control than would be possible in a more static system.

The chemistry and biochemistry of these storage lipids is discussed entirely within Chapter 4. Chapter 5 deals mainly with the dietary lipids of mammals and their assimilation by the body. The emphasis is on human nutrition.

As the foods we eat originate from living things, so, therefore our dietary lipids are derived from the structural and storage lipids of plants and animals. These lipids are digested in the alimentary tract by hydrolytic enzymes. The digestion products are taken up into the intestinal absorbing cells and resynthesized into complex lipids. Lipids absorbed and resynthesized in this way are transported to sites of storage or metabolism depending on the animal's current energy needs. The route of transport is the bloodstream, which is essentially an aqueous fluid in which small molecules are dissolved and macromolecules and cells are suspended. This poses a problem when fatty substances, insoluble in water, have to be accommodated. The solution to the problem is the conjugation of lipids with proteins to form a wide range of transport particles called lipoproteins.

Lipids transported as lipoproteins are taken up into tissues where they may be stored as energy reserves in the adipose tissue, incorporated into the structural lipids of membranes or oxidized to supply energy, depending on the nutritional and physiological state of the organism at the time. Much of the lipid in the body can be synthesized by body tissues. However, some fatty acids (the essential fatty acids) and the fat-soluble vitamins have to be supplied in the diet because, although vital for the animal, they cannot be synthesized endogenously.

Deficiencies in the diet or genetically determined errors in metabolism give rise to disease states. Lipids may be important in the aetiology of these diseases or play a part in therapy. These aspects of health and disease are addressed in Chapter 5.

1.3.3 Lipids in metabolic control: storage and barrier functions make use of the bulk properties of lipids. At the level of individual molecules, lipids participate as chemical messengers and are involved in the control of metabolism

Lipids not only contribute to the structure of cells and provide an energy store, they also participate in the transmission of chemical messages in living organisms.

In Chapters 3, 5 and 7, while discussing the biosynthesis and breakdown of specific types of lipids and the way in which their own metabolism is controlled, we allude to their specialized physiological roles. Thus, in Chapter 3, we discuss the conversion of specific polyunsaturated fatty acids into a whole range of oxygenated fatty acids, the eicosanoids, which have a variety of potent physiological effects at extremely low concentrations. These effects, which include stimulation of muscle contraction and stimulation or inhibition of platelet aggregation, are indeed so powerful that the eicosanoids need to be produced at or near the site of their action and quickly destroyed. They are, in effect, local hormones. The way in which our understanding of how nutrition influences their production and the way in which they act is explored further in Chapter 5, while their role in the regulation of the immune system is discussed in Chapter 8.

The formation of eicosanoids is an example of tightly controlled enzymic peroxidation of lipids. Peroxidation reactions can also occur chemically and unless controlled by the presence of natural antioxidant systems can give rise to massive cell damage, disintegration and disease. The chemistry involved is discussed in Chapter 3, while the pathological implications are described in Chapter 8.

A variety of lipid molecules take part in diverse aspects of metabolism and its control. Polyunsaturated fatty acids and their metabolites have been discussed above. Others are the fat-soluble vitamins, retinol (vitamin A) and tocopherol (vitamin E) (Chapters 5 and 8). Sterols, such as cholesterol, regulate membrane function and act as precursors for a range of molecules with diverse metabolic activities: cholecalciferol (vitamin D), which is metabolized further to hydroxylated derivatives that regulate calcium metabolism and other aspects of cellular function (Chapters 5 and 7); bile acids, which are involved in lipid absorption (Chapters 4 and 7) and steroid hormones (Chapter 7).

Perhaps the greatest interest in recent years has been the development of ideas about the role of the inositol phospholipids whose breakdown products act as **second messengers** in cell signalling. This story begins in Chapter 7 in relation to the metabolism of the inositol phospholipids and continues in Chapter 8 with a discussion of their role in cells. Indeed, we have collected together in Chapter 8 a diversity of material which has a common theme, namely, specific physiological functions of lipids. In addition to those described above, the chapter covers adaptive changes in membranes, the use of lipid vesicles for clinical purposes such as drug delivery, the roles of lipids particularly the sphingolipids in membrane receptors. It also discusses the involvement of lipids in skin diseases, in the so-called lipid storage diseases and their role in pulmonary surfactant.

This book is about the chemical nature of the many types of water-insoluble substances known as lipids: how they are synthesized and broken down; how they are incorporated into and released from living cells and tissues; how they contribute to the structural elements of cells (Chapters 3–7). It is also concerned with the normal biological control of these metabolic processes and the implications for the organisms when the normal metabolic control mechanisms are faulty or when environmental conditions create a stress that requires a process of adaptation. It is about the roles of lipids in biological structures, the means by which they provide metabolic energy and about their important functions in the foods we eat. Although it is probably true to say that most students of biochemistry are interested in the metabolism of man and other mammals, it should not be forgotten that, because of the vast quantities of plants and microorganisms distributed over the earth's surface, the plant kingdom accounts for much the largest proportion of the world's lipids. Consequently, this book will still deal very extensively with plant lipids even though there has been increased emphasis on human lipid biochemistry compared with previous editions.

REFERENCES

Farquar, J.W., Insull, W., Rosen, P., Stoffel, W. and Ahrens, E.H. (1959) Nomenclature of fatty acids. *Nutrition Reviews*, **17**, Suppl.

Gunstone, F.D. (1967) *An Introduction to the Chemistry and Biochemistry of Fatty Acids and their Glycerides*, Chapman and Hall, London.

IUPAC–IUB Commission on Biochemical Nomenclature (1978) *Biochemical Journal*, **169**, 11–14.

Nomenclature Committee of the International Union of Biochemistry (1984) *Enzyme Nomenclature*, Academic Press, London.

2 Isolation, separation and detection of lipids

Since lipids are characterized generally by their ability to dissolve in water-immiscible organic solvents, advantage is taken of this property at many stages of analysis.

2.1 EXTRACTION OF LIPIDS FROM NATURAL SAMPLES

Extraction of lipids from natural samples, like that of many biological molecules, is best accomplished as soon as possible to minimize the degradative changes which would otherwise take place. When storage has to take place it is best done at as low a temperature as possible (say −20°C or less) and under an inert nitrogen atmosphere to minimize oxidation of groups such as double bonds. If great care is not taken during the extraction process, many lipids will be partly or completely lost. For example, the polyphosphoinositides (section 8.7, 8.8) are extremely labile due to the very active degradative enzymes present in many animal tissues. Similar degradative enzymes can pose particular problems in plant tissues since they are active at very low temperatures (certainly at −20°C) and also retain activity (or may even be activated) in organic solvents. Such enzymes are obviously best inactivated as quickly as possible by, for example, brief exposure of the tissue to steam or boiling water or by a prior extraction with hot isopropanol – all measures which inactivate the lipases.

The actual extraction method depends on the type of tissue and also the lipids it is desired to analyse. However, few lipids can be extracted by a single solvent and binary mixtures are usually used. One of the components should have some water solubility and hydrogen bonding ability because lipid–protein complexes such as those encountered in membranes have to be split.

A very common method is that of Bligh and Dyer. These workers used a mixture of chloroform and methanol in a ratio with tissue water (1:2:0.4) to form a one-phase system. Homogenization of tissues in this mixture efficiently extracts most lipids. More chloroform and methanol are then added to give two phases, the upper (aqueous) one containing non-lipid impurities; the lower (chloroform) phase can then be removed, washed with fresh upper phase and finally evaporated to dryness. Residual water can be removed with anhydrous sodium sulphate or by filtration through Sephadex columns. In some cases it may be desirable to use salt or dilute acid solutions in the upper phase to prevent losses of polar lipids. Finally after removing solvent by vacuum evaporation, the crude lipid residue should be protected from oxidation by inert nitrogen gas. In fact, before storage for any length of time, lipids are best redissolved in a small amount of solvent containing an antioxidant such as 2,6-di-*tert*-butyl-*p*-cresol(BHT) before storage at −20°C or less in the dark under nitrogen.

2.2 LIKELY COMPONENTS OF THE CRUDE LIPID EXTRACT

Since the initial extraction has been based on solubility properties, the crude lipid extract will contain any molecule which dissolves preferentially in the organic solvents used. Thus, significant quantities of non-lipids, e.g. hydrophobic proteins, may be present at this stage. The mixture of lipids (Table 2.1) will depend on the nature of the sample extracted.

Table 2.1 Major components of typical lipid extracts from different tissues

Erythrocytes	Liver	Leaves	Cyanobacteria	Gram⁻Bacteria
P. Lipid	P. Lipid	P. Lipid	P. Lipid	P. Lipid
Sphingolipid	Sphingolipid	—	—	—
—	—	Glycosyl-glycerides	Glycosyl-glycerides	—
Sterols	Sterols	—	—	—
—	TAG	—	—	—
		Others*‡	Others*	Others†

TAG = triacylglycerols, P. Lipid = phosphoglycerides; * pigments; † lipopolysaccharide; ‡ waxes, cutin.

2.3 GENERAL FEATURES OF LIPIDS IMPORTANT FOR THEIR ANALYSIS

Most of the lipids cited as major components in different tissues contain esterified fatty acids. These are termed acyl lipids. This is important during analysis for several reasons:

1. the nature of the fatty acids determines much of the physical and biological properties of the lipid and, therefore, it is of importance to analyse their properties;
2. the type of fatty acid influences the stability of the sample;
3. because mixtures of fatty acids are found in any given lipid class, the latter contains a number of molecular species.

Total fatty acid analysis is usually performed by forming volatile derivatives (such as methyl esters) for gas chromatography. Special techniques have to be used sometimes to avoid destroying unusual functional groups such as cyclopropene rings. Moreover, the danger of auto-oxidation means that analysis of samples containing polyunsaturated fatty acids has to be especially careful. For complete identification of individual acyl groups, degradation or derivatization usually has to be employed – for example, when double bond positions have to be assigned.

In order to determine the positional distribution of acyl groups on, say, the glycerol backbone, enzymic cleavage is usually employed. For example, phospholipases are available which have a specificity for either the *sn*-1 or the *sn*-2 position. Use of such enzymes will release fatty acids from one position and these can be separated from the partly deacylated product. Analysis of fatty acids and deacylated lipid will then reveal which fatty acids are present at each glycerol carbon.

Molecular species of lipids can be separated on the basis of size and/or unsaturation. In the past it has often been necessary to block or remove the polar part of the lipid making analyses time-consuming. For example, because the charge on the head-group of phosphatidylcholine is large in relation to differences in acyl unsaturation, it was usually necessary to degrade such phosphoglycerides to diacylglycerols before analysis by chromatography. Modern methods of HPLC have, however, rendered such methods unnecessary provided that adequate methods of detection are available.

Single lipid classes can be separated from each other by methods which make use of differences in their size and charge. They can often be provisionally identified by co-chromatography with authentic standards in various systems. Important constituent groups will be revealed by spectroscopic techniques or with specific colour reagents. However, unambiguous identification may require that the various products of hydrolysis be isolated, characterized and quantified. When enzymes are used to cause hydrolysis, their action (or otherwise) may also provide information about the stereochemistry of particular linkages.

2.4 CHROMATOGRAPHIC TECHNIQUES FOR SEPARATING LIPIDS

2.4.1 The principles of chromatography are based on distribution between two phases, one moving, the other stationary. The two phases can be arranged in a variety of ways

A chromatogram (so-named by its Polish inventor, Tswett, because he used the technique to separate plant pigments) consists of two immiscible phases. One phase is kept stationary by either being held on an inert microporous support or being itself a microporous or particulate adsorbent solid; the other phase is percolated continuously through the stationary phase. Various phase pairs are possible although liquid–solid and gas–liquid are the most common.

If we take any single substance and mix it with any of these phase pairs, it will distribute itself between the two phases, the ratio of the concentrations in the two phases (at equilibrium) being known as the partition coefficient.

The partition coefficient is a physical constant dependent on the nature and magnitude of solute–solvent interactions in the two phases. Let us consider two substances A and B and imagine that, at equilibrium, substance A distributes itself between the two phases so that 90% is in the stationary phase and 10% in the moving phase. Substance B, however, distributes so that 10% is in the stationary phase and 90% in the moving phase. Then a mixture of A and B dissolved in a small volume of moving phase and applied to the chromatogram will begin to separate when the moving phase is added and washes them through the system. They will distribute themselves independently of one another: B will move as a zone at 9/10ths of the velocity of the moving phase and A at 1/10th of the velocity. Clearly the two substances will rapidly move apart, the original rectangular profile of the zones changing to the shape (ideally) of a Gaussian error curve because of diffusion. The substances can either be visualized **directly on the system** by colour sprays or be eluted as pure components.

There are basically only two types of chromatogram geometry. (1) **The column** consists of a metal, glass or even plastic tube with a ratio of length to diameter of at least 10 : 1 and packed with either an adsorbent solid (silica gel, alumina etc.) or an inert solid, such as Kieselguhr, of large surface area, that can hold, by surface tension, a liquid as one member of the phase pair. Gas chromatograms can afford a much greater ratio of length to diameter because of their inherent lower resistance to flow than can liquid columns. (2) **The plate or strip** consists of the stationary phase support arranged as a flat surface. Mixtures can be spotted and dried on the surface and when the bottom of the plate is immersed in the moving phase in a closed vessel, capillarity ensures that the liquid will move through the porous material (paper, or a porous solid held to a glass or metal surface). The fact that both phases exist as relatively thin films means that the solutes have only short distances to move as they pass from phase to phase. Very refined separations can thus be obtained rapidly. The thin layer chromatogram is particularly useful for lipid separations.

In general, compounds are not eluted from flat plate chromatograms; instead, the development (i.e. the movement of the liquid phase) is stopped when the front has reached the end of the strip, the strip is dried to remove solvent and the position of the zones revealed by spraying. The sprays can be of a destructive type such as dilute sulphuric acid followed by heating (this produces black spots by carbonization where there is an organic material) (Figure 2.1) or a non-destructive type such as dichloro- or dibromo-fluoresceins that show a changed fluorescence where there is a zone. In the latter case the compounds can be recovered by scraping the adsorbent from the plate where the spray indicates a zone to be present, followed by extraction with a suitable solvent. Under standard conditions it will be found that a given substance will move relative to a standard substance to a constant ratio or relative R_F (Figure 2.1). This relative R_F is a useful confirmation of structure of an unknown substance but it is unwise to use it as an absolute indicator.

Only infrequently can every component of a complex mixture be resolved with one solvent system. However, by using two-dimensional development, i.e. first by one solvent system in one direction, then after drying the plate by running a different solvent at right angles, refined separations can be achieved.

2.4.2 Gas–liquid chromatography is a particularly useful method for volatile derivatives of lipids

With this method the moving phase is a permanent gas and columns contain either an inert support (e.g. Celite) on whose surface is the stationary liquid phase or else are themselves narrow tubes on whose wall is a thin layer of the stationary phase. The latter are known as wall-coated-open-tubular (WCOT) or capillary columns. Because no inert support is present, the flow of gas is relatively unimpeded and very long thin columns (say 100 m long and only 0.25 mm internal diameter) capable of quite remarkable separations are possible. Columns containing inert support may be either

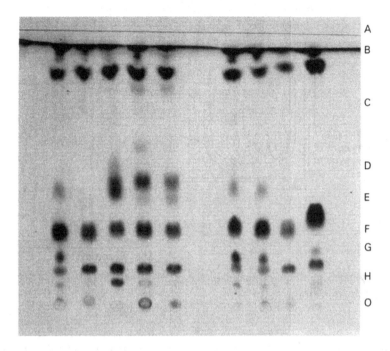

Figure 2.1 Thin layer chromatogram of plant lipids.

Developed in chloroform–methanol–acetic acid–water (85 : 15 : 10 : 4, by vol.), on silica gel G. Compounds detected by spraying with 50% sulphuric acid and charring. The source of the lipids is as follows:

    1. *Chlorella vulgaris.* 2. *Anacystis nidulans.* 3–5. Nitrogen-fixing blue-green algae. 6. Spinach leaves. 7. *Chlorella vulgaris.* 8. *Anacystis nidulans.* 9. Spinach chloroplasts. Identification of the lipids: A. Neutral lipids. B. Monogalactosyl diacylglycerol. C. Sterol glycoside. D. In 1, 6 and 7, Phosphatidyl ethanolamine. In 3–5, a mixture of glycosides of a fatty alcohol. E. Digalactosyl diacylglycerol + phosphatidyl glycerol. F. Phosphatidyl choline. G. Sulphoquinovosyl diacylglycerol. H. In 1, 6 and 7, Phosphatidyl inositol. In remainder: unknown. O. Origin of application.

The R_F of, for example, spot B relative to spot C is the ratio of the distances of the centres of spots B and C from the origin O. This depends only on the solvent, the nature of B and C, and the temperature. Reproduced with kind permission of Dr B. W. Nichols.

packed or support-coated-open-tubular (SCOT). Packed columns are the 'work horses' of gas chromatography and glass or stainless steel columns are usual. Glass columns are better because they are almost completely inert and any breaks or deterioration in the packing material can be seen easily. Of course, they are fragile – as many students know to their cost!

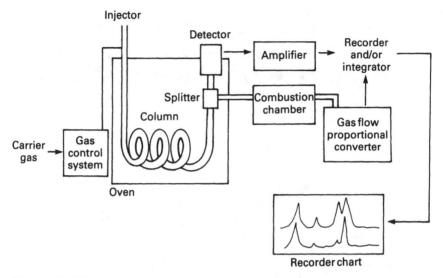

Figure 2.2 Diagrammatic representation of a gas chromatograph coupled to a radioactivity detector.

In a typical gas chromatograph (Figure 2.2) the column is held in an oven either at a fixed temperature (isothermal operation) or at a temperature which increases during the separation (temperature programming). The latter method is particularly useful when the mixture to be analysed contains components with a wide range of molecular weights. At the end of the column is the detector which quantifies the eluted components. For lipids the flame ionization detector is usually used. Eluates are burned in a flame of hydrogen and air to produce ions which are detected and measured by an electrical device capable of detecting about 10^{-12} g/ml. Flame ionization detectors respond to almost all organic compounds, have a good signal-to-noise ratio and are robust.

The mixture to be separated is injected through a flexible septum at the top of the column. The injection port is usually heated to a temperature about 25°C above the operating temperature of the column although, for certain purposes, unheated injectors may be used. If the sample is not very volatile, a flash-heater is used so that all components enter the column as a vapour. However, on-column injection – where the sample, in a volatile solvent, is injected directly into the packing material of the column – is preferred in most circumstances. At the far end of the column the effluent is quantified by the use of a detector. In addition, a T-piece can be used either for collecting fractions or for separate radioactivity measurements (in, for example, a gas-flow proportional counter).

At a fixed temperature (isothermal use) and with constant flow conditions, the time taken for individual components to emerge from a given column will be constant. However, because of the difficulties in keeping conditions exactly constant and also the slow loss of stationary phase from the column, it is better to use **relative retention volume** (or time) to provisionally identify particular components. By the use of different column packings, confirmation of these identities can be made although for complete analysis some degradative techniques must be used also. A rather convenient method uses a combination of gas chromatography and mass spectrometry. Here, the effluent is split after emerging with part going to the detector and the remainder entering a mass spectrometer for (fragmentation) analysis.

In some cases the gas chromatograph can be used to predict the elution behaviour of an unknown compound. In the absence of an authentic standard, this allows provisional identifications to be made. For example, members of a homologous series show a straight line relationship between the logarithm of the absolute retention time and the chain length (Figure 2.3). Thus, the retention time of a higher or lower homogene can be predicted from the behaviour of a few members of the series.

Gas–liquid chromatography can, of course, only be used for compounds which can be volatilized without decomposition. In practice this normally limits its application to molecules with molecular mass of less than 800 Da. Even so many lipids, such as phosphoglycerides, are not amenable to analysis as intact molecules.

Perhaps the most important use for the lipid biochemist is in the analysis of the fatty acid composition of lipids or tissues. In addition, gas chromatography may be used for steroid, glycerol, sphingosine, inositol and carbohydrate analysis. Molecular species of partial glycerides or, on highly thermostable silicone liquid phases, even of triacylglycerols may be analysed also.

2.4.3 Adsorption column chromatography is used for the separation of large amounts of lipids

In this technique a moving liquid phase elutes lipids selectively from a solid support. Traditionally the most useful columns for lipid separations have been those containing silicic acid. Polar lipids tend to be tightly adsorbed onto these and can be separated easily from less-polar materials such as sterol (-esters) or triacylglycerols. By an appropriate choice of elution solvents, good separations may also be obtained within a single lipid class – say phospholipids. Two simple methods are summarized in Table 2.2. In method A, glycolipids are eluted with acetone, which although rather polar, fails to elute phospholipids because of their poor solubility in this solvent. In

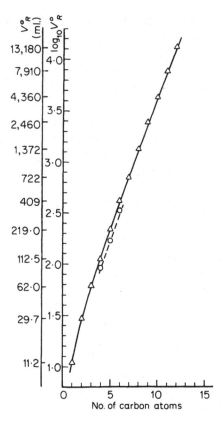

Figure 2.3 The linear relationship between corrected retention volume and chain length for short chain fatty acids at 137°. Stationary phase: silicone DC 550 containing stearic acid.

—△—   straight chain acid
—O—   *iso*-branched chain acid

method B, increasingly polar mixes of methanol and chloroform cause some separation between phospholipid classes. Other packing materials may be used instead of silicic acid. For example, columns containing DEAE-cellulose can be used to separate lipids carrying a net negative charge from uncharged or zwitterionic molecules.

High performance liquid chromatography (HPLC) has recently been developed sufficiently for it to be used for many lipid separations. Stainless steel columns packed with microspherules of silicic acid are most often used. Excellent separations of a host of different lipids can be made but often the lack of suitable universal detectors for lipids limits use of the method. If only lipids had convenient absorbances then HPLC would become the method of choice for practically every occasion. Even so, it offers particular merits for

Table 2.2 Separation of complex lipids on silicic acid columns

Fraction	Compound eluted	Solvent	Column volume
Method A			
1	Simple lipids	Chloroform	10
2	Glycolipids (and traces of acidic phospholipids)	Acetone	40
3	Phospholipids	Methanol	10
Method B (for fraction 3 separated above)			
1	DPG + PA	Chloroform/methanol (95 : 5)	10
2	PE + PS	Chloroform/methanol (80 : 20)	20
3	PI + PC	Chloroform/methanol (50 : 50)	20
4	Sphingomyelin + lyso PC	Methanol	20

DPG = diphosphatidylglycerol; PA = phosphatidic acid; PE = phosphatidylethanolamine; PS = phosphatidylserine; PI = phosphatidylinositol; PC = phosphatidylcholine.
Adapted from Christie (1984) *Lipid Analysis*, 2nd edn, Pergamon Press.

many analyses and is being used very frequently for separation of molecular species of lipid classes. Perhaps the best detector is the differential refractometer, which senses minute differences in the refractive index of an eluant brought about by material eluting from the column. However, it can only be used easily with a solvent of constant composition. Recently efforts have been made to use a sampling device in the form of a moving belt to collect a constant portion of the eluant and analyse it with a flame ionization detector. In theory, this should be an excellent method but mechanical and sampling problems continue to bedevil the technique as they did its forerunner the moving wire. Another method lipid biochemists use is to derivatize the lipids for analysis so that they absorb in the UV or visible region and can, therefore, be readily measured.

2.4.4 Thin layer adsorption chromatography can achieve very good separation of small lipid samples

In this technique a thin (usually 0.25 mm) layer is spread on a glass plate and separations are achieved by allowing appropriate solvents to rise up the plate by capillary action. Silica gel is the most usual adsorbant and samples, which are applied close to the edge of the plate, are fractionated due to their different adsorptions and solubility in the solvent. Very good separations can be achieved although the thinness of the layer means that only small amounts of lipids are usually analysed. When the plates have been run, they are dried and can be sprayed with various reagents to reveal individual lipid classes. If the spray reagent is non-destructive (e.g. 2',7'-dichloro-fluorescein which causes all lipids to fluoresce) then the lipids can be recovered by scraping the appropriate area off the plate and eluting with a suitable solvent.

Complex lipid mixtures cannot always be separated by one-dimensional chromatography. In that case, either a preliminary fractionation on a column (to give, say, neutral, glycolipid and phospholipid fractions) or two-dimensional thin layer chromatography can be used. An example of the separation of lipid extracts from a variety of tissues by TLC is given in Figure 2.1.

TLC offers a number of advantages over column chromatography. It is more rapid and sensitive, gives better resolution and is usually much quicker. Moreover, the apparatus required is minimal and, especially if plates are made in the laboratory, the technique is inexpensive. By incorporating various chemicals into the thin layer, special lipid separations can be made. For example, silver nitrate allows fatty acids (or more complex lipids) to be separated on the basis of their unsaturation. Silicone oil–silica gel TLC works on the basis of reverse-phase separation and can be used to fractionate fatty acid mixtures based on their hydrophobicity, with shorter chain or unsaturated components migrating faster. Boric acid impregnation allows separation of *threo-* or *erythro-* isomers of vicinal diols or fractionation of molecular species of ceramides.

A number of useful chemical methods are available for detecting or quantifying lipid classes. In some cases, the specificity of the colours developed can be used to help with identifications. Also when a non-destructive fluorescent reagent is used, lipids can be quantitated on the plates by a scanning fluorimeter – though, because different lipids have different fluorescence values, calibration curves must be made.

Specific chemical methods can be used to reveal phospholipids, glycolipids, sterols or their esters as well as compounds with quaternary nitrogens or vicinal diols. Particularly useful for many membrane extracts is the reaction of ammonium molybdate with inorganic phosphate released from phospholipids. The phosphomolybdic acid so produced is then reduced to give an intense blue colour. The method can be adapted to a spray reagent or, more often, used to detect as little as 1 μg of phosphorus in scraped samples.

In order to increase the sensitivity of TLC and also to allow the rapid quantification of a variety of separated lipids, efforts have been made to apply flame ionization detection systems. The most successful method involves separating small amounts of lipid mixtures on quartz rods to which a thin silica gel has been fused. After separating the mixtures, the rods are placed in a rack and pass in turn through a flame ionization detector. The act of passing the rod through the flame regenerates it and each rod can be used up to a hundred times. Unfortunately, different lipids give varying responses (often non-linear) and separations are poor compared to regular TLC. The expense of the machine also limits the usefulness of this new development.

2.5 OTHER USEFUL METHODS

A detailed discussion of the wide variety of methods which are available to the lipid biochemist for identification and analysis is beyond the scope of this book. However, a few of the more-widespread techniques are listed in Table 2.3.

Table 2.3 Some other methods for lipid analysis

Method	Use
I.R. spectroscopy	Identification of organic bases in phospholipids or *trans*-double bonds
N.M.R. spectroscopy	Widely used for lipid structure determination particularly identification and location of double bonds in fatty acids, functional groups (e.g. hydroxyl) on fatty acids and preliminary identification of glycerides, glycolipids and phospholipids
U.V. spectroscopy	Analysis of conjugated double bond systems, especially those formed by oxidation reactions
Mass spectrometry	Often linked to a gas chromatograph. Has a wide variety of uses from the estimation of molecular weights, to the location of double bonds in aliphatic chains or identification of functional groups. Different ionization methods (chemical ionization, fast atom bombardment, electron impact) are used for various purposes
Enzyme degradation	Used to show nature of specific bonds and substituents in lipids. For example, the positional distribution of acyl groups can be determined or the presence of individual sugars or bases revealed. The stereo-specificity of most enzymes also allows the configuration of the target linkage to be demonstrated

2.6 SUMMARY

Lipids can usually be extracted easily from tissues by making use of their hydrophobic characteristics. However, such extractions yield a complex mixture of different lipid classes which have to be purified further for quantitative analysis. Moreover, the crude lipid extract will be contaminated by other hydrophobic molecules, e.g. by intrinsic membrane proteins.

Of the various types of separation processes, thin layer and column chromatography are most useful for intact lipids. High performance liquid chromatography (HPLC) is also rapidly becoming more popular, especially for the fractionation of molecular species of a given lipid class.

The most powerful tool for quantitation of the majority of lipids is gas–liquid chromatography (GLC). The method is very sensitive and, if adapted with capillary columns, can provide information with regard to such subtle features as the position or configuration of substitutions along acyl chains. By coupling GLC or HPLC to a radioactivity detector, then the techniques are also very useful for metabolic measurements.

Although research laboratories use generally sophisticated analytical methods such as GLC to analyse and quantify lipid samples, chemical derivatizations are often used in hospitals. For these methods, the lipid samples are derivatized to yield a product which can be measured simply and accurately – usually by colour. Thus, total triacylglycerol, cholesterol or phospholipid-phosphorus can be quantitated conveniently without bothering with the extra information of molecular species, etc. which might be determined by more thorough analyses.

REFERENCES

Christie, W.W. (1982) *Lipid Analysis*, 2nd edn, Pergamon Press, Oxford.

Christie, W.W. (1987) *High-Performance Liquid Chromatography and Lipids*, Pergamon Press, Oxford.

Christie, W.W. (1989) *Gas Chromatography and Lipids*, The Oily Press, Ayr, Scotland.

Gunstone, F.D., Harwood, J.L. and Padley, F.B. (eds) (1986) *The Lipid Handbook*, Chapman and Hall, London.

Kates, M. (1986) *Techniques of Lipidology*, 2nd edn, Elsevier, Amsterdam.

3 *Fatty acid structure and metabolism*

3.1 STRUCTURE AND PROPERTIES

In lipid biochemistry – as in most other fields of science – various trivial names and shorthand nomenclatures have come into common usage. Thus, the poor student is faced with a seemingly endless series of illogical names and symbols for the various new compounds to be learnt. Fatty acid names are no exception and there are trivial names for most commonly occurring fatty acids as well as numerical symbols (as a shorthand) for given structures. The latter derive from the use of gas–liquid chromatography (section 2.4.2) where separations are achieved due to both carbon chain length as well as unsaturation. The shorthand nomenclature consists of two numbers separated by a colon. The number before the colon gives the carbon number and the figure after denotes the number of double bonds. Thus a saturated fatty acid such as palmitic would be 16 : 0 while a monounsaturated acid such as oleic would be 18 : 1. Chain branching or substitution is denoted by a prefix thus: br-16 : 0 for a branched chain hexadecanoic acid or HO-16 : 0 for an hydroxy-palmitic acid. This shorthand has the merit of not attributing a more precise identification than the gas chromatogram alone can give.

Where additional information, such as the exact position and configuration of double bonds, is known then this knowledge can be incorporated into both the shorthand nomenclature as well as the systematic names. For example, linoleic acid could be written as *cis*(Δ-)9, *cis*(Δ-) 12–18 : 2 or (*cis,cis*)9,12-octadecadienoic acid to indicate that it is an 18-carbon fatty acid with *cis* double bonds 9 and 12 carbons from the carboxyl end (see Chapter 1 for conventions for geometrical isomerism). Occasionally, because of their metabolic connections, it is useful to number the double bonds from the methyl end. In that case linoleic acid would become (*cis,cis*)n-6,9-octadecadienoic acid with the n (ω in older literature) showing that numbering has been from the methyl end.

3.1.1 Saturated fatty acids

Most saturated fatty acids are straight chain structures with an even number of carbon atoms. Acids from C_2 to longer than C_{30} have been reported but the most common lie in the range C_{12}–C_{22}. Some of the more important naturally occurring straight chain saturated acids are shown in Table 3.1 together with further information. In general, fatty acids do not exist as free carboxylic acids because of their marked affinity for many proteins. (One result of this is an inhibitory action on many enzymes.) In fact, where free acids are reported as major tissue constituents they are usually artefacts due to cell damage which allows lipases to break down the endogenous acyl lipids. Exceptions to this statement are the albumin bound fatty acids of mammalian blood. Free acids are often referred to as FFA (free fatty acids) or, preferably, NEFA (non-esterified fatty acids).

The configuration of a typical saturated chain is shown in Figure 3.1(a). Because of continuous thermal motion in living systems and the free rotation about the carbon–carbon bonds, the fatty acids are capable of adopting a huge number of possible configurations but with a mean resembling an extended straight chain. Steric hindrance and interactions with other molecules in Nature will, of course, restrict the motion of non-esterified fatty acids and the acyl chains of complex lipids. The physical properties of acyl lipids are obviously affected by their individual fatty acids – a most obvious one being the melting point. As a general rule, membranes are incapable of operating with lipids whose acyl chains are crystalline. For mammals, this means about 37°C and for poikilotherms (organisms which cannot regulate their own temperature) anything between about −10° and over 100°C. Although it is possible for the membranes of, for example, a mammal to contain lipids with acyl chains whose melting points are slightly above 37°C, the presence of other lipid types ensures that the mixture is in fact semi-liquid.

Table 3.1 Some naturally occurring straight chain saturated acids

No. of carbon atoms	Systematic name	Common name	M. pt (°C)	Occurrence
2	*n*-Ethanoic	Acetic	16.7	As alcohol acetates in many plants, and in some plant triacylglycerols. At low levels widespread as salt or thiolester. At higher levels in the rumen as salt

Table 3.1 *Continued*

No. of carbon atoms	Systematic name	Common name	M. pt (°C)	Occurrence
3	n-Propanoic	Propionic	−22.0	At high levels in the rumen
4	n-Butanoic	Butyric	−7.9	At high levels in the rumen, also in milk fat of ruminants
6	n-Hexanoic	Caproic	−8.0	Milk fat
8	n-Octanoic	Caprylic	12.7	Very minor component of animal and plant fats. Major component of many milk triacylglycerols
10	n-Decanoic	Capric	29.6	Widespread as a minor component. Major component of many milk triacylglycerols
12	n-Dodecanoic	Lauric	42.2	Widely distributed, a major component of some seed fats (e.g. palm kernel oil)
14	n-Tetradecanoic	Myristic	52.1	Widespread; occasionally as a major component
16	n-Hexadecanoic	Palmitic	60.7	The most common saturated fatty acid in animals, plants and microorganisms
18	n-Octadecanoic	Stearic	69.6	Major component in animals and some fungi, minor constituent in plants (but major in a few, e.g. cocoa butter)
20	n-Eicosanoic	Arachidic	75.4	Widespread minor component, occasionally a major component
22	n-Docosanoic	Behenic	80.0	Fairly widespread as minor component in seed fat triacylglycerols and plant waxes
24	n-Tetracosanoic	Lignoceric	84.2	Fairly widespread as minor component in seed fat triacylglycerols and plant waxes
26	n-Hexacosanoic	Cerotic	87.7	Widespread as component of plant and insect waxes
28	n-Octacosanoic	Montanic	90.9	Major component of some plant waxes

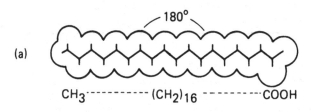

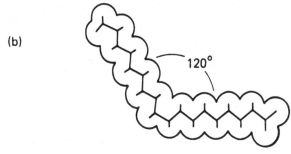

Figure 3.1 Preferred conformation of (a) a saturated (stearic) and (b) a mono-unsaturated (oleic) fatty acid.

While most natural fatty acids are even-numbered (due to their mode of biosynthesis) odd-numbered acids do occur. The formation of both types is discussed in section 3.2.2.

3.1.2 Branched chain fatty acids

Although branched chain fatty acids are usually also saturated, they are discussed separately here. Two distinct series which are often found in bacteria, are the *iso-* series where the terminal group is:

$$\begin{array}{c} CH_3 \\ | \\ CH_3{-}CH{-} \end{array}$$

and the *anteiso-* series where the terminal group is:

$$\begin{array}{c} CH_3 \\ | \\ CH_3{-}CH_2{-}CH{-} \end{array}$$

However, branch points can also be found in other positions. The presence of the side-chain has a similar effect on fluidity as the presence of a *cis* double bond (i.e. lowers the melting temperature). Branched chain acids occur widely but mainly at low concentrations in animal fats and some marine oils. They are rarely found in plant lipids. Butter fats, bacterial and skin lipids contain significant amounts. In the latter class, the uropygial (preen) gland of birds is a major source. Branched chain fatty acids are major components of the lipids of gram-positive bacteria and more complex structures with several branches may be found in the waxy outer coats of mycobacteria.

3.1.3 Unsaturated fatty acids

(a) Monoenoic acids

Over one hundred naturally occurring monoenoic acids have been identified but most of these are extremely rare. In general, the more common compounds have an even number of carbon atoms, a chain length of 16–22C and a double bond with the *cis* configuration. Often the *cis* bond begins at the Δ9 position. *Trans* isomers are rare but do exist, one of the most interesting being *trans*-3-hexadecenoic acid, a major fatty acid esterified to phosphatidylglycerol in the photosynthetic membranes of higher plants and algae.

The presence of a double bond causes a restriction in the motion of the acyl chain at that point. Furthermore, the *cis* configuration introduces a kink into the average molecular shape (Figure 3.1(b)) while the *trans* double bond ensures that the fatty acid has properties nearer to that of an equivalent chain length saturated acid (Tables 3.1, 3.2). Because the *cis* forms are less stable thermodynamically than the *trans* forms, they have lower melting points than the latter or their saturated counterparts.

In addition to the normal **ethylenic** double bonds, some fatty acids possess **acetylenic** bonds. These occur in a number of rare seed oils and a few mosses. Little is known about their biosynthesis.

(b) Polyunsaturated fatty acids

All dienoic acids are derived from monoenoic acids, the position of the second double bond being a function of the biochemical system. Thus, mammals have desaturases which are capable of removing hydrogens only from carbon atoms between an existing double bond and the carboxyl group. Because of this, further desaturations may need to be preceded by chain elongation. Higher plants on the other hand carry out desaturation between the existing double bond and the terminal methyl group. The double bonds are almost invariably separated from each other by a methylene grouping.

Table 3.2 Some naturally occurring monoenoic fatty acids

No. of carbon atoms	Systematic name	Common name	M. pt (°C)	Occurrence
16	*trans*-3-hexadecenoic		53	Plant leaves; eukaryotic algae; specifically as component of phosphatidylglycerol
16	*cis*-5-hexadecenoic			Ice plant, Bacilli
16	*cis*-7-hexadecenoic			Algae, higher plants, bacteria
16	*cis*-9-hexadecenoic	palmitoleic	1	Widespread: animals, plants, microorganisms. Major component in some seed oils
18	*cis*-6-octadecenoic	petroselenic	33	Found in umbelliferous seed oils
18	*cis*-9-octadecenoic	oleic	16	Most common monoenoic fatty acid in plants and animals. Also found in microorganisms
18	*trans*-9-octadecenoic	elaidic	44	Ruminant fats, hydrogenated margarines
18	*trans*-11-octadecenoic	*trans*-vaccenic	44	Found in rumen fats via biohydrogenation of polyunsaturated fatty acids
18	*cis*-11-octadecenoic	vaccenic	15	*E. coli* and other bacteria
20	*cis*-11-eicosenoic	gondoic	24	Seed oil of rape: fish oils
22	*cis*-13-docosenoic	erucic	24	Seed oil of Cruciferae (rape, mustard, etc.)

3.1.4 Cyclic fatty acids

These acids are rather uncommon but, nevertheless, examples of them are important metabolic inhibitors. The ring structures are usually either cyclopropyl or cyclopentyl. Cyclopropane and cyclopropene fatty acids are produced by many bacteria and are also found in some plants and fungi (Table 3.4).

3.1.5 Oxy acids

A great range of keto, hydroxy and epoxy acids has been identified in recent years. The most widely occurring epoxy acid is vernolic acid. Hydroxy acids do not occur very extensively although they occur generally in some sphingolipids. They are major components of surface waxes, cutin and suberin of plants (Table 3.4).

3.1.6 Conjugated unsaturated fatty acids

The major polyunsaturated fatty acids all contain *cis* methylene interrupted sequences and for years it was thought that most conjugated systems were artefacts of isolation. However, many such acids have now been firmly identified and are found in sources as diverse as seed oils, some micro-organisms and some marine lipids (especially sponges). An example of one such acid would be α-eleostearic acid (Table 3.3).

3.1.7 Fatty aldehydes and alcohols

Many tissues contain appreciable amounts of fatty alcohols or aldehydes whose chain length and double bond patterns reflect those of the fatty acids from which they can be derived. Sometimes the alcohols form esters with fatty acids and these 'wax esters' are important in marine waxes such as sperm whale oil or its plant equivalent jojoba wax.

3.1.8 Some properties of fatty acids

The short chain fatty acids (i.e. of chain lengths up to 8C) are poorly water-soluble although in solution they are associated and do not exist as single molecules. Indeed, the actual solubility (particularly of longer chain acids) is often very difficult to determine because it is influenced very much by the pH and also because the tendency for fatty acids to associate together leads to monolayer or micelle formation. Micelle formation is characteristic of many lipids. McBain and Salmon many years ago introduced the concept of micelles, heavily hydrated charged aggregates, based on their demonstration that a 1.0 N 'solution' of potassium stearate has an osmotic activity equivalent to only a 0.42 N solution of an undissociated salt and equivalent conductivity of that of the same concentration of potassium acetate. The most striking evidence for the formation of micelles in aqueous solutions of lipids lies in extremely rapid changes in physical properties over a limited

Table 3.3 Some naturally occurring polyunsaturated fatty acids

No. of carbon atoms	Systematic name	Common name	M. pt. (°C)	Occurrence
Dienoic acids				
18	cis, cis-6,9-octadecadienoic		−11	Minor component in animals
	cis, cis-9,12-octadecadienoic	linoleic	−5	Major component in plant lipids In animals it is derived only from dietary vegetables, and plant and marine oils
Trienoic acids (methylene interrupted)				
16	all-cis-7,10,13-hexadecatrienoic			Higher plants and algae
18	all-cis-6,9,12-octadecatrienoic	γ-linolenic		Minor component in animals and some algae. Important constituent of some plants
	all-cis-9,12,15-octadecatrienoic	α-linolenic	−11	Higher plants and algae, especially as component of galactosyl diacylglycerols

Trienoic acids (conjugated)				
18	*cis*-9, *trans*-11, *trans*-13-octadecatrienoic	eleostearic	49	Some seed oils, especially Tung oil
Tetraenoic acids				
16	all-*cis*-4,7,10,13-hexadecatetraenoic			*Euglena gracilis*
20	all-*cis*-5,8,11,14-eicosatetraenoic	arachidonic	−49.5	A major component of animal phospholipids. Major component of marine algae and some terrestrial species
Pentaenoic acids				
20	all-*cis*-5,8,11,14,17-eicosapentaenoic			Major component of marine algae, fish oils
22	all-*cis*-7,10,13,16,19-docosapentaenoic	clupanodonic		Animals, especially as phospholipid component. Abundant in fish
Hexaenoic acids				
22	all-*cis*-4,7,10,13,16,19-docosahexaenoic			Animals, especially as phospholipid component. Abundant in fish

Table 3.4 Examples of some substituted fatty acids

Structure	Name	Notes
Cyclic fatty acids		
$CH_3(CH_2)_5CH-CH(CH_2)_9COOH$ with CH_2 bridge	Lactobacillic	Produced by Lacto-bacilli, *Agrobacterium tumefaciens*
$CH_3(CH_2)_7C=C(CH_2)_7COOH$ with CH_2 bridge	Sterculic	Found in plants of the Malvales family
$CH_2(CH_2)_{11}COOH$ on cyclopentene ring	Chaulmoogric	Used for leprosy treatment
Epoxy-fatty acid		
$CH_3(CH_2)_4CH-CHCH_2CH=CH(CH_2)_7COOH$ with O epoxy	Vernolic	Epoxy derivative of oleate
Hydroxy-fatty acids		
$CH_3CH_2(CH_2)_{12}CHCOOH$ with OH	α-hydroxy palmitate	Found in galactosyl-cerebrosides
$HOCH_2(CH_2)_{14}COOH$	ω-hydroxy palmitate	Constituent of suberin coverings of plants
$CH_3(CH_2)_5CHCH_2CH=CH(CH_2)_7COOH$ with OH	Ricinoleic	Represents >90% of the total acids of castor bean oil

range of concentration, the point of change being known as the **critical micellar concentration** or CMC and exemplifies the great tendency of lipids to self-associate rather than stay as single molecules. The CMC is not a fixed value but a small range of concentration and is markedly affected by the presence of other ions, neutral molecules etc. The value of CMC can be conveniently measured by following the absorbance of a lipophilic dye such as Rhodamine in the presence of increasing 'concentrations' of the lipid.

Fatty acids are easily extracted from solution or suspension by lowering the pH to form the uncharged carboxyl group and extracting with a non-polar solvent such as light petroleum. In contrast, raising the pH increases solubility because of the formation of alkali metal salts which are the familiar soaps. Soaps have important properties as association colloids and are surface-active agents.

The influence of fatty acid structure on its melting point has already been mentioned with branch chains and *cis* double bonds lowering the melting points of equivalent saturated chains. In addition, the melting point of fatty acids depends on whether the chain is even- or odd-numbered (Table 3.5).

Saturated fatty acids are very stable but unsaturated acids are susceptible to oxidation; the more double bonds the greater the susceptibility. Unsaturated fatty acids, therefore, have to be handled under an atmosphere of inert gas (e.g. nitrogen) and kept away from (photo) oxidants or substances giving rise to free radicals. Anti-oxidant compounds have frequently to be used in the biochemical laboratory just as organisms and cells have to utilize similar compounds to prevent potentially harmful attack of acyl chains *in vivo* (section 8.11).

Table 3.5 The melting points of a series of saturated fatty acids

Fatty acid	Chain length	M. pt. (°C)
Butanoic	C_4	−8
Pentanoic	C_5	−35
Hexanoic	C_6	−8
Heptanoic	C_7	17
Octanoic	C_8	13
Nonanoic	C_9	32
Decanoic	C_{10}	29
Undecanoic	C_{11}	44
Dodecanoic	C_{12}	42
Tridecanoic	C_{13}	55
Tetradecanoic	C_{14}	52
Pentadecanoic	C_{15}	63
Hexadecanoic	C_{16}	61

3.1.9 Quantitative and qualitative fatty acid analysis

In Chapter 2 it was pointed out that chromatographic methods cannot only be used to separate closely related compounds but may also help to define their structures and, in combination with suitable detection processes, will give quantitative data. For fatty acids, gas–liquid chromatography has proved particularly useful (section 2.4.2).

The preparation of the samples and the choice of column packing material are rather important. For example, it is customary to form methyl esters of fatty acids by refluxing the sample with methanolic-HCl (methylation lowers the boiling point of the derivative sufficiently for injection onto the gas chromatograph). However, if this technique is used for cyclic fatty acids they arc destroyed. These acids would have to be transferred from lipids to methanol by base-catalysed methanolysis (with sodium methoxide).

No single stationary phase is capable of separating every acid – although capillary-gas chromatography (where the columns have dimensions such as 50 m by 0.25 mm interior diameter) usually separates most components including positional isomers. Generally two different types of column are necessary to resolve all the individual components.

Columns having saturated paraffin hydrocarbons (Apiezon L grease) or silicone greases separate largely on the basis of molecular weight; unsaturated acids emerge from the column earlier than the corresponding saturated acids. Branched chain acids emerge before the saturated fatty acids of corresponding carbon number. Thus, these acids will overlap with unsaturated fatty acids.

Columns containing polar materials such as polyethylene glycol adipate give separations based on number of double bonds as well as on molecular weight (i.e. chain length). The order of emergence from this type of column will, therefore, be 16:0, 16:1, 18:0, 18:1, 18:2 etc. (Figure 3.2). Branched chain fatty acids may, in consequence, overlap highly unsaturated fatty acids of shorter chain length.

In order to quantify the amounts of individual fatty acids in a particular sample, it is usual to include an internal standard. This will consist of a known amount of a fatty acid which does not appear in the mixture being analysed. Frequently, odd-chain length acids such as 17:0 or 21:0 are used (Figure 3.2) although again care must be taken that the peak of the internal standard does not obscure unsaturated components of the mixture.

(a) Determination of the structure of an unknown acid

Merely running an unknown fatty acid on a column of known properties can give some information about its structure. Cochromatography with authentic standards on different types of columns will serve to identify a

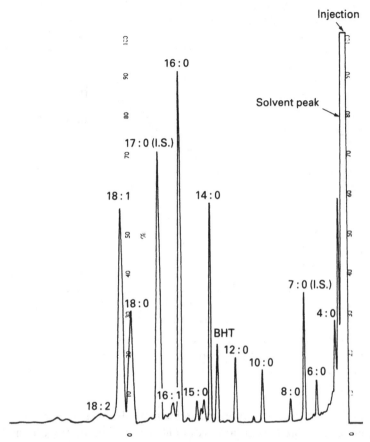

Figure 3.2 Separation of fatty acid methyl esters by gas–liquid chromatography on a 2 m × 4 mm column. Stationary phase: 10% polyethylene glycol adipate (stabilized with orthophosphoric acid) on Diatomite C-AAW. Carrier gas: argon. Flow rate 50 ml/1 min. The oven temperature was programmed from 65°C to 185°C at 8°C per min. The sample consists of fatty acids derived from a milk diet for calves. Heptanoic and heptadecanoic acids were added as 'internal standards' in concentrations of 1.55 mg 7 : 0 and 4.94 mg 17 : 0 in 10 ml sample solution and the computer was programmed to print out the concentration in mg/10 ml of each fatty acid in the mixture. IS = internal standard; BHT = butylated hydroxytoluene (antioxidant). Reproduced with kind permission of Mr J.D. Edwards-Webb.

fatty acid provisionally and the chain length of an unknown substance can be estimated from identified peaks because there is a simple linear relationship between the log of the retention volume (time) and the carbon number for each particular type of acid (Figure 3.3) on a given column. Samples of a given component can be collected from the gaseous effluent of the column by allowing the gas stream to pass through a wide glass tube loosely packed

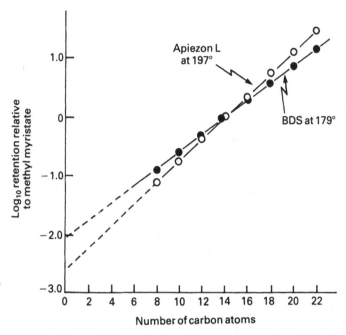

Figure 3.3 Plot of log relative retention time versus carbon chain length for saturated fatty acid methyl esters. Two liquid phases were used: Apiezon L and butanediol succinate.

with defatted cotton or glass wool wetted with a solvent such as methanol or petroleum ether. Such an isolated component can then be run on another type of column or subjected to specific chemical reactions. For example, the identity of an unsaturated fatty acid can be confirmed by H_2 reduction to the corresponding saturated acid and by oxidation with permanganate or ozone. With oxidation, the unsaturated acid is split into fragments at its double bonds so that analysis of the products gives information about the position of the double bond (e.g. oleic acid, n-9 18 : 1 gives oxidation products consisting of a C9 dicarboxylic acid and nonanoic (9 : 0) acid). Infra-red spectroscopy can be used to differentiate between *cis* and *trans* double bonds. Instead of collecting a sample of an individual component the gas chromatograph can also be connected directly to a mass spectrometer which allows the identification of many lipid components including most fatty acids.

(b) Thin layer chromatography

This is a very useful ancillary technique since the inclusion of $AgNO_3$ in the absorbant allows selective retardation of unsaturated acids by reversible complexing with the Ag^+ ion. The larger the number of double bonds, the greater the retardation (Figure 3.4); *trans* acids are retarded less than their *cis* counterparts and positional isomers can be separated in some cases.

Preparative TLC can be used to isolate groups of acids defined by the number of double bonds. After elution from the plate and removal of Ag^+ ions, the fatty acids can then be analysed further. Study of such a group by GLC and, after single component isolation, oxidative degradation followed by identification of the fragments by GLC, often allows the complete structure of a fatty acid to be defined.

Silver nitrate TLC is also useful for separation of monohydroxy acids. Dihydroxy or tetrahydroxy acids are best dealt with by TLC using borate or arsenate-impregnated silica gel.

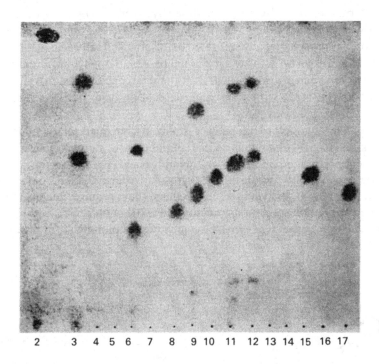

Figure 3.4 Thin layer chromatography of isomeric octadecenoates on silver nitrate–silica gel G (30 : 70). The position of the double bonds is indicated by the sample number, samples being 2, 3, 6, 9, 11 and 12 *trans*-octadecenoates, 3, 6, 8, 9, 10, 11, 12 and 15 *cis*-octadecenoates and the vinyl compound 17-octadecenoate (in mixtures, *trans* runs ahead of *cis*). The plate was developed at − 25°C, three times with toluene. The spots were located with chlorosulphonic acid–acetic acid (1 : 2) and charring. Reproduced with kind permission of Dr L.J. Morris (1987) and the Elsevier Science Publishers, BV, from *Journal of Chromatography*, **31**, p. 74, Figure 2(a).

3.2 THE BIOCHEMISTRY OF FATTY ACIDS

3.2.1 Conversions of fatty acids into metabolically active thiol esters is often a prerequisite for their further metabolism

For most of the metabolic reactions in which fatty acids take part, whether they be anabolic (synthetic) or catabolic (degradative), thermodynamic considerations dictate that the acids be 'activated'. For these reactions thiol esters are generally utilized. The active form is usually the thiol ester of the fatty acid with the complex nucleotide, *coenzyme A* (CoA) or the small protein known as acyl carrier protein (ACP) (Figure 3.5). These molecules contain a thiol ester and, at the same time, render the acyl chains water soluble.

The formation of acyl-CoAs is catalysed by several acyl-CoA synthetases. The overlapping of their chain length specificities and their tissue distribution are such that any saturated or unsaturated fatty acid in the range 2–22C or more can be activated in animal tissues, though at different rates. These enzymes can also activate branch chain fatty acids, dicarboxylic acids and carboxylic acids with unusual groups. Bile acids require separate, specific enzymes, however.

Acetyl-CoA synthetase has been crystallized from heart mitochondria. In heart, kidney and skeletal muscle it is confined to the mitochondrial matrix but in adipose tissue and mammary gland it is also present in the cytosol. In liver, it is only found in the cytosol. Where tissues contain the extramitochondrial enzyme, it is because they require acetate to be activated as the first step in fatty acid synthesis (section 3.2.2).

Acetyl-CoA synthetase (which also activates propionate at slower rates),

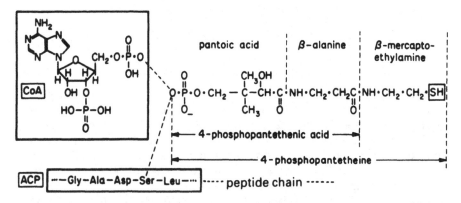

Figure 3.5 The structures of coenzyme A and acyl carrier protein.

along with other short chain acyl-CoA synthetases, is important in ruminants. The rumen microorganisms in these animals generate huge amounts of short chain acids including acetate. Other sources of acetate in animals include the oxidation of ethanol by the combined action of alcohol and aldehyde dehydrogenases and from β-oxidation. Under fasting conditions or in diabetes the ketotic conditions cause a rise in the production of ketone bodies and of acetate (via the action of acetyl-CoA hydrolase). The absence of acetyl-CoA synthetase in liver mitochondria allows acetate to be transported from this tissue and activated and oxidized in mitochondria of other tissues.

Butyryl-CoA synthetase is active with acids in the range 3–7C and has been isolated from heart mitochondria. It is a mitochondrial enzyme in other tissues but has not been found in liver where a special propionyl-CoA synthetase has been demonstrated. The latter enzyme may be especially important in ruminants where propionate is formed by rumen micro-organisms and is an important substrate for gluconeogenesis.

The medium chain acyl-CoA synthetase, again isolated from heart mitochondria, has its highest activity with heptanoate. It is active with acids in the range 4–12C, the lowest Km being for octanoate. Liver mitochondria also contain a second enzyme activating medium chain fatty acids including branched or hydroxy fatty acids but which also functions with benzoate and salicylate. Thus, its purpose seems to lie in the formation of glycine conjugates of aromatic carboxylic acids such as benzoate and salicylate.

The long chain acyl-CoA synthetases are firmly membrane bound and can only be solubilized by the use of detergents. Within the cell, activity has been detected in endoplasmic reticulum and the outer mitochondrial membrane with small amounts in peroxisomes (when the latter are present). There is some dispute as to whether the activity present in mitochondrial and microsomal fractions is due to the same enzyme. Because long chain fatty acid activation is needed for both catabolism (β-oxidation) and for synthesis (acylation of complex lipids) it would be logical if the long chain acyl-CoA synthetases of mitochondria and the endoplasmic reticulum formed different pools of cellular acyl-CoA. This compartmentation has been demonstrated with yeast mutants where it plays a regulatory role in lipid metabolism (section 3.2.7) and, perhaps, in other organisms.

The long chain acyl-CoA synthetases have equal activity for substrates between 10 and 18C and have decreasing activity with shorter or longer chains. Unsaturated fatty acids are activated at higher rates than saturated acids. In intact cells, the activity of the synthetases is regulated both by substrate (free fatty acids, CoA) availability and by product inhibition by acyl-CoA.

The reaction mechanism of acyl-CoA synthetases has been studied using heart acetyl-CoA and butyryl-CoA synthetase. The reaction proceeds in two stages:

$$\text{carboxylic acid} + \text{ATP} \xrightleftharpoons{\text{Mg}^{2+}} \text{Acyl-AMP} + \text{PP}_i \tag{3.1}$$

$$\text{Acyl-AMP} + \text{CoA} \rightleftharpoons \text{Acyl-CoA} + \text{AMP} \tag{3.2}$$

Both reactions are freely reversible but the overall reaction is driven to the right by pyrophosphatase activity and/or by the re-phosphorylation of AMP. The short chain acyl-AMP is thought to be enzyme-bound. In the medium and long chain acyl-CoA synthetases, however, acyl-AMP is probably only formed as a side-product or with unphysiological or conformationally changed enzyme forms. This would explain why intermediary enzyme complexes were formed when medium and long chain synthetases were incubated with ATP and medium and long chain fatty acids but that acetyl-AMP could not be identified when the short chain synthetase was incubated with acetate.

Although most activating enzymes are ATP-dependent, some acyl-CoA synthetases have been found which are GTP-dependent. These enzymes have low activities and their physiological significance is uncertain.

In some microorganisms, CoA may be esterified to short chain fatty acids (up to 5C) by a different mechanism. The enzyme responsible, thiophorase, first forms an acetyl-phosphate and then catalyses the transfer of the acetyl group to CoA:

$$\text{CH}_3\text{COO}^- + \text{ATP} \xrightleftharpoons{\text{Mg}^{2+}} \text{CH}_3\text{COPO}_3^- + \text{ADP} \tag{3.3}$$

$$\text{CH}_3\text{COPO}_3^- + \text{CoA} \rightleftharpoons \text{CH}_3\text{COCoA} + \text{H}_2\text{PO}_4^- \tag{3.4}$$

Acetyl-CoA can then react with the fatty acid to form an acyl-CoA.

$$\text{CH}_3\text{COCoA} + \text{RCOO}^- \rightleftharpoons \text{RCOCoA} + \text{CH}_3\text{COO}^- \tag{3.5}$$

3.2.2 Biosynthesis of saturated fatty acids: *de novo* synthesis by acetyl-CoA carboxylase and fatty acid synthetase and elongation reactions for pre-formed acids

Most naturally occurring fatty acids have even numbers of carbon atoms. Therefore, it was natural for biochemists to postulate that they were formed by condensation from two-carbon units. This suggestion was confirmed in 1944 when Rittenberg and Bloch isolated fatty acids from tissues of rats which had been fed acetic acid labelled with ^{13}C in the carboxyl group and ^{2}H in the methyl group. The two kinds of atoms were located at alternate positions along the chain, showing that the complete chain could be derived from acetic acid.

This stimulated interest in the mechanism of chain elongation and when the main details of the β-oxidation pathway (the means by which fatty acids are broken down two carbon atoms at a time, section 3.3.1) were worked out in the early 1950s it was natural for many biochemists to ask the question: can β-oxidation be reversed in certain circumstances to synthesize fatty acids instead of breaking them down?

The study of fatty acid biosynthesis began in the laboratories of Gurin in the USA who studied liver and Popjak in London, studying mammary gland. Several discoveries soon indicated that the major biosynthetic route to long chain fatty acids was distinctly different from β-oxidation. In the first place, a pyridine nucleotide was involved; but the reduced form, NADPH, not NAD^+ as in β-oxidation (Figure 3.20 and Table 3.7). Second, a requirement for bicarbonate or carbon dioxide was noticed by Wakil and his colleagues who were studying fatty acid biosynthesis in pigeon liver and by Brady and Klein studying rat liver and yeast respectively.

(a) Acetyl-CoA carboxylase

The study of a biosynthetic pathway usually progresses in well-defined steps: first the overall pathway is demonstrated in the living animal (*in vivo*); then the specific organ where the reaction takes place is located and the reaction demonstrated in the isolated organ or in slices of the organ; then in a cell-free homogenate; next in subcellular components; and finally, each step is studied at the level of the isolated, purified enzyme. Fatty acid biosynthesis was no exception and in pigeon liver the enzymes of fatty acid synthesis were found to be in the supernatant fraction which could be split into two protein fractions by ammonium sulphate precipitation; both of these fractions added together were necessary to produce fatty acids from acetate, and the cofactor biotin was found to be bound to one of the protein fractions. At that time biotin was already known to be involved in carboxylation reactions and when Wakil discovered soon afterwards that malonate was an intermediate in fatty acid synthesis, the essential features of the pathway began to emerge. The first step in fatty acid synthesis is the carboxylation of the 2-carbon fragment acetate to malonate. The acetate must be in an 'activated form' as its CoA thiol ester (section 3.2.1 and Figure 3.5). The carboxylation is catalysed by the enzyme, acetyl-CoA carboxylase (acetyl-CoA: carbon dioxide ligase [ADP]) which could be identified with the protein fraction from pigeon liver supernatant to which biotin was bound. One of the first suggestions that biotin was the prosthetic group of acetyl-CoA carboxylase came from the observation that the carboxylation step was inhibited by avidin. This protein was already well known as a potent inhibitor of biotin, for it is the component of raw egg white which can bind to biotin causing a vitamin deficiency known as 'egg white injury'. The active carbon dioxide is attached to one of the ureido nitrogens of the biotin ring.

carboxyl biotinyl enzyme

The carboxylation reaction of acetyl-CoA carboxylase proceeds in two main steps:

$$\text{BCCP-biotin} + \text{ATP} + \text{HCO}_3^- \rightleftharpoons \text{E-biotin-CO}_2 + \text{ADP} + \text{P}_i \qquad (3.6)$$

$$\text{BCCP-biotin-CO}_2 + \text{acetyl-CoA} \rightleftharpoons \text{E-biotin} + \text{malonyl-CoA} \qquad (3.7)$$

In *E. coli*, acetyl-CoA carboxylase consists of three dissociable components. One protein contains biotin, has a molecular mass of 22.5 kDa and is the **biotin carboxyl carrier protein** (BCCP). A second component (102 kDa) consists of two identical subunits, catalyses reaction (3.6) and is the **biotin carboxylase**. The final component (130 kDa) contains two pairs of non-identical subunits of 30 and 35 kDa. It catalyses reaction (3.7) and is the **carboxyltransferase**.

Although *E. coli* acetyl-CoA carboxylase can be separated into three components, crude cell extracts contain complexes of the BCCP and biotin carboxylase proteins. Furthermore, the acetyl-CoA carboxylase from *P. citronellolis* which has a subunit structure similar to that from *E. coli* can be isolated as a complex of 250 kDa from high salt solutions. Therefore, bacterial acetyl-CoA carboxylases may function as enzyme complexes *in vivo* (Table 3.6).

Evidence from a variety of studies (protein purification, immunology, genetics) indicates that the acetyl-CoA carboxylases from yeasts, higher plants, birds and mammals consist of a single multifunctional protein. The molecular masses of these proteins are rather similar between organisms with, for example, values of 189–230 kDa for the yeasts *Saccharomyces cerevisiae* and *Candida lipolytica*, 220–260 kDa for rat and chicken liver, rat and rabbit mammary gland and goose uropygial gland and 240 kDa for rape seed. Thus all eukaryotic acetyl-CoA carboxylases have a single subunit which carries all the catalytic activities with the possibility also of a regulatory domain in some cases (Table 3.6).

Acetyl-CoA carboxylase from animals requires citrate or isocitrate for activity and this activation is accompanied by an increase in the molecular weight form of the enzyme. Electron microscopic and light scattering studies have shown that the large molecular form has a mass of 4–11×10^6 Da. In the absence of citrate or isocitrate enzyme activity is lost and a small molecular form is present which is usually a mixture of the monomeric subunit

Table 3.6 Properties of different acetyl-CoA carboxylases

	Animal	Higher plants	Yeast	E. coli
Nature of enzyme	Polymeric filaments (4–11×10^6 Da) of protomers (220–260 kDa) of single multifunctional protein	Multifunctional protein (200–250 kDa) possibly active as a dimer	Multifunctional protein (190–230 kDa)	Multienzyme complex of 3 proteins
Regulation	Activated by citrate Inhibited by acyl-CoAs	*In vivo* activity regulated by pH, Mg^{++}, ATP/ADP	Activated but not polymerized by citrate	Inhibited by (p)ppGpp

Table 3.7 Reactions of fatty acid synthesis in *E. coli*

1 Malonyl transacylase	$HOOC \cdot CH_2 \cdot \overset{O}{\overset{\|}{C}} \cdot S \cdot CoA + ACP\!-\!SH \rightleftharpoons$ $HOOC \cdot CH_2 \cdot \overset{O}{\overset{\|}{C}}S\text{-}ACP + CoA\!-\!SH$	Specific for malonate: not a saturated acyl-CoA. Malonyl-S-pantetheine is also a substrate. Both enzymes (1 and 2) were characterized by (i) the amount of ^{14}C-acetate or malonate transferred to ACP or (ii) paper chromatography of acetyl or malonyl hydroxamate. Intermediate is acyl-S-enzyme
2 Acetyl transacylase	$CH_3\overset{O}{\overset{\|}{C}} \cdot S\!-\!CoA + ACP\!-\!SH \rightleftharpoons$ $CH_3 \cdot \overset{O}{\overset{\|}{C}} \cdot S \cdot ACP + CoA\!-\!SH$	There is now doubt about the importance of this reaction because acetyl-CoA can be used directly in reaction 7 to form acetoacetyl-ACP
3 β-ketoacyl-ACP synthetase	$CH_3 \cdot \overset{O}{\overset{\|}{C}} \cdot S \cdot ACP + HOOC \cdot CH_2 \cdot \overset{O}{\overset{\|}{C}} \cdot S \cdot ACP \rightleftharpoons$ $CH_3 \cdot \overset{O}{\overset{\|}{C}} \cdot CH_2 \cdot \overset{O}{\overset{\|}{C}} \cdot S \cdot ACP + CO_2 + ACP \cdot SH$	Assayed by coupling with the next reaction and following NADPH oxidation spectrophotometrically. SH-enzyme; inhibited by iodoacetamide, etc. Protected by preincubation with acetyl-ACP not malonyl-ACP
4 β-ketoacyl-ACP reductase	$CH_3 \cdot \overset{O}{\overset{\|}{C}} \cdot CH_2\overset{O}{\overset{\|}{C}} \cdot S \cdot ACP + NADPH + H^+ \rightleftharpoons$ $D(-)CH_3 \cdot \overset{OH}{\overset{\|}{C}H} \cdot CH_2 \cdot \overset{O}{\overset{\|}{C}} \cdot S \cdot ACP + NADP^+$	Also reacts with CoA, pantetheine esters but much more slowly. Stereospecific for D(−) isomer

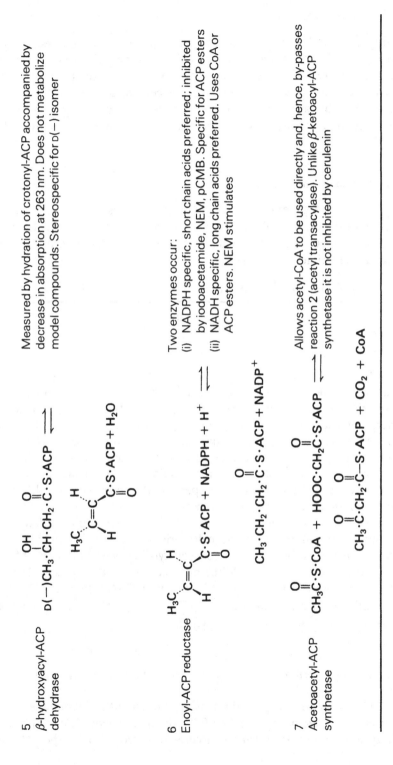

5
β-hydroxyacyl-ACP dehydrase

$$D(-)CH_3 \cdot CH \cdot CH_2 \cdot C \cdot S \cdot ACP \rightleftharpoons$$

with OH on the CH and O on the C

$$\underset{H}{\overset{H_3C}{\diagdown}} C = \underset{H}{\overset{}{C}} \diagup C \cdot S \cdot ACP + H_2O$$

Measured by hydration of crotonyl-ACP accompanied by decrease in absorption at 263 nm. Does not metabolize model compounds. Stereospecific for D(−) isomer

6
Enoyl-ACP reductase

$$\underset{H}{\overset{H_3C}{\diagdown}} C = \underset{H}{\overset{}{C}} \diagup C \cdot S \cdot ACP + NADPH + H^+ \rightleftharpoons$$

$$CH_3 \cdot CH_2 \cdot CH_2 \cdot C \cdot S \cdot ACP + NADP^+$$

Two enzymes occur:
(i) NADPH specific, short chain acids preferred; inhibited by iodoacetamide, NEM, pCMB. Specific for ACP esters
(ii) NADH specific, long chain acids preferred. Uses CoA or ACP esters. NEM stimulates

7
Acetoacetyl-ACP synthetase

$$CH_3C \cdot S \cdot CoA + HOOC \cdot CH_2C \cdot S \cdot ACP \rightleftharpoons$$

$$CH_3 \cdot C \cdot CH_2 \cdot C - S \cdot ACP + CO_2 + CoA$$

Allows acetyl-CoA to be used directly and, hence, by-passes reaction 2 (acetyl transacylase). Unlike β-ketoacyl-ACP synthetase it is not inhibited by cerulenin

(230 kDa) and its dimer. Biotin is needed for the polymerization because the apoenzyme fails to aggregate even in the presence of citrate. There is evidence for the existence of the large molecular form of acetyl-CoA carboxylase *in vivo* and, moreover, that the cellular content of this form is controlled by the ratio of the cytoplasmic contents of citrate and acyl-CoAs. The regulation of acetyl-CoA carboxylase is discussed more fully in section 3.2.7.

(b) Fatty acid synthetase

The malonyl-CoA generated by acetyl-CoA carboxylase forms the source of nearly all of the carbons of the fatty acyl chain. In most cases, only the first two carbons arise from a different source – acetyl-CoA – and the latter is known as the 'primer' molecule.

$$CH_3CH_2\,|CH_2CH_2CH_2CH_2CH_2CH_2CH_2CH_2CH_2CH_2CH_2CH_2CH_2COOH$$

from   from malonyl-CoA
'primer'
acetyl-CoA

The fatty acid chain then grows in a series of reactions illustrated for *E. coli* in Table 3.7. The basic chemistry is similar in all organisms although the organization of the enzymes is not. In some cases acetyl-CoA is not the usual primer. For example, in mammalian liver and mammary gland, butyryl-CoA is more active. In goat mammary gland there are enzymes which can catalyse the conversion of acetyl-CoA into crotonyl-CoA by what is, essentially, the reverse of β-oxidation (section 3.3.1). Finally the use of propionyl-CoA or branched primers permits the formation of odd-chain length or branch chain fatty acids, respectively.

Most of the animal cell's acetyl-CoA is derived from the oxidation of pyruvate in mitochondria. A difficulty arises in that the mitochondrial membrane is relatively impermeable to acetyl-CoA while the sites of fatty acid synthesis are mainly outside the mitochondria in the cytosol. Acetyl-CoA may be formed by direct esterification with CoA catalysed by acetyl-CoA synthetase ((3.8) and section 3.2.1) or from citrate by a reaction catalysed by citrate cleavage enzyme (3.9).

$$\text{acetate} + \text{ATP} + \text{HS-CoA} \rightleftharpoons \text{Acetyl-CoA} + \text{AMP} + \text{PP}_i \qquad (3.8)$$

$$\text{citrate} + \text{ATP} + \text{HS-CoA} \rightleftharpoons \text{Acetyl-CoA} + \text{oxaloacetate} + \text{ADP} + \text{P}_i \quad (3.9)$$

Acetyl-CoA may be transported across the mitochondrial membrane in the four ways illustrated in Figure 3.6. A fuller discussion of the role of carnitine in fatty acid metabolism will be found in section 3.3.1. The relative

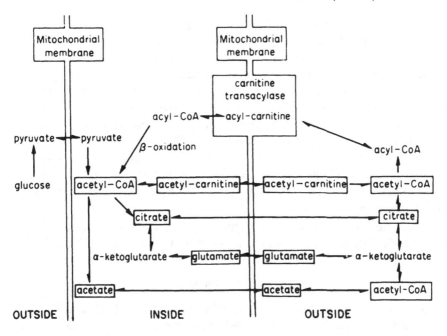

Figure 3.6 Formation of extramitochondrial acetyl-CoA.

importance of these four ways of obtaining acetyl-CoA outside mito-chondria is still unclear.

In plants, where fatty acid synthesis *de novo* is concentrated in the plastid (chloroplasts in green tissues), the source of acetyl-CoA is also not entirely clear. Pyruvate dehydrogenase activity has been demonstrated in some chloroplasts but thioesterase action on mitochondrial acetyl-CoA followed by movement of acetic acid into the plastid and reactivation by acetyl-CoA synthetase has been suggested to be important, at least in species such as spinach.

The overall reaction of fatty acid synthesis can be summarized:

$$CH_3CO\text{---}CoA + 7HOOC.CH_2.CO\text{---}CoA + 14NADPH + 14H^+$$

$$\rightleftharpoons CH_3(CH_2)_{14}COOH + 7CO_2 + 8CoASH + 14NADP^+ + 6H_2O$$

(3.10)

Six separate enzyme activities are involved and the steps have been elucidated mainly from studies of *E. coli*, yeast, the tissues of various animals and some higher plants. The group of enzymes are known collectively as **fatty acid synthetase** and it is now clear that, although the overall biochemistry is the same in all cases, the organization of the synthetases is very different.

Fatty acid synthetases can be divided mainly into Type I and Type II enzymes (Table 3.8). Type I synthetases are multifunctional proteins in which the proteins catalysing the individual partial reactions are discrete domains. This type includes the animal synthetases and those from higher bacteria and yeast. Type II synthetases contain enzymes which can be separated, purified and studied individually. This system occurs in lower bacteria and plants and has been studied most extensively in *E. coli*. In addition, Type III synthetases – occurring in different organisms – catalyse the addition of C_2 units to preformed acyl chains and are also known as elongases. Although, historically, the reactions of the yeast synthetase were unravelled first, we shall start by describing the separate reactions catalysed by the enzymes of *E. coli*.

The Americans, Goldman and Vagelos, discovered that the product of the first condensation reaction of fatty acid synthesis in *E. coli* was bound to a protein through a thiol ester linkage:

Acetoacetyl-*S*-protein

When they incubated this product with the coenzyme NADPH and a crude protein fraction containing the synthetase, they obtained butyryl-*S*-protein and when malonyl-CoA was included in the reaction mixture, long chain fatty acids were produced. Subsequently, intermediates at all steps of the reaction were shown to be attached to this protein, which was therefore called acyl carrier protein (ACP). It is a small molecular mass protein (about 8.8 kDa), has one SH-group per mole of protein and is stable to heat and acid pH. Vagelos and his colleagues worked out the structure of the prosthetic group to which the acyl moieties were attached, in the following way. They made acyl-ACP derivatives in which the acyl group was labelled with radioactive carbon atoms (such as 2-[^{14}C]-malonyl-ACP) and then hydrolysed the acyl protein with proteolytic enzymes. This yielded small radioactive peptides whose structure was fairly easy to determine. The prosthetic group turned out to be remarkably similar to coenzyme A: the acyl groups were bound covalently to the thiol group of 4'-phospho-pantetheine, which in turn was bound through its phosphate group to a serine hydroxyl of the protein (Figure 3.5). To study the biosynthesis of ACP, Vagelos made use of a strain of *E. coli* that had to be supplied with pantothenate in order to grow. When the bacteria were fed with radioactive pantothenate, the labelled substance was incorporated into the cell's ACP and CoA. However, whereas the ACP level remained constant under all conditions the level of CoA was very much dependent on the concentration of pantothenate. When the cells were grown in a medium of high pantothenate concentration, the level of CoA produced was also high; but if these cells were washed free of pantothenate and transferred to a medium with no pantothenate, then the CoA level dropped rapidly while that of

Table 3.8 Types of fatty acid synthetases in different organisms

Source	Subunit types	Subunit (mol. mass)	Native (mol. mass)	Major products
Type I—Multicatalytic polypeptides				
Mammalian, avian liver	α	220–270×10^3	450–550×10^3	C_{16} free
Mammalian mammary gland	α	200–270×10^3	400–550×10^3	C_4–C_{16} free
Goose uropygial gland	α			2,4,6,8-tetramethyl C_{10} 2,4,6-trimethyl C_{12}
M. smegmatis	α	$290{,}000$	2×10^6	C_{16}-, C_{24}-CoA
S. cerevisiae	α, β	$185{,}000$	2.3×10^6	C_{16}-, C_{18}-CoA
Dinoflagellates	α	$180{,}000$ $180{,}000$	4×10^5	
Type II—Freely dissociable enzymes				
Higher plant chloroplasts	Separate enzymes	—	—	C_{16}-ACP, C_{18}-ACP
E. gracilis chloroplast	Separate enzymes	—	—	C_{12}-, C_{14}-, C_{16}-, C_{18}-ACP
E. coli	Separate enzymes	—	—	C_{16}-, $C_{18:1}$-ACP

ACP remained constant. In other words, ACP was synthesized at the expense of CoA. It seems that ACP biosynthesis is under tight control and this may be yet another factor involved in the overall control of fatty acid biosynthesis.

Two enzymes are involved in the turnover of the prosthetic group of ACP. Both enzymes are highly specific and even small modifications of the ACP structure prevents activity. The first enzyme catalyses the transfer of 4'-phosphopantetheine from CoA:

$$\text{apo-ACP} + \text{CoA} \underset{\text{holo-ACP synthetase}}{\overset{Mg^{2+}}{\rightleftharpoons}} \text{holo-ACP} + 3'5'\text{-ADP} \qquad (3.11)$$

$$\text{holo-ACP} + H_2O \underset{\text{ACP hydrolase}}{\rightleftharpoons} \text{apo-ACP} + 4'\text{-phosphopantethcine} \qquad (3.12)$$

In vivo both enzymes have been demonstrated in *E. coli* and in mammalian synthetases the turnover of the 4'-phosphopantetheine prosthetic group is an order of magnitude faster than the rest of the synthetase, implying the presence of holo-ACP synthetase and ACP hydrolase. In plants, also, the holo-ACP synthetase has been detected.

After the discovery and characterization of ACP, the different steps in the reaction sequence were quickly elucidated and enzymes purified as detailed in Table 3.7. The *E. coli* synthetase does not only produce saturated fatty acids but also makes monounsaturated (mainly *cis*-vaccenic) acids. It does this because of the presence of a β-hydroxydecanoyl-ACP-β,γ-dehydrase which produces *cis*-3-decenoyl-ACP (precursor of unsaturated acids) instead of *trans*-2-decenoyl-ACP (converted to saturated acids; section 3.2.4). In addition, two forms of β-ketoacyl-ACP synthetase (designated 1 and 2) have been identified in fatty acid biosynthetic mutants of *E. coli*. The two enzymes were purified and shown to have different properties – particularly in their substrate specificity. In particular, β-ketoacyl-ACP synthetase 2 was much better at utilizing palmitoleoyl-ACP as a substrate. This β-ketoacyl-ACP synthetase 2 is thought to play a particular role in the thermal regulation of membrane fatty acids by normally catalysing the elongation of palmitoleoyl-ACP to *cis*-vaccenoyl-ACP. The *E. coli* fatty acid synthetase produces acyl-ACPs as products and these can serve as acyl donors for complex lipid synthesis. The chain length of the product is thought to be determined by the specificities of the two β-ketoacyl-ACP synthetases.

Very recently a third condensing enzyme has been reported in *E. coli*. This condensing enzyme is distinctly different from the other β-ketoacyl-ACP synthetases in *E. coli* in that it is: (a) cerulenin-insensitive; (b) specific for very short chain acyl-ACPs; and (c) prefers acetyl-CoA over acetyl-ACP. It has been termed acetoacetyl-ACP synthetase. A similar enzyme has also been found in some plant tissues. Therefore, the Type II FAS complexes may be able to by-pass the very slow acetyl transacylase step which would not, therefore, be rate-limiting for overall fatty acid synthesis.

The discovery that the enzymes of the Type I synthetases were apparently in a tight complex presented difficulties in studying the individual enzymic steps. In *E. coli*, the individual intermediates could be isolated, purified, characterized and used as substrates to study the enzymology of each reaction. In yeast this was not possible because the intermediates remained bound to the enzyme all the time. The German biochemist, Lynen, solved this problem elegantly by synthesizing model substrates – acyl derivatives of *N*-acetyl-cysteamine – in the hope that they would have enough affinity for the enzyme to be able to demonstrate the reactions. This was not always the case, but in certain reactions the model substrates had enough affinity (though very small) for the reaction to be measured. It was partly for this work that Lynen was awarded the Nobel Prize in 1964. Lynen visualized the yeast synthetase as a complex of seven distinct enzymes catalysing reactions analogous to those in Table 3.7 but including a final step in which the product – a palmitoyl or stearoyl group – is transferred to coenzyme A from the enzyme. However, it was perplexing that the complex could not be dissociated into functioning monomers. Studies with mutants were again responsible for a major advance in elucidating this problem. Schweizer and his colleagues in Germany found three types of mutants which required fatty acids for growth. One type lacked acetyl-CoA carboxylase whereas the other two (*fas*-1 and *fas*-2) were devoid of fatty acid synthetase activity. Initial studies of these mutants showed that one mutant was deficient in protein-bound pantetheine while the second class of mutant lacked FMN. (In yeast, the enoyl reductase uses FMN and this proved a fortunate feature which allowed the easy initial classification of mutants.) Further investigation showed that three different fatty acid synthetase functions, i.e. β-ketoacyl-ACP synthetase, β-ketoacyl-ACP reductase and the binding of pantetheine (ACP function), were associated with *fas*-2. The remaining four activities were assigned to *fas*-1. It was, therefore, proposed that yeast fatty acid synthetase contained two dissimilar multifunctional peptides α and β. Subsequently, these genetic results were confirmed by purifying the enzyme and dissociating it into α and β subunits by acylation with dimethyl maleic anhydride. Each subunit could then be assayed for the partial reactions in order to confirm the genetic assignments. Moreover, *in vitro* complementation of appropriate pairs of mutant FAS proteins has been carried out by deacylation techniques to restore activity.

	β-ketoacyl- ACP synthase	ACP	β-ketoacyl- ACP reductase

α-chain (*fas*-1) (185 kDa)

Acetyl transferase	Enoyl reductase	Dehydratase	Malonyl/Palmitoyl (Acyl) transferase

β-chain (*fas*-2) (180 kDa)

A model for the minimal functional entity of yeast FAS contains dimers of the two chains (i.e. α_2,β_2). The model is fully consistent with data from *fas* mutants as well as experiments with cross-linking agents in mammalian FAS where a dimer is also the minimal functional unit. The way in which yeast FAS is thought to function is shown in Figure 3.7, where only one half of each dimer is functional at a given time. Hydrophobic interaction of the palmitoyl group with palmitoyl transferase at the end of the cycle of 2C additions causes a conformational change in the β-subunit. This facilitates transfer of the palmitoyl group to the palmitoyl transferase and, at the same time reduces its transacylation to β-ketoacyl-ACP synthetase. Inactivation of one half of the $\alpha_2\beta_2$ dimer leads simultaneously to activation of the other. Although the minimal functional entity of yeast FAS is $\alpha_2\beta_2$, the full yeast complex as purified normally has a molecular mass of about 2.2×10^6 Da and, therefore, consists of a $\alpha_6\beta_6$ structure.

Animal fatty acid synthetases (FAS) have been purified from a variety of sources. One of the best sources is the goose uropygial gland which is a highly specialized organ producing large amounts of fatty acids. FAS comprises up to 30% of the total soluble protein in this gland. The immunological cross-reactivity between FAS from different sources has been checked and the results indicated that there were very few common antigenic determinants on the avian and mammalian enzymes. In contrast, mammalian enzymes from various tissues and species showed some common antigenic determinants.

Animal FAS complexes consist of homodimers of molecular mass 450–550 kDa. For some years there has been considerable debate as to whether the animal enzymes are heterodimers like yeast or homodimers. The evidence that the latter is the case has come from several lines of research.

1. When dissociated into monomers, only a single peptide band is obtained on electrophoresis.
2. 4'-Phosphopantetheine was found associated with the peptide and there were approximately 2 moles per dimer.
3. A thioesterase domain could be released by partial proteinase digestion and there were 2 domains per dimer.

Studies with specific ligands and inhibitors have succeeded in giving

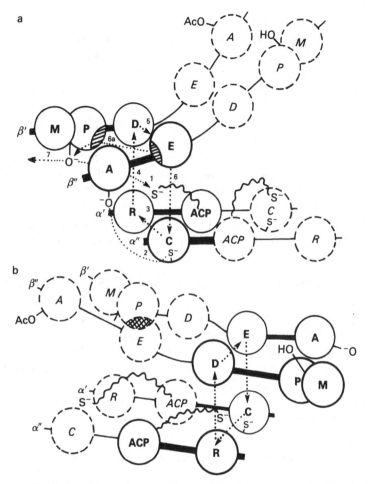

Figure 3.7 Model of intermolecular fatty acid synthetase mechanism in the $\alpha_2\beta_2$ protomer of yeast. A, acetyl transferase; E, enoyl reductase; D, dehydratase; P, palmitoyl transferase; M, malonyl transferase; C, β-ketoacyl synthase; R, β-ketoacyl reductase; ACP, acyl carrier protein. Dotted lines and arrows delineate the route taken by intermediates when sequentially processed on different FAS domains. Numbers indicate the reaction sequence. Catalytically active domains, at a specific moment, are marked by bold lines. Shaded areas on E and P domains potentially interact by hydrophobic attraction in the presence of palmitate (b). On the protomer depicted in (a) fatty acyl chain elongation occurs in one half of the $\alpha_2\beta_2$ protomer. In (b) chain termination is induced by hydrophobic interaction between E-bound palmitate and P. Subsequently, palmitate is transferred to its *O*-ester binding · site on P. Inactivation of the left half of $\alpha_2\beta_2$ simultaneously activates its right half (b). Redrawn from Schweizer (1984) with permission of the author and Elsevier Science Publishers, BV. From *Fatty Acid Metabolism and its Regulation* (1984) (ed. S. Numa), p. 73, Figure 7.

considerable information about the sites for the partial reactions along the FAS molecule. The work has been carried out in a number of laboratories but those of Kumar, Porter and Wakil have made major advances. It is known, for example, that there is considerable sequence homology for the ACPs from *E. coli* and barley and the ACP part of the multifunctional protein of rabbit. The sequence around the active-site serine of the acyl transferase of rabbit FAS is similar to that of the malonyl- and acetyl-transferases of yeast FAS and the thioesterase of goose FAS. These observations have led to the conclusion that the multifunctional forms of FAS have arisen by gene fusion. Thus, in simple terms the genes for Type II FAS enzymes would fuse to give two genes which could code for a yeast-type FAS. The latter would then fuse to give a single gene coding for the mammalian Type I FAS enzyme. However, the fusion must have occurred by independent events (rather than by the simple scheme described above) for several reasons.

1. The mass of the two yeast FAS proteins combined is 50% greater than the mammalian FAS.
2. The termination mechanisms for yeast FAS (transfer to CoA) and mammalian FAS (liberation of free acid) are different.
3. The second reductase of yeast FAS is unique in using FMN.
4. In mammalian FAS, a single active site transfers the acetyl- and malonyl-residues whereas in yeast FAS there are two.
5. Recent studies on the site of the partial reactions on mammalian FAS have shown the following arrangement for the dimer:

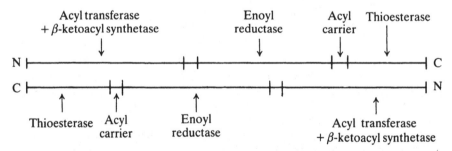

This arrangement places the enoyl reductase between β-ketoacyl synthase and ACP in mammalian FAS whereas the reductase is on a separate protein (β-unit) from the other two functions (α-unit) in yeast FAS.

The FAS from *M. smegmatis* may, however, have originated from fusion of the genes for the yeast subunits. It has a molecular mass of about 290 kDa (Table 3.8) and, like the yeast enzyme, uses FMN for its enoyl reductase and transfers the products to CoA.

Overall FAS activity in mammals is catalysed only by the dimeric form of the enzyme and not by the monomers. However, the monomeric species are capable of catalysing all of the partial reactions except the β-ketoacyl-ACP synthetase reaction. By the use of bifunctional reagents, it has been demonstrated that this reaction needs two essential thiol groups, the second coming from the acyl carrier domain of the second monomer.

(c) Termination

The typical end product of animal FAS enzymes is free palmitic acid. The cleavage of this acid from the complex is catalysed by a thioesterase which, as discussed above, is an integral part of the enzyme. However, in some mammalian tissues products of 4C–14C are released. Several factors have been found to influence chain termination.

1. In rat mammary gland, where the milk triacylglycerols contain large quantities of $C_{8:0}$ and $C_{10:0}$ acids, Smith found a second thioesterase responsible for the release of medium chain acids. A similar thioesterase II has also been found in rabbit mammary gland.
2. In contrast, goat mammary gland which also produced short and medium chain fatty acids did not contain thioesterase II. In this case a transacylase was found in the FAS and when incubated in combination with microsomes and appropriate cofactors rapid transfer of medium chain acids into triacylglycerols is found.
3. The uropygial gland FAS from a number of birds produces medium branch chain fatty acids as products. These products are released by a specific hydrolase which is absent from those birds whose end products are long branched chain fatty acids.
4. For *E. coli* early experiments showed that the specificity of the β-keto-acyl-ACP synthetase was such that palmitoyl- and vaccenoyl-ACP could not act as primers. However, additional evidence from Cronan's laboratory suggests chain elongation can continue in cells if they are starved of glycerol 3-phosphate so that the fatty acid products are not transferred into membrane phospholipids.
5. In *Mycobacterium smegmatis* termination involves transacylation of 16C–24C fatty acids to CoA. This transacylation is stimulated by polysaccharides which seem to act by increasing the diffusion of the acyl-CoA esters from FAS rather than promoting acyl transfer from ACP to CoA.

There may be other mechanisms for controlling the chain length of the fatty acids produced by the various FAS complexes. Certainly, there are plenty of other theories which have been proposed and evidence has been obtained, in some cases, *in vitro*. Moreover, there are numerous cases where unusual distributions of fatty acid products are found but about which we know very little of the mechanism of termination.

(d) Elongation

In contrast to the *de novo* formation of fatty acids by the Type I and Type II FAS enzymes, the chain lengthening of preformed fatty acids (either formed endogenously or originating from the diet) is catalysed by the Type III synthetases or elongases.

In one such system, discovered by Wakil in rat liver mitochondria, elongation occurs by addition of acetyl-CoA units with either NADPH or NADH as the reducing coenzyme. Neither CO_2 nor malonyl-CoA is involved in this pathway. Another example is the reversal of the pathway by which fatty acids are degraded (β-oxidation, section 3.3.1). The formation of the butyryl-CoA primer has already been discussed. In addition, the German biochemist, Seubert, has demonstrated the virtual reversal of β-oxidation in soluble extracts of animal mitochondria. The activities of thiolase, β-hydroxyacyl-CoA dehydrogenase and enoyl-CoA hydratase are reversed, while a fourth enzyme, NADPH-specific enoyl-CoA reductase replaces the NADH-specific acyl-CoA dehydrogenase of β-oxidation. The activity is highest with 10C substrates and is the major mechanism for fatty acid synthesis in heart muscle where the rate may be governed by the ratio of NADH to NADPH.

The principal reactions for the elongation of longer chain fatty acids are found in the membranes of the endoplasmic reticulum that are isolated as the **microsomal fraction** by ultracentrifugation of tissue homogenates. This reaction involves acyl-CoAs as primers, malonyl-CoA as the donor of 2C units and NADPH as the reducing coenzyme. An example of the microsomal elongation system is in the nervous system, where large amounts of 22C and 24C saturated fatty acids are constituents of myelin sphingolipids (Chapters 6 and 7). Before myelination begins, the activity of stearoyl-CoA elongase is hardly measurable but it rises rapidly during myelination. The mutant 'quaking mouse' is deficient in myelination and has proved to be a useful model for studies of the elongation process. The rate of elongation of $C_{18:0}$-CoA to $C_{20:0}$-CoA is normal and that of $C_{16:0}$-CoA to $C_{18:0}$-CoA is rather lower than normal in this mutant. The elongation of $C_{20:0}$-CoA, however, is very much reduced, suggesting that there are at least three elongases in this tissue.

One of the most important functions of elongation is in the transformation of dietary essential fatty acids to the higher polyunsaturated fatty acids (section 5.2.2(a)). The starting point is linoleoyl-CoA which is first desaturated to a trienoic acid. This is followed by a sequence of alternate elongations and desaturations that are described in further detail in section 5.2.2(a).

The end product of the Type II FAS of plants is palmitoyl-ACP and this serves as the substrate for an elongation system (palmitate elongase). This is soluble and only appears to differ from the Type II FAS forming palmitate in having a specific condensing enzyme, β-ketoacyl-ACP synthetase II. Because the palmitate elongase is able to chain-lengthen preformed palmitate it can be regarded as an elongase but since it usually functions as part of the *de novo* system for fatty acid production it can also be considered part of the Type II FAS by analogy with the two different condensing enzymes present in *E. coli*.

Very long chain fatty acids in plants are made by membrane-bound enzyme systems utilizing malonyl-CoA as the source of 2C units in similar fashion to the animal elongases. Acyl-CoAs have been shown to be the substrates in some of these systems and various elongases have been demonstrated which have different chain-length specificities. The production of very long chain ($> 18C$) fatty acids is required for the formation of the surface-covering layers, cutin and suberin (section 6.6.1), as well as for seed oil production in commercially-important crops such as rape and jojoba (sections 4.3 and 4.7).

(e) Branched chain fatty acids

The formation of branched chain fatty acids by the Type I FAS of the sebaceous (uropygial) glands of waterfowl has already been mentioned (section 3.2.2(b) and Table 3.8). These acids arise because of the use of methylmalonyl-CoA rather than malonyl-CoA which is rapidly destroyed by a very active malonyl-CoA decarboxylase. The utilization of methylmalonyl-CoA results in the formation of products such as 2,4,6-trimethyl lauric acid and 2,4,6,8-tetramethyl decanoic acid as major products.

A high proportion of odd chain and of various polymethyl-branched fatty acids occurs in the adipose tissue triacylglycerols of sheep and goats when they are fed diets based on cereals such as barley. Cereal starch is fermented by bacteria in the rumen to form propionate, and when the animals' capacity to metabolize propionate via methylmalonyl-CoA to succinate is overloaded, propionyl- and methylmalonyl-CoA accumulate. Garton and his colleagues showed that methylmalonyl-CoA can take the place of malonyl-CoA in fatty acid synthesis and that with acetyl- or propionyl-CoA as primers, a whole range of mono-, di- and tri-methyl branched fatty acids can be produced.

The major fatty acids in most gram-positive and some gram-negative genera are branched chain *iso* or *anteiso* fatty acids. The Type II FAS enzymes present in these bacteria make use of primers different from the usual acetyl-CoA. For example, *Micrococcus lysodeikticus* is rich in 15C acids of both the *iso* type, 13-methyl-C_{14} or the *anteiso* type, 12-methyl-C_{14}. These have been shown to originate from leucine and isoleucine respectively (Figure 3.8).

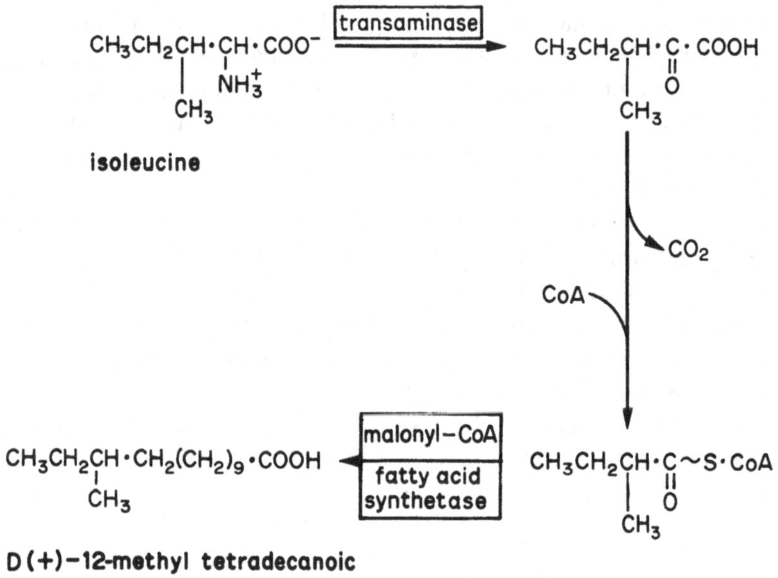

$$CH_3CH_2CH\cdot CH\cdot COO^- \xrightarrow{\boxed{transaminase}} CH_3CH_2CH\cdot C\cdot COOH$$

with NH_3^+ and CH_3 below the left structure, and O (double bond) and CH_3 below the right structure.

isoleucine

CO_2

CoA

$$CH_3CH_2CH\cdot CH_2(CH_2)_9\cdot COOH \xleftarrow[\substack{\boxed{malonyl-CoA}\\ \boxed{fatty\ acid\\ synthetase}}]{} CH_3CH_2CH\cdot C\sim S\cdot CoA$$

with CH_3 below left structure; CH_3 and O below right structure.

D(+)-12-methyl tetradecanoic

Figure 3.8 Production of an *anteiso* branched chain fatty acid in bacteria.

Another common branched chain fatty acid is 10-methyl stearic acid, tuberculostearic acid, a major component of the fatty acids of *Mycobacterium phlei*. In this case, the methyl group originates from the methyl donor *S*-adenosyl methionine, while the acceptor is oleate esterified in a phospholipid. This is an example, therefore, of fatty acid modification taking place while the acid is in an *O*-ester rather than the *S*-esters of CoA or ACP. The formation of tuberculostearic acid takes place in two steps: the intermediate being 10-methylene stearic acid which is then reduced to 10-methyl stearic acid (Figure 3.9).

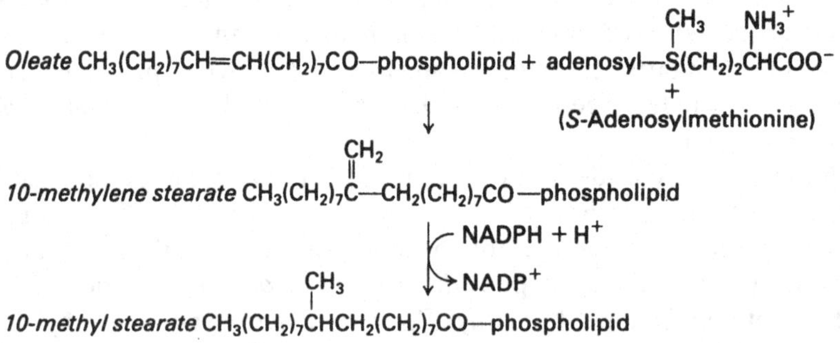

Oleate $CH_3(CH_2)_7CH=CH(CH_2)_7CO$—phospholipid + adenosyl—$S(CH_2)_2\overset{\overset{CH_3}{|}}{\underset{\underset{NH_3^+}{|}}{C}}HCOO^-$

(*S*-Adenosylmethionine)

10-methylene stearate $CH_3(CH_2)_7\overset{\overset{CH_2}{\|}}{C}$—$CH_2(CH_2)_7CO$—phospholipid

NADPH + H^+

NADP$^+$

10-methyl stearate $CH_3(CH_2)_7\overset{\overset{CH_3}{|}}{C}HCH_2(CH_2)_7CO$—phospholipid

Figure 3.9 Production of tuberculostearic acid in *Mycobacterium phlei*.

3.2.3 The biosynthesis of hydroxy fatty acids results in hydroxyl groups in different positions along the fatty chain

Hydroxy fatty acids are formed as intermediates during various metabolic pathways (e.g. fatty acid synthesis, β-oxidation) and also because of specific hydroxylation reactions. Usually the hydroxyl group is introduced close to one end of the acyl chain. However, mid-chain hydroxylations are also found – a good example being the formation of ricinoleic acid (D-12-hydroxyoleic acid; Table 3.4). This acid accounts for about 90% of the triacylglycerol fatty acids of castor oil and about 40% of those of ergot oil, the lipid produced by the parasitic fungus, *Claviceps purpurea*. In developing castor seed, ricinoleic acid is synthesized by hydroxylation of oleoyl-CoA in the presence of molecular oxygen, NADH and iron. Although the enzyme has many of the properties of a *mixed function oxidase*, the mechanism contrasts with ω-oxidation (section 3.3.3) in requiring an acyl thiol ester substrate and in not involving cytochrome P_{450} as a cofactor. In contrast to the method of hydroxylation in castor seed, the pathway in *Claviceps* involves hydration of linoleic acid under anaerobic conditions. Thus the hydroxyl group in this case comes from water and not from molecular oxygen.

α-Oxidation systems producing α-hydroxy (2-hydroxy) fatty acids have been demonstrated in microorganisms, plants and animals. In plants and animals these hydroxy fatty acids appear to be preferentially esterified in sphingolipids. α-Oxidation is described more fully in section 3.3.2.

ω-Oxidation involves a typical mixed-function oxidase. The major hydroxy fatty acids of plants have an ω-OH and an in-chain OH group (e.g. 10,16-dihydroxypalmitic acid). Their synthesis seems to involve ω-hydroxylation with NADPH and O_2 as cofactors, followed by in-chain hydroxylation with the same substrates. If the precursor is oleic acid then the double bond is converted to an epoxide which is then hydrated to yield 9,10-hydroxy groups. These conversions involve CoA esters. ω-Oxidation is discussed further in section 3.3.3.

3.2.4 The biosynthesis of unsaturated fatty acids can use anaerobic or aerobic mechanisms

(a) Anaerobic desaturation

Basically, there are two completely different pathways by which unsaturated fatty acids are produced. In an earlier section, we mentioned that the fatty acid synthetase of *E. coli*, in contrast to the mammalian and yeast synthetases, produced unsaturated as well as saturated acids. An idea of how this might arise was first put forward by Bloch, who was studying the biosynthesis of unsaturated fatty acids in the anaerobic bacterium *Clostridium butyricum*. This organism, like almost all other bacteria, produces only monounsaturated fatty acids. Two pairs of monoene isomers are produced: Δ7 and Δ9 hexadecenoic and Δ9 and Δ11 octadecenoic acids. The occurrence of these isomers can be explained if there are branch

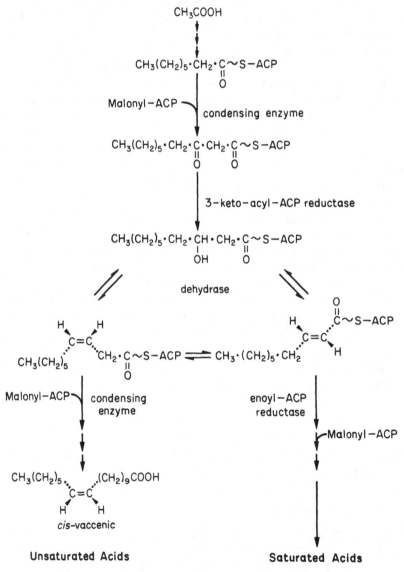

Figure 3.10 The anaerobic pathway for unsaturated fatty acid synthesis.

points in fatty acid synthesis at $C_{8:0}$ and $C_{10:0}$ according to the scheme in Figure 3.10. At these points, the normal *trans*-2-enoyl-ACP is isomerized to *cis*-3-enoyl-ACP which is not a substrate for the enoyl-ACP reductase, but is capable of elongation. Thus the *trans*-isomer formed at the branch point gives rise to a long chain saturated acid while the *cis*-isomer yields a long chain *cis*-unsaturated fatty acid. The positions of the double bonds in the final products, therefore, depend only on the positions of branch points.

Subsequent studies on the mechanism of the branching process have been done using *E. coli*, one of the most extensively studied organisms in the biochemical world. The fatty acid synthetase contains a β,γ-dehydrase which has activity with β-OH-decanoyl-ACP and, therefore, produces *cis*-3-decenoyl-ACP. This β,γ-dehydrase was first revealed when mutants were discovered which could synthesize saturated fatty acids but required unsaturated fatty acids for growth. The enzyme can also catalyse the interconversion of *cis*-3 to *trans*-2-decenoate and can be purified and separated from the synthetase by making use of the fact that the *dehydrase* is stable to heating at 50°C for 10 min, whereas after 2 min of this treatment the synthetase is completely denatured.

(b) Aerobic desaturation

Monounsaturated fatty acids
The anaerobic mechanism for monounsaturated fatty acid synthesis is by no means the most important. Indeed, it has only been observed in a limited number of bacteria – the Eubacteriales in particular. By far the most widespread pathway is by an oxidative mechanism, also discovered by Bloch's team, in which a double bond is introduced directly into the preformed saturated long chain fatty acid with O_2 and a reduced compound (such as NADH) as cofactors. This pathway is almost universal and is used by bacteria, yeasts, algae, higher plants, protozoa and animals. Apparently, the two pathways are mutually exclusive because no organism has yet been discovered which contains aerobic and anaerobic mechanisms of desaturation. Most of the acids produced have a $\Delta 9$ double bond. Exceptions are $\Delta 7$ bonds in some algae, $\Delta 5$ and $\Delta 10$ monoenoic acids in Bacilli and a $\Delta 6$ acid (petroselenic) in some plants.

The pathway was first demonstrated in yeast. Cell-free preparations could catalyse the conversion of palmitic into palmitoleic acid (hexadec-9-enoic acid, $\Delta 9$-$C_{16:1}$) only if both a particulate fraction (microsomes) and the supernatant fraction were present. The membrane fraction alone could perform the dehydrogenation provided that the substrate was the acyl-CoA thiolester. The supernatant contained the acid : CoA ligase to activate the fatty acid. More recently another protein fraction in the soluble cytoplasm has been found to stimulate desaturation. This is probably a fatty acid binding protein which regulates the availability of fatty acid or fatty acyl-CoA for lipid metabolizing enzymes. Bloch found that the cofactors for the desaturation were NADH or NADPH and molecular oxygen which suggested to him a mechanism similar to many mixed-function oxygenase reactions. These cofactor requirements are now known to be almost universal for oxidative desaturation reactions.

It has been particularly difficult and slow to obtain a detailed understanding of the biochemistry of the desaturase enzymes. Not only are they bound in an insoluble form to membranes, but the substrates are micellar (section 3.1.8) at concentrations that are suitable for studies *in vitro*. It was only when methods for solubilizing membranes with detergents were developed that the stearoyl-CoA desaturase enzyme was purified and a better understanding of the enzymic complex emerged. Work by Sato's group in Japan and Holloway in the USA, has identified three component proteins of the complex: a flavoprotein, NADH-cytochrome b_5 reductase (I in Figure 3.11); a haem-containing protein, cytochrome b_5 (II in Figure 3.11) and the desaturase itself (III in Figure 3.11) which, because of its inhibition by low concentrations of cyanide, is sometimes referred to as the **cyanide-sensitive factor** (CSF). Strittmatter's group in the USA have purified this latter protein (and identified its gene), which is a single polypeptide chain of 53 kDa containing one non-haem iron atom per molecule of enzyme. The iron can be reduced in the absence of stearoyl-CoA, by NADH and the electron transport proteins, and when it is removed, enzymic activity is lost. A possible function of the electron transport proteins may be to transfer electrons from NADH to the non-haem iron of the desaturase protein. The reduced iron would then bind molecular oxygen which, after activation by a mechanism that is by no means understood, brings about the desaturation of enzyme-bound stearoyl-CoA.

In Figure 3.11 we have illustrated how the complex may fit into the lipid bilayer of the endoplasmic reticulum membrane. Proteins I and II are each folded to form a globular hydrophilic domain containing the active centre and a hydrophobic tail that serves to anchor the single polypeptide chain of the protein in the phospholipid bilayer of the endoplasmic reticulum. For example, cytochrome b_5 has a catalytic hydrophilic region of 85 residues (including the NH_2-terminal) and a hydrophobic COOH-terminal tail of approximately 40 amino acids. The greater difficulty of solubilization of the terminal desaturase protein suggests that this protein is more intimately associated with the membrane than either of the electron transport proteins.

Although the desaturation reaction has all the characteristics of a mixed-function oxygenation, nobody has ever successfully demonstrated an hydroxylation as an intermediate step in double bond formation. In spite of our lack of knowledge of the mechanism, certain details have emerged

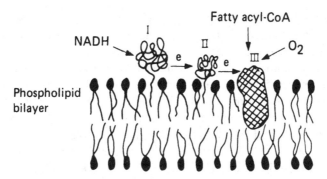

Figure 3.11 Diagrammatic representation of the Δ9 desaturase complex. The three proteins, NAD-cytochrome b reductase (I), cytochrome b (II) and the cyanide-sensitive factor (III) are shown in the phospholipid bilayer of the endoplasmic reticulum. Reproduced with kind permission of Dr R. Jeffcoat (1977) and The Biochemical Society from *Biochemical Society Transactions*, **5**, Figure 1, p. 813.

concerning the stereochemistry of the dehydrogenation. Schroepfer and Bloch in the USA and James and coworkers in the UK have demonstrated that the D-9 and D-10 (*cis*) hydrogen atoms are removed by animal, plant and bacterial systems. Experiments with deuterium-labelled stearate substrates showed isotope effects at both the 9 and 10 positions which are consistent with the concerted removal of hydrogens rather than a mechanism involving a hydroxylated intermediate followed by dehydration.

Studies on the chain length specificity of Δ9 desaturases from rat liver show that there are two enzymes. One (which is widely distributed in other mammalian tissues) has maximum rates with a 18C substrate. The chain length effectiveness is in the order: 18C > 17C > 16C > 15C > 14C. The activity with 19C and 20C fatty acids is low but perceptible. The second enzyme has a maximum velocity at 14C, the order in this case being: 14C > 13C > 12C > 11C. In all cases the double bond was introduced at carbon atoms 9 and 10 counting from the carboxyl end, so that clearly the initial reaction between substrate and enzyme is at the carboxyl end of the fatty acid chain.

The Δ9 desaturases in plants and algae (and *Euglena gracilis*) appear to use stearoyl-ACP as substrate. The enzyme from developing safflower seeds has been partly purified in Stumpf's laboratory in California by the use of affinity chromatography on Sepharose columns to which ACP had been linked. The enzyme can use either NADPH or photosystems I and II with ferredoxin acting as the intermediate electron carrier. The safflower enzyme is soluble and is absolutely specific for stearoyl-ACP – two properties which contrast with those of the Δ9 desaturases of animals.

The naturally occurring cyclopropenoid fatty acid, sterculic acid (Table 3.4) when supplied to the diet of laying hens, gives rise to a condition known as pink-white disease. The effect is believed to be due to an accumulation of stearic acid-containing lipids in the membrane of the egg (instead of oleic acid-containing lipids) that alters the permeability of the membrane so that one of the pink ferroproteins passes from the yolk into the white. The only known biochemical effect of sterculic acid is to block the conversion of stearic into oleic acid in animals and yeasts – hence the change in the membrane lipids. Jeffcoat has now shown that the active inhibitor is sterculoyl-CoA and not sterculic acid itself. However, we have to explain how sterculic acid can by synthesized in plants without apparently blocking the stearoyl desaturase. We know that in higher plant preparations *in vitro*, added long chain fatty acids are converted into CoA thiolesters but not into ACP thiolesters and this is why added stearic acid is never converted into oleic acid. Added oleic acid, however, is known to be converted into sterculic acid and hence this reaction is presumed to involve a CoA thiolester. Sterculoyl-*S*-CoA would not (by analogy) be expected to be an inhibitor of the stearoyl desaturase in higher plants since this reaction involves stearoyl-*S*-ACP as substrate. Alternatively, sterculate may be synthesized outside the chloroplast and, hence, not inhibit the plastid-localized stearate desaturase.

An unusual monounsaturated fatty acid, specifically linked to phosphatidyl glycerol and found in chloroplasts, is *trans*-3-hexadecenoic acid (Table 3.2). Palmitic acid is its precursor and oxygen is required as co-factor. When labelled *trans*-3-hexadecenoic acid is incubated with chloroplast preparations, it is not specifically esterified in phosphatidylglycerol but is either randomly esterified in all chloroplast lipids or reduced to palmitic acid. These results could be explained if the direct precursor of *trans*-3-hexadecenoic acid were palmitoyl-phosphatidylglycerol and not palmitoyl-*S*-CoA or palmitoyl-*S*-ACP. However, the precise details of the reaction have yet to be worked out.

Polyunsaturated fatty acids
Although most bacteria are incapable of producing polyunsaturated fatty acids, other organisms, including many cyanobacteria and all Eukaryotes, can. These acids usually contain methylene interrupted double bonds, i.e. they are separated by a single methylene group. Animal enzymes normally introduce new double bonds between an existing double bond and the carboxyl group (Figure 3.12); plants normally introduce the new double bond between the existing double bond and the terminal methyl group (cf. Figure 3.13). Some primitive organisms, intermediate in the evolutionary scale (such as the phytoflagellate *Euglena*), have the ability to desaturate in either direction.

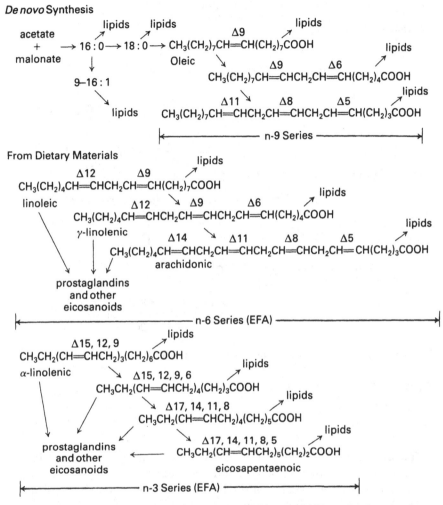

Figure 3.12 Important pathways for unsaturated fatty acid formation in animals.

We shall describe polydesaturations in plants first. The reason for this is that one of the most abundant polyunsaturated acids produced by plants, linoleic acid (*cis, cis*-9, 12–18 : 2) cannot be made by animals* yet this acid is necessary to maintain animals in a healthy condition. For this reason linoleic acid must be supplied in the diet from plant sources, and in order to discuss adequately the metabolism of polyunsaturated fatty acids in animals, it is first necessary to understand their formation in plants. Acids of the linoleic family are known as essential fatty acids and will be discussed in later sections.

---

* It is always dangerous in biochemistry to make dogmatic statements and, indeed, some protozoa and a few species of insects are capable of forming linoleic acid.

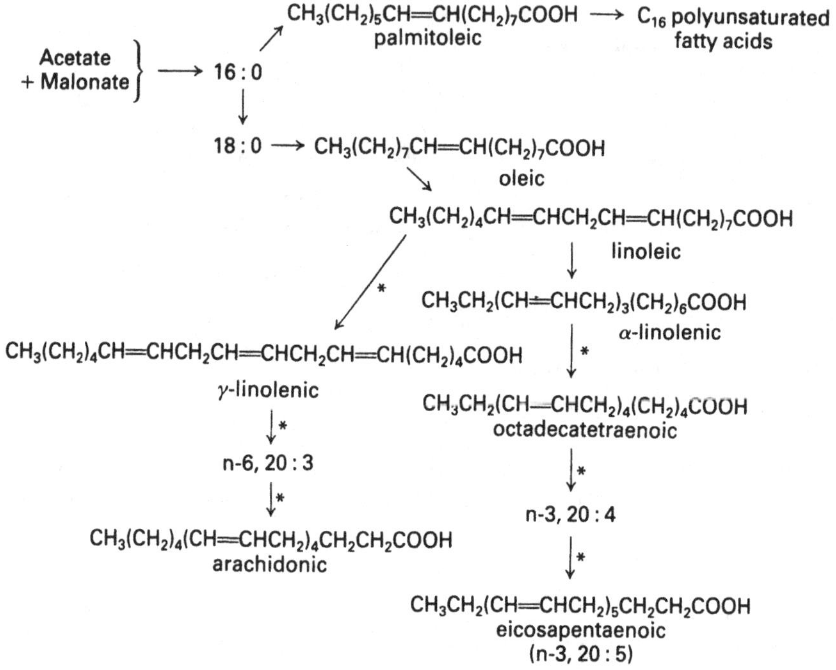

$$CH_3(CH_2)_5CH{=}CH(CH_2)_7COOH \longrightarrow C_{16} \text{ polyunsaturated}$$
palmitoleic   fatty acids

Acetate + Malonate $\longrightarrow$ 16 : 0

$$18:0 \longrightarrow CH_3(CH_2)_7CH{=}CH(CH_2)_7COOH$$
oleic

$$CH_3(CH_2)_4CH{=}CHCH_2CH{=}CH(CH_2)_7COOH$$
linoleic

$$CH_3CH_2(CH{\doteq}CHCH_2)_3(CH_2)_6COOH$$
α-linolenic

$$CH_3(CH_2)_4CH{=}CHCH_2CH{=}CHCH_2CH{=}CH(CH_2)_4COOH$$
γ-linolenic

$$CH_3CH_2(CH{-}CHCH_2)_4(CH_2)_4COOH$$
octadecatetraenoic

n-6, 20 : 3

n-3, 20 : 4

$$CH_3(CH_2)_4(CH{=}CHCH_2)_4CH_2CH_2COOH$$
arachidonic

$$CH_3CH_2(CH{=}CHCH_2)_5CH_2CH_2COOH$$
eicosapentaenoic
(n-3, 20 : 5)

* Indicates a pathway found in high levels in marine algae but less commonly in other algae or plants.

Figure 3.13 Major pathways for polyunsaturated fatty acid synthesis in plants and algae.

The precursor for polyunsaturated fatty acid formation in plants and algae is oleate. The next double bond is introduced at the 12,13-position (Δ12 desaturase) to form linoleate followed by desaturation at the 15,16-position (Δ15 desaturase) to form α-linolenic acid (all *cis*-9,12,15–18:3) as summarized in Figure 3.13. With the exception of some Cyanobacteria, α-linolenic acid is the most common fatty acid found in plants and fresh-water algae. In marine algae, highly unsaturated 20C acids are predominant, the principal of which (arachidonic and eicosapentaenoic) are made by the pathways shown in Figure 3.13.

The possibility that desaturation could occur in fatty acyl chains esterified in complex lipids was first suggested by experiments with the phyto-flagellate, *Euglena gracilis*. This organism can live as a plant or an animal and can synthesize both plant and animal types of polyunsaturated fatty acids. The animal type of fatty acids accumulate in the phospholipids and the plant types in the galactolipids. It proved impossible to demonstrate plant-type desaturations *in vitro* when acyl-CoA or acyl-ACP thiolesters were incubated with isolated cell fractions. Either the desaturase enzymes were labile during the fractionation of the plant cells, or the substrates needed to be incorporated into the appropriate lipids before desaturation could take place. The next series of experiments were done with *Chlorella vulgaris*, a green alga that produces a very simple pattern of plant type lipids. When cultures of the algae were labelled with ^{14}C-oleic acid as a precursor of the 18C polyunsaturated fatty acids, labelled linolenic and linoleic acids were produced and the label was only located in the phosphatidylcholine fraction. Next, synthetic ^{14}C-labelled oleoyl-phosphatidylcholine was tested as a substrate for desaturation: the only product formed was linoleoyl-phosphatidylcholine. An important loophole that had to be closed was the possibility that during the incubation, labelled oleic acid might be released, activated to the CoA or ACP thiolester, desaturated as a thiolester and then re-esterified to the same complex lipid very rapidly. Appropriate control experiments eliminated this possibility.

Desaturations involving lipid-bound fatty acids have also been shown to occur in the mould *Neurospora crassa*, various yeasts such as *Candida utilis* and *Candida lipolytica*, higher plants and several animal tissues.

One of the best studied systems has been that in the leaves of higher plants. It will be recalled that synthesis of fatty acids *de novo* in plants occurs predominantly in the plastids. Fatty acid synthetase forms palmitoyl-ACP which is elongated to stearoyl-ACP and then desaturated to oleoyl-ACP. The latter can then be hydrolysed and re-esterified by the chloroplast envelope to oleoyl-CoA. Although oleoyl-CoA can be used as a substrate for *in vitro* systems and will be desaturated rapidly to linoleate in the presence of oxygen and NADH, both substrate and product accumulate in phosphatidylcholine. Indeed, careful experiments by Roughan and Slack in New Zealand, Stymne and Stobart (Sweden, UK) and others have provided considerable evidence that the actual desaturase substrate is *sn*-1-acyl-2-oleoyl-phosphatidylcholine. Although phosphatidylcholine may also be the substrate for linoleate desaturation in a few systems, for leaf tissue the final desaturation appears usually to utilize another lipid, monogalactosyl-diacylglycerol, as a substrate. There is indirect evidence *in vivo* for this desaturation and also a direct demonstration of the desaturation of linoleoyl-monogalactosyldiacylglycerol by chloroplasts. These pathways are indicated in Figure 3.14.

One difficulty with the scheme is the necessity for movement of oleate out of and linoleate back into the plastid. The transport of oleate and its esterification into phosphatidylcholine is solved by the participation of oleoyl-CoA which is water-soluble. An explanation for the movement of linoleate back into the chloroplast was found when Yamada in Tokyo noted that incubation of isolated plastids with ^{14}C-linoleoyl-phosphatidylcholine allowed desaturation to linolenate when a soluble phospholipid exchange

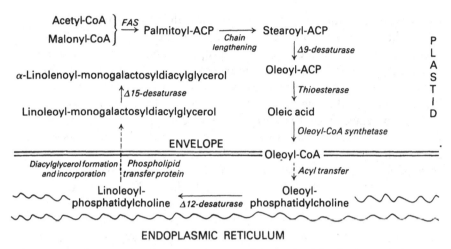

Figure 3.14 Proposed pathway for the overall formation of α-linolenate in leaves from plants operating the 'eukaryotic pathway' of lipid synthesis.

protein (isolated from plants) was also added. It is envisaged that such proteins allow movements of complex lipids within cells – including animal as well as plant cells.

The pathway for polyunsaturated fatty acid formation in plants which was discussed above (Figure 3.14) is probably that used by the majority of plants. It has been termed the **eukaryotic pathway** because it involves the participation of extra-chloroplastic compartments and particularly because 18C fatty acids are esterified in the *sn*-2 position of participating lipids (as they would be in other eukaryotes like animals). By contrast, desaturation (and formation of chloroplast lipids) continues within the chloroplast in some plants and such mechanisms are termed **prokaryotic**. For the latter desaturations, monogalactosyldiacylglycerol is used as substrate – allowing the formation of α-linolenate and, also, hexadecatrienoate (16:3) at its *sn*-2 position. An example of a plant operating the prokaryotic pathway would be spinach (see Table 6.6). However, the most important point to stress is that for all plants, polyunsaturated fatty acids are made on complex lipid substrates.

Although complex lipid substrates have been studied best in plants, the conversion of eicosatrienoyl-phosphatidylcholine into arachidonoyl-phosphatidylcholine is an example of a similar reaction in animals (e.g. rat liver). Interestingly, some yeasts have been shown to contain oleoyl-CoA as well as oleoyl-phosphatidylcholine desaturases.

The inability of animals to desaturate oleic acid towards the methyl end of the chain gives rise to distinct families of polyunsaturated fatty acids that are not interconvertible. In general, polyunsaturation in animals is accomplished by three separate desaturases, designated as Δ4, Δ5 and Δ6 because they introduce double bonds between carbon atoms 4–5, 5–6 and 6–7 respectively (Figures 3.12 and 5.1). None of these enzymes has been studied in as much detail as the Δ9 desaturase, but there is evidence for the involvement of cytochrome b_5 and NADH-cytochrome b_5 reductase. They also require molecular oxygen, a reduced nicotinamide nucleotide, appear to be sensitive to cyanide and usually use acyl-CoA substrates. The most important substrates for the first polydesaturation are oleic acid (either produced by the animal or coming from the diet), linoleic and α-linolenic acids (only from the diet). The structural relationships between the families of fatty acids that arise from these three precursors are most easily recognized by using the system that numbers the double bonds from the methyl end of the chain. Hence, oleic acid gives rise to a series of (n-9) fatty acids, the linoleic family is (n-6) and the α-linolenic family is (n-3). The first desaturation is at Δ6 and the sequence is one of alternate elongations (Type III synthetase, section 3.2.2(d)) and desaturations in the order Δ6, Δ5, Δ4, (Figures 3.12 and 5.2). The rate limiting step in this pathway is the Δ6 desaturase and some physiological consequences of this are discussed in section 5.2.2(a).

Interconversions within a family can occur not only by chain elongation, but by a chain shortening by two-carbon units. This is called retroconversion and is a limited form of β-oxidation (section 3.3.1).

Further discussion of the role of polyunsaturated fatty acid synthesis in the formation of prostaglandins, thromboxanes and prostacyclins will be found in section 3.4.

3.2.5 Biohydrogenation of unsaturated fatty acids takes place in rumen microorganisms

Desaturation of an acyl chain is a reaction widespread in Nature. The reverse process, namely the hydrogenation of double bonds, is found in only a few organisms. These organisms are commonly found in the rumens of cows, sheep and other ruminant animals. Linoleic acid, for example, can be hydrogenated by rumen flora (anaerobic bacteria and protozoa) to stearic acid by the series of reactions shown in Figure 3.15.

First of all the substrate fatty acids must be released from leaf complex lipids by the action of acyl hydrolases (section 7.2.5). The first reaction of the unesterified linoleic acid involves isomerization of the *cis*-12,13 double bond to a *trans*-11,12 bond which is then in conjugation with the *cis*-9,10 double bond. The enzyme responsible for this isomerization has been partially purified from the cell envelope of *Butyrivibrio fibrisolvens* and can

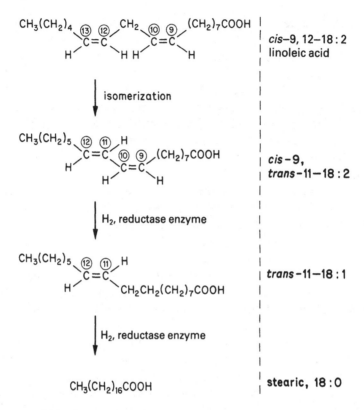

Figure 3.15 Biohydrogenation.

act on α-linolenic as well as linoleic acid. Next, hydrogen is added across the *cis*-9,10 bond to form *trans*-vaccenic acid which is further reduced to stearic acid. Analogous reactions occur with fatty acids other than linoleic acid but the positions of the *cis*- and *trans*- bonds will, of course, be different.

In spite of the high activity of rumen microorganisms (including the further breakdown of the fatty acids by oxidation) ruminants do not appear to suffer from essential fatty acid (EFA) deficiency. The amount of unchanged EFA passing through the rumen (up to 4% of dietary intake) is sufficient for the needs of the animal (section 5.4). Hydrogenation can, however, be reduced by giving ruminants 'protected fats', thereby enriching their tissues with polyunsaturated fatty acids (section 5.1.2).

An interesting example of hydrogenation occurs in *Bacillus cereus* which can reduce oleic to stearic acid. This reductase is induced by an increase in growth temperature and seems to be involved in the overall control of membrane fluidity (section 8.2).

3.2.6 The biosynthesis of cyclic acids provided one of the first examples of a complex substrate for fatty acid modifications

The only ring structures we shall discuss are the cyclopropanes and cyclopropenes.

The methylene group in cyclopropane acids originates from the methyl group of methionine in *S*-adenosyl methionine ('active methionine'). This is the same methyl donor which is involved in the formation of 10-methylene stearic and 10-methyl stearic acid from oleic acid (section 3.2.2(e)). The acceptor of the methyl group is likewise an unsaturated fatty acid. Thus, *cis*-vaccenic acid gives rise to *lactobacillic acid*, while oleic acid yields *dihydrosterculic acid* (see Table 3.4). These reactions occur in a number of bacteria and in certain families of higher plants, *Malvaceae* and *Sterculaceae*.

When Law and his colleagues purified cyclopropane synthetase from *Clostridium butyricum*, they found that the enzyme would catalyse the formation of cyclopropane fatty acids from ^{14}C-labelled methionine only if phospholipids were added in the form of micellar solutions. They discovered that the real acceptor for the methylene group was not the free monounsaturated fatty acid or its CoA or ACP thiol ester, but phosphatidyl ethanolamine – the major lipid of the organism (see Figure 3.16).

The biosynthesis of cyclopropane and the related cyclopropene acids in higher plants has been studied by experiments with radioactive precursors. In this method, a supposed precursor for the compounds being studied is supplied to the plant and its incorporation into more complex molecules and/or conversion to products is studied at successive time intervals. The sequence in which the radiolabel appears in different compounds can be used to deduce the pathways by which they are made. Thus it has been deduced that oleic acid gives rise to the cyclopropane derivative of stearic acid (dihydrosterculic acid). The latter can either be shortened by

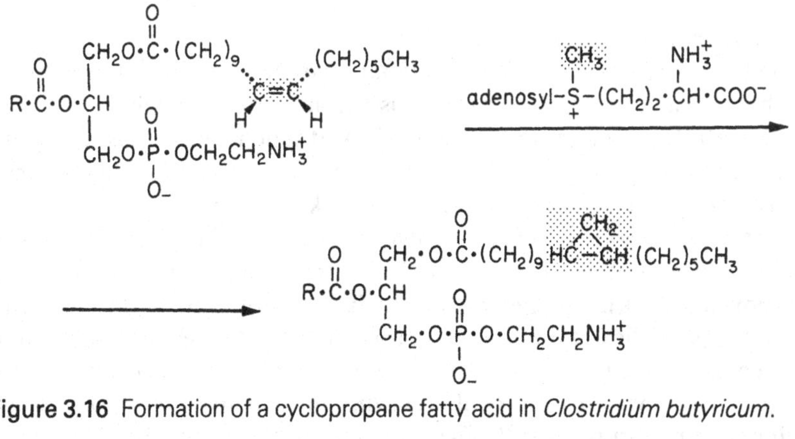

Figure 3.16 Formation of a cyclopropane fatty acid in *Clostridium butyricum*.

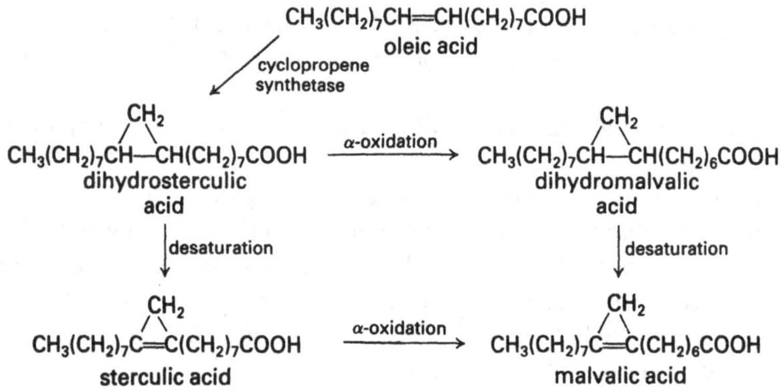

$$CH_3(CH_2)_7CH\!=\!CH(CH_2)_7COOH$$

oleic acid

cyclopropene
synthetase

$$\overset{\displaystyle CH_2}{\overset{\displaystyle /\backslash}{CH_3(CH_2)_7CH\!-\!CH(CH_2)_7COOH}}$$
dihydrosterculic
acid

α-oxidation →

$$\overset{\displaystyle CH_2}{\overset{\displaystyle /\backslash}{CH_3(CH_2)_7CH\!-\!CH(CH_2)_6COOH}}$$
dihydromalvalic
acid

↓ desaturation

↓ desaturation

$$\overset{\displaystyle CH_2}{\overset{\displaystyle /\backslash}{CH_3(CH_2)_7C\!=\!C(CH_2)_7COOH}}$$
sterculic acid

α-oxidation →

$$\overset{\displaystyle CH_2}{\overset{\displaystyle /\backslash}{CH_3(CH_2)_7C\!=\!C(CH_2)_6COOH}}$$
malvalic acid

Figure 3.17 Synthesis of sterculic and malvalic acids by plants.

α-oxidation or desaturated to give sterculic acid (Figure 3.17). Desaturation of the α-oxidation product (dihydromalvalic acid) similarly yields malvalic acid (8,9-methylene-8–17 : 1).

3.2.7 The control of fatty acid synthesis can take place at a number of enzyme steps

(a) Acetyl-CoA carboxylase in animals

It is now accepted generally that the control of acetyl-CoA carboxylase activity plays a key role in the overall regulation of fatty acid synthesis, especially in animal tissues. In support of this view is the fact that tissue concentrations of malonyl-CoA vary in parallel with the amount of inhibition by the overall end products (fatty acyl-CoAs). The enzyme's activity can be regulated in a number of ways (Table 3.9).

Activation and inhibition
Acetyl-CoA carboxylase from animal tissues requires a hydroxy tricarboxylic acid, such as citrate or isocitrate, for its catalytic activity. Because citrate is a precursor of acetyl-CoA, it functions as a positive feed-forward activator. Fatty acyl-CoAs which can be regarded as end products of fatty acid synthesis, act as negative feedback inhibitors.

Citrate acts on both of the partial reactions of the carboxylase. However, at low concentrations (about 10 μM) citrate acts primarily on the uncarboxylated form of the enzyme to induce polymerization. In turn, carboxylation of the enzyme's biotin causes a marked decrease in the affinity of the enzyme for citrate. Thus, the main role of citrate in the catalysis is to keep the carboxylated form of the enzyme in its active conformation by shifting the equilibrium from the inactive to the active species of the enzyme.

Table 3.9 Methods for regulating acetyl-CoA carboxylase activity in animals

Mediator	Comments
Hydroxytricarboxylic acid (e.g. citrate)	Polymerization induced Activity much increased
Long chain acyl-CoAs	Negative feedback inhibitors Active at very low concentrations
Phosphorylation/dephosphorylation	Dephosphorylated enzyme form much more active
Enzyme synthesis degradation	Used for long term regulation. Enzyme protein amounts decreased by diabetes or by fasting and increased by re-feeding low-fat diets. Protein translation rates generally more important than catabolism

Long chain acyl-CoAs inhibit mammalian acetyl-CoA carboxylases. The inhibition is competitive towards citrate and non-competitive towards the three substrates acetyl-CoA, ATP and bicarbonate. One mole of palmitoyl-CoA was found completely to inhibit one mole of enzyme. The inhibition constant for this is 5.5 nM which is three orders of magnitude lower than the critical micellar concentration (CMC) for palmitoyl-CoA. Saturated fatty acyl CoAs of 16–20C are most effective and unsaturated acyl-CoAs are considerably less inhibitory. (For a discussion of CMC and its significance, see section 7.2.1.)

It is, of course, important to consider whether the cellular concentration of the putative regulator molecules – citrate and acyl-CoAs – are sufficient to allow the physiological control of acetyl-CoA carboxylase. Indeed, cellular concentrations of these molecules under various metabolic conditions are consistent with their role in regulation. For example, it has been found that in liver 50–75% of the cellular citrate is in the cytoplasm, which is where the acetyl-CoA carboxylase is. The estimated concentration of 0.3–1.9 mM is close to the concentration needed for half maximal activation. Furthermore, glucagon or (dibutyryl)cAMP, which lower fatty acid synthesis, also reduce the cytoplasmic concentration of citrate. The cytoplasmic concentration of acyl-CoAs is difficult to measure because they bind to intracellular proteins and membranes. However, total cellular concentrations up to 150 μM have been reported, which are more than sufficient completely to inhibit the carboxylase.

Phosphorylation/dephosphorylation
Acetyl-CoA carboxylase can be isolated from animals that have been

injected with ^{32}P-orthophosphate and found to contain radioactivity. Furthermore, if the purification is carried out from tissues in different physiological states, those preparations with the highest specific activity contain the lowest amounts of phosphate . mg^{-1} protein. In addition phosphorylation of the purified enzyme with a cAMP-dependent protein kinase leads to the incorporation of additional phosphate per subunit and a severe decrease in activity.

Highly phosphorylated acetyl-CoA carboxylase can be dephosphorylated with appropriate protein phosphatases. The release of phosphate is accompanied by a large increase in activity. Although up to 6 moles of phosphate can be present per mole of enzyme, experiments on the release of different phosphate molecules have shown that probably only one (or at most two) of the phosphates are involved in regulation. The regulatory domain appears to be at one end of the polypeptide chain.

Synthesis and degradation

Long-term regulation of acetyl-CoA carboxylase in animals can be due to changes in enzyme amounts. The tissue concentration of the carboxylase protein has been shown to vary with the rate of fatty acid synthesis under a variety of nutritional, hormonal, developmental or genetic conditions. Measurements with specific antibodies have shown that, for example, fasted rats have only one quarter the normal levels of the liver enzyme while genetically obese mice contain four times as much compared to control animals. The amount of an enzyme protein accumulating is due to the net rates of synthesis and degradation. Depending on the specific trigger for changes in acetyl-CoA carboxylase levels, one or both of these factors may be altered. For example, the increase in enzyme content in fasted/re-fed rats is due only to changes in the rate of its synthesis (Table 3.10). By contrast,

Table 3.10 Relative acetyl-CoA carboxylase levels and rate of enzyme synthesis in different conditions

	Content	Enzyme synthesis
Rat		
Normal	1	1
Fasted	0.25	0.50
Re-fed	4	4
Alloxan diabetic	0.50	0.50
Mouse		
Normal	1	1
Obese mutant	4	3

the decrease in enzyme content in fasted animals is due both to diminished synthesis and to accelerated breakdown. It appears that a metabolite responsible for a major control in the synthetic rate of the enzyme is unesterified fatty acid.

(b) Acetyl-CoA carboxylase regulation in other organisms

The acetyl-CoA carboxylase from yeast is also inhibited by acyl-CoAs but unlike the animal enzyme is unaffected by citrate. In some mutant strains an activation by fructose-1,6-bisphosphate has been demonstrated. Like the mammalian carboxylase, the enzyme from yeast is also regulated by changes in amount. An interesting example of the role of fatty acyl-CoA in mediating the rate of synthesis of acetyl-CoA carboxylase has been demonstrated in fatty acid mutants of *C. lipolytica*. These mutants contained no apparent fatty acyl-CoA synthetase and, hence, were unable to grow on exogenous

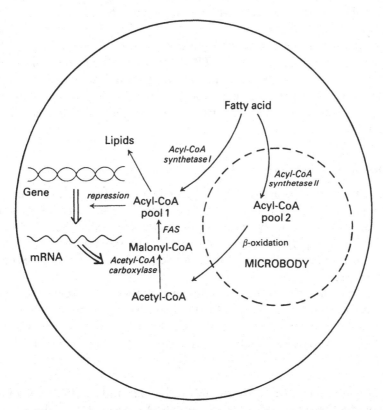

Figure 3.18 Interaction of acyl-CoA pools with fatty acid metabolism in the hydro-carbon-utilizing yeast *Candida lipolytica*. Re-drawn from Numa (*1981*). Reproduced with kind permission of Dr S. Numa (1981) and Elsevier Trends Journals from *Trends in Biochemical Science*, p. 115, Figure 2.

fatty acids when their own fatty acid synthesis was blocked by inhibitors. However, further examination showed that the mutants did have one type of fatty acyl-CoA synthetase (called II) in common with normal cells but lacked fatty acyl-CoA synthetase I which was needed for membrane lipid synthesis. Fatty acyl-CoA synthetase II is used for activating fatty acids destined for β-oxidation. Thus, these two acyl-CoA synthetases are responsible for generating two pools of acyl-CoAs in different parts of the cell. The acyl-CoAs formed by acyl-CoA synthetase I are in the cytosol and these cause repression of acetyl-CoA carboxylase (Figure 3.18).

Regulation of *E. coli* acetyl-CoA carboxylase, on the other hand, seems to be by a completely different system. In this bacterium (and others such as *P. citronellolis*) highly phosphorylated guanosine nucleotides seem to be important. These compounds (Guanine-5'-diphosphate-3'diphosphate, ppGpp and Guanine-5'-triphosphate-3-diphosphate,pppGpp) act by inhibiting the carboxyltransferase.

In plants, acetyl-CoA carboxylase levels seem to play an important role in the overall rates of fatty acid synthesis *de novo*. The purified enzyme is unaffected by citrate or a whole range of organic acids. There is evidence that *in vivo* it may be controlled by a combination of factors. Thus, the change in chloroplast pH, Mg^{2+}, ATP and ADP levels which occur during illumination have been shown by Hawke to cause an increase of about 24-fold in the activity of the maize enzyme *in vitro*. This is almost exactly the same as the observed increase in fatty acid synthesis *in vivo* when leaves are illuminated.

(c) Regulation of fatty acid synthetase

In the same way as the actual levels of acetyl-CoA carboxylase protein can be changed in animals under dietary influence so can that of fatty acid synthetase (Table 3.11). Up to 20-fold differences have been observed in the levels of, for example, liver fatty acid synthetase between starved and re-fed animals. Alterations in both synthesis and degradation of the enzyme seem to be involved.

It should be noted that dietary factors which affect fatty acid synthetase levels do not affect all tissues equally. While the liver is highly influenced by such regulation, the fatty acid synthetase of brain is unaffected. This is just as well for it would be extremely disadvantageous for a young animal to have its brain development influenced dramatically by its day-to-day nutritional state!

In addition, a number of hormones including insulin, glucocorticoids, glucagon, theophylline and estradiol have also been found to produce acute alterations in levels of various mammalian fatty acid synthetases (Table 3.11). Longer term factors can also play a role in determining enzyme levels. Thus, for example, the fatty acid synthetase increases in mammary gland

Table 3.11 Effect of nutritional state or hormones on liver FAS levels

Cause increase	Cause decrease
Re-feeding	Starvation
Insulin	Alloxan-diabetes
β-Estradiol	Glucagon*
Hydrocortisone	Glucocorticoids†
Growth hormone	

* Glucagon can induce enzyme synthesis in embryonic liver but will reduce levels of FAS in adult animals.
† Glucocorticoids decrease adipose tissue FAS but have no effect on the liver enzyme.

during mid to late pregnancy and early in lactation. In brain development, on the other hand, the synthetase is highest in fetal and neonatal rats and decreases with maturity. Like acetyl-CoA carboxylase, the concentration of fatty acid synthetase is also much higher in tissues of certain genetic mutants such as the obese hyperglycemic mouse. The enzyme in the latter mutant is less subject to control by dietary factors, such as fasting, when compared to normal mice.

These changes in animal fatty acid synthetase (FAS) levels have now been studied at the subcellular level. FAS mRNA has been isolated from different tissues after various hormonal or dietary manipulations and translated *in vitro*. Recombinant plasmids have been used to ascertain the size of FAS mRNA and also the amounts of this mRNA in tissues. By these means it has been shown that differences in FAS activity are caused by changes in enzyme levels rather than its intrinsic activity. These alterations are themselves changed by the balance of enzyme synthesis and degradation, mediated by the quantity of FAS mRNA present in the cell.

A common way in which the rate of a particular metabolic reaction can be controlled is through the supply of substrate. In the case of animal fatty acid synthetase, this regulation has been examined with regard to NADPH. However, it appears that NADPH production is adjusted to cope with the altering demands of fatty acid synthesis rather than the other way around. In contrast, supply of malonyl-CoA by acetyl-CoA carboxylase activity (section 3.2.7(a)) is considered to be a major factor regulating overall fatty acid (and lipid) formation under many conditions.

The above mechanisms probably apply also to organisms other than animals. Thus, in developing seeds, where there is a spectacular rise in fat accumulation during a particular stage of maturation, the increase in fatty acid synthesis has been correlated with increases in levels of various synthetase proteins including acyl carrier protein. Conversely, the rise in activity of fatty acid synthetase in photosynthetic tissues on illumination undoubtedly requires the simultaneous production of NADPH substrate by photo-system I.

(d) Control of animal desaturases

The discussion so far has centred on the control of the production of saturated fatty acids by fatty acid synthetase. Unsaturated fatty acids are present in all living cells and are important in regulating the physical properties of lipoproteins and membranes (Chapter 8) as well as acting as regulators of metabolism in cells (see above) or as precursors for physiologically active compounds (sections 3.4 and 5.2.2(a)). The control of their production must therefore be important, yet our knowledge of the subject is sparse. Research has lagged behind that on fatty acid synthetase, because of the inability to obtain purified enzymes with which to raise antibodies to study induction of new enzyme protein or to study regulatory control by small molecules.

Dietary changes that result in the induction of synthesis of fatty acid synthetase produce similar rises in activity of $\Delta 9$ desaturase but have little effect on $\Delta 4$, $\Delta 5$ or $\Delta 6$ desaturases. Several hormones have been reported to influence desaturase activity, notably insulin. In the rat, made diabetic by poisoning the pancreas with streptozotocin, liver $\Delta 9$ desaturase activity is low and can be restored by administration of insulin. However, dietary fructose, glycerol or saturated fatty acids can also restore enzyme activity, so that the role of insulin may be indirect. Little further progress can be made until purified desaturases are available so that short- and long-term regulation can be studied directly.

3.3 DEGRADATION OF FATTY ACIDS

The main pathways of fatty acid breakdown involve oxidation at various points on the acyl chain or lipoxidation at certain double bonds of specific unsaturated fatty acids. The main forms of oxidation are termed α-, β- and ω-. They are named depending on which carbon of the acyl chain is attacked:

$$CH_3(CH_2)n\ CH_2\ CH_2COO^-$$

$$\omega \qquad\qquad \beta \quad \alpha$$

Of these oxidations, β-oxidation is the most general and prevalent.

3.3.1 β-Oxidation is the most common type of biological oxidation of fatty acids

Long chain fatty acids, combined as triacylglycerols, provide the long-term storage form of energy in the adipose tissues of animals. In addition, many plant seeds contain triacylglycerol stores. These fats are degraded principally by the liberation of two-carbon (acetyl-CoA) fragments in β-oxidation. The mechanism was originally proposed over 80 years ago (1904) by Knoop. He synthesized a series of phenyl-substituted fatty acids with odd-numbered or even-numbered carbon chains. He found that the odd-numbered substrates gave rise to phenylpropionate while the even-numbered substrates were metabolized to phenylacetate. Knoop was using the phenyl group in the same way that modern biochemists would use a ^{14}C-radiolabel. At around the same time the proposed intermediates were isolated by Dakin and, therefore, the basic information about β-oxidation was available 50 years before the enzymological reactions were demonstrated.

When Leloir and Munoz (1944) showed that β-oxidation could be measured in cell-free preparations from liver, it was not long before a number of its important features were revealed. Lehninger found that ATP was needed to initiate the process, which seemed particularly active in mitochondria. Following the isolation of coenzyme A by Lipmann, Lynen was able to demonstrate that the active acetate was acetyl-CoA and Wakil and Mahler showed that the intermediates were CoA-esters. With the availability of chemically-synthesized acyl-CoA substrates, Green in Wisconsin, Lynen in Munich and Ochoa in New York could study the individual enzymes involved in detail.

(a) Cellular site of β-oxidation

Up until recently it had always been considered that β-oxidation was confined to mitochondria. Although animal mitochondria do contain all the enzymes necessary and are a major site for β-oxidation, other subcellular sites, such as the microbodies, are implicated. Peroxisomes or glyoxysomes, together are often referred to as **microbodies**. They contain a primitive respiratory chain where energy released in the reduction of oxygen is lost as heat. The presence of an active β-oxidation pathway in microbodies was first detected in glyoxysomes from germinating seeds by de Duve in 1976. Since that time the various enzymes involved have been purified and characterized for microbodies from rat liver.

Microbodies occur in all major groups of eukaryotes including yeasts, protozoa, plants and animals. The contribution of these organelles to total β-oxidation in a given tissue varies considerably. In animals microbodies are particularly important in liver and kidney. In fact, in liver it seems that mitochondria and microbodies collaborate in overall fatty acid oxidation. Thus, microbodies oxidize long chain fatty acids to medium chain products which are then transported to mitochondria for complete breakdown. In this way very long chain fatty acids, such as erucate, which are poor substrates for mitochondria, can be catabolized.

In contrast, the glyoxysomes from germinating seeds are capable of the complete breakdown of fatty acids to acetyl-CoA. They also integrate this metabolism with the operation of the glyoxylate cycle which allows plants (in contrast to animals) to synthesize sugars from acetyl-CoA. Leaf tissues also contain peroxisomes and recent work indicates that β-oxidation in leaves is confined to peroxisomes with no detectable activity in mitochondria.

(b) Transport of acyl groups to the site of oxidation: the role of carnitine

Fatty acids are transported between organs in animals either in the form of non-esterified fatty acids bound to serum albumin or as triacylglycerols associated with lipoproteins (especially chylomicrons and very low density lipoproteins: section 5.3.5). Triacylglycerol is hydrolysed on the outer surface of cells by lipoprotein lipase and fatty acids have been shown to enter liver, adipose and heart tissue cells by saturatable and non-saturatable mechanisms.

Once inside cells, fatty acids can be activated to acyl-CoAs by various ligases. Most of the activating enzymes are ATP-dependent acyl-CoA synthetases which act in a 2-step reaction (section 3.2.1).

Both non-esterified fatty acids and acyl-CoAs are capable of binding to distinct cytosolic proteins known as fatty acid binding proteins (FABP). The best known of these is the Z-protein of liver. These small molecular weight (about 14 kDa) FABPs have been suggested to function for intracellular transport or to provide a temporary binding site for potentially damaging compounds such as acyl-CoAs (section 5.3.3).

Because the inner mitochondrial membrane is impermeable to CoA and its derivatives, fatty acyl-CoAs formed in the cytosol cannot enter the mitochondria directly for oxidation. The observation, by Bremer and others, that carnitine could stimulate the oxidation of fatty acids *in vitro* led to the idea that long chain fatty acids could be transported in the form of carnitine esters.

$$(CH_3)_3\overset{+}{N}CH_2CHCH_2COO^-$$
$$|$$
$$OH$$

carnitine

The theory really began to take shape when an enzyme (carnitine : palmitoyl transferase, CPT) was discovered which would

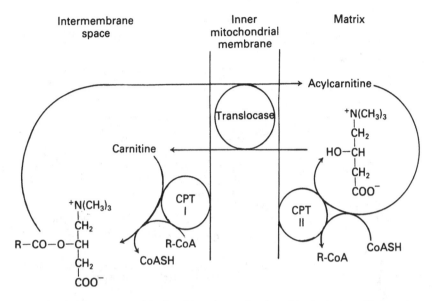

Figure 3.19 Movement of acyl residues into mitochondria via the carnitine carrier.

transfer long chain acyl groups from CoA to carnitine. Two forms of the enzyme have been identified which have different chain length specificities. Both are associated with the inner mitochondrial membrane but one form, CPT I (or CPT A), faces the intermembrane space whereas the other, CPT II (or CPT B), is directed toward the matrix. The acylcarnitines cross the membrane via a carnitine : acylcarnitine translocase which causes a one to one exchange, thus ensuring that the mitochondrial content of carnitine remains constant (Figure 3.19). Fatty acids of less than ten carbons can be taken up by mitochondria as free acids independent of carnitine. This is in spite of the fact that a third carnitine acyltransferase, which transfers acyl groups of 2–10 carbon atoms, has been found in mitochondria. The function of this enzyme is uncertain although it has been suggested to be useful in regenerating free CoA within the mitochondrial matrix.

(c) The overall pathway of β-oxidation in mitochondria

A summary of the sequential reactions of the β-oxidation cycle and the relationship of the overall process to energy metabolism in mitochondria is shown in Figure 3.20. Two carbon units (as acetyl-CoA) are removed from the fatty acyl-CoA substrate by the successive action of four enzymes – acyl-CoA dehydrogenase, enoyl hydratase, a second dehydrogenase and a thiolase. These enzymes are detailed in Table 3.12.

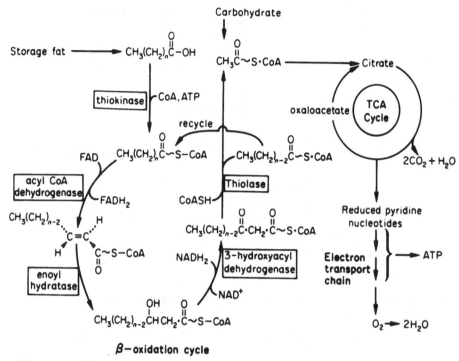

β-oxidation cycle

Figure 3.20 β-Oxidation and energy metabolism in mitochondria.

(d) Enzymes of β-oxidation

The first step in the β-oxidation cycle proper is the introduction of a *trans-α,β* double bond into the hydrocarbon chain of the activated fatty acid, catalysed by the flavoprotein enzyme, acyl-CoA dehydrogenase.

Acyl-CoA dehydrogenase
The acyl-CoA dehydrogenases are unique among flavoproteins for a number of reasons. The binding of substrate and enzyme is so tight as to be virtually non-dissociable. The flavin moiety is very stable and not oxidized in the presence of molecular oxygen, ferricyanide or other elecron acceptors. For this reason, the oxidation of the reduced form of the enzyme must be catalysed by a specific flavoprotein – the electron transport flavoprotein (ETF) discovered in 1956 by the American biochemists Crane and Beinert. This seems to be a unique case where an oxidation-reduction enzyme requires another specific protein for its reoxidation.

Enoyl-CoA hydratase
The hydratase catalyses the addition of the elements of water across the

Table 3.12 Enzymes of β-oxidation cycle. (The reaction scheme is given in Figure 3.21)

Full name	Trivial name	Description	Assay
1 Acyl-CoA dehydrogenase (EC 1.3.2.2)		Introduces a *trans*-2,3-double bond into the activated acyl chain. Chain-length specificities: 'green flavoprotein': C_4—C_8 'yellow flavoprotein' (medium chain): C_8—C_{12} 'yellow flavoprotein' (long chain): C_8—C_{18} All have two moles FAD/mole protein. The green F.P. has no copper, colour probably due to a complex of the prosthetic group with a chemical group of the protein. In addition, two acyl-CoA dehydrogenases specific for metabolites of branched-chain amino acids have been isolated and purified	Reversible reaction (i) Couple reaction to reduction of electron acceptor in presence of E.T.F. (ii) Measure reduction of unsaturated acyl-CoA (reverse reaction) by certain dyes
2 Enoyl-CoA hydratase (EC 4.2.1.17)	Enoyl hydrase or Crotonase	Two enzymes have been identified in heart mitochondria: one is crotonase or short-chain enoyl-CoA hydratase, the second is a long-chain enoyl-CoA hydratase. Crotonase activity is so high in some tissues that the second enzyme may have little function in β-oxidation there (e.g. in liver). However, in most tissues they probably co-operate in fatty acid degradation	Measure (i) Change in absorption of $\Delta2$-unsaturated (α,β-unsaturated) bond at 232 nm (ii) Couple the reaction with the next enzyme of the sequence (3-hydroxyacyl-CoA dehydrogenase) and measure change in absorption of NAD^+/NADH at 340 nm

Table 3.12 *Continued*

Full name	Trivial name	Description	Assay
3 L-3-hydroxyacyl-CoA dehydrogenase (EC 1.1.1.30)		Converts L(+)-3-hydroxyacyl-CoA into 3-ketoacyl-CoA. Completely specific for L-configuration and NAD$^+$. Less specificity for chain length or CoA. An example of an enzyme which can be studied with Lynen model compounds. At pH 7 equilibrium favours OH acid; at high pH equilibrium shifts in favour of keto acid. 100% yield by trapping keto acid with Mg^{2+}. Main dehydrogenase is soluble but a second enzyme, with a preference for long chain substrates, is located on the inner mitochondrial membrane	Measure appearance or disappearance of NADH absorption at 340 nm

4
Acyl-CoA: Acetyl-CoA
C-acyltransferase
(EC 2.3.1.16)

Thiolase

Catalyses a thiolytic cleavage of the keto acid in which —SH group of CoA displaces an acetyl-CoA moiety. One enzyme (purified and crystallized) for C_4–C_{18}. Purified protein catalyses 2 reactions:

Thiolytic cleavage:

$R \cdot CH_2 \cdot CO \cdot CH_2 \cdot COS \cdot CoA + HS\text{-}E \rightleftharpoons$

$R \cdot CH_2 \cdot COS\text{-}E + CH_3 COS \cdot CoA$

Acyl transfer:

$R \cdot CH_2 \cdot COS\text{-}E + CoASH \rightleftharpoons$

$R \cdot CH_2 \cdot COS \cdot CoA + HS\text{-}E$

Overall thiolase:

$R \cdot CH_2 \cdot CO \cdot CH_2 COS \cdot CoA + CoASH \rightleftharpoons$

$R \cdot CH_2 \cdot COS\text{-}CoA + CH_3 COS \cdot CoA$

In addition, an acetoacetyl-CoA thiolase is found in mitochondria. This is involved in ketone body formation and degradation

(i) Measure appearance or disappearance of CoA groups by nitroprusside reaction

(ii) Measure appearance or disappearance of acetoacetyl-CoA band at 303 nm

(iii) Link reaction to citrate synthesis in presence of acyl-CoA, malate dehydrogenase and condensing enzyme. Measure change in absorption of NADH at 340 nm

trans double bond of the unsaturated acyl-CoA to form 3-hydroxyacyl-CoA (β-hydroxyacyl-CoA). It is fairly well established that a *trans* double bond of the unsaturated acyl-CoA is transformed into L(+)-3-hydroxyacyl-CoA (the hydroxy acid having the absolute configuration of the L-series and positive optical rotation). In the past, controversies about the absolute configuration of this intermediate arose because no less than five enzymic activities are possessed by the same highly purified crystalline enzyme.

L-3-Hydroxyacyl-CoA dehydrogenase

The next dehydrogenation step, which converts the L-hydroxy-compound into the corresponding keto-compound, is completely specific for the L-absolute configuration and for NAD^+ as cofactor, but most preparations also contain an activity specific for the D-configuration which tends to give the overall appearance of a racemase.

Acyl-CoA : acetyl-CoA acyltransferase (β-ketothiolase)

The last enzyme of the sequence is generally known as thiolase and catalyses a thiolytic cleavage of the keto acid in which the –SH group of CoA displaces an acetyl-CoA moiety (Table 3.12). The remaining fragment is thus an acyl-CoA two carbon atoms shorter than the molecule which began the cycle and the process can be repeated over and over again until the carbon chain has been completely oxidized to C_2 fragments. Lynen obtained highly purified and crystalline thiolase and studied its properties. One enzyme appears to catalyse the cleavage of the whole range of keto acids from 4C to 18C. The protein has a molecular mass of about 170 kDa but in the presence of 5 M-urea dissociates reversibly into four subunits of about 42 kDa.

In addition to the 3-ketoacyl-CoA thiolase which functions in β-oxidation, another enzyme, acetoacetyl-CoA thiolase, is present in mitochondria and is believed to be important in ketone body synthesis and degradation.

(e) Other fatty acids containing branched chains, double bonds and an odd number of carbon atoms can also be oxidized

So far we have assumed that the fatty acid being oxidized is a straight chain, fully saturated compound. This is not necessarily the case and the ease with which other compounds are oxidized depends on the position along the chain of the extra group or the capacity of the cell for dealing with the end products. From acids of odd chain length, one of the products is propionic acid and the ease with which the organism oxidizes such fatty acids is governed by its ability to oxidize propionate. Liver, for example, is equipped to oxidize propionate and therefore deals with odd chain acids quite easily; heart, on the other hand, cannot perform propionate oxidation and degradation of odd chain acids grinds to a halt. The end product of propionate oxidation, succinyl-CoA, arises by a mechanism involving the B_{12} coenzyme:

$$\text{CH}_3\text{CH}_2\text{COSCoA} + \text{CO}_2 + \text{ATP} \xrightleftharpoons[\text{biotin propionyl-CoA carboxylase}]{}$$

$$\underset{\text{methylmalonyl-CoA}}{\text{CH}_3\text{CH(COOH)COSCoA} + \text{ADP} + \text{P}_i} \xrightleftharpoons[\text{B}_{12}]{} \underset{\text{succinyl-CoA}}{\text{HOOCCH}_2\text{CH}_2\text{COSCoA}}$$

Similarly, branched chain fatty acids with an even number of carbon atoms may eventually yield propionate, while the oxidation of the odd numbered branched chain acids proceeds by a different route involving 3-hydroxy-3-methylglutaryl-CoA (HMG-CoA) (Figure 3.21).

Most natural fatty acids are unsaturated. In addition, as discussed before (section 3.1.3) most double bonds are *cis* and, in polyunsaturated fatty acids, are methylene interrupted (three carbons apart). When unsaturated fatty acids are β-oxidized, two problems may be encountered – the unsaturated acids have *cis* double bonds and these may be at the wrong position for β-oxidation. The German biochemist, Stoffel, has shown that an isomerase exists to convert the *cis*-3 compound into the necessary *trans*-2-fatty acyl-CoA. The isomerase will also act with *trans*-3 substrates, though at lower rates. Once over this obstacle, β-oxidation can again continue to eliminate a further two carbons and then dehydrogenation to produce a 2-*trans*,4-*cis*-decadienoyl-CoA from linoleoyl-CoA. The discovery of a 2,4-dienoyl-CoA

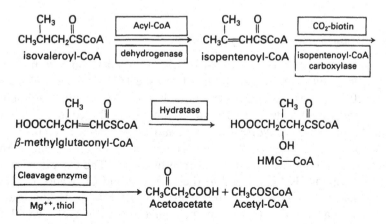

Figure 3.21 Oxidation of odd numbered branched chain fatty acids.

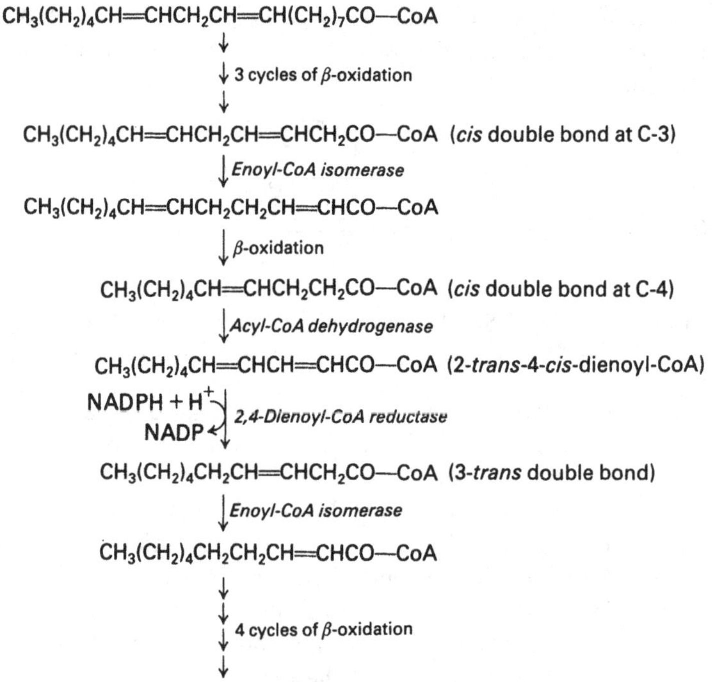

$$CH_3(CH_2)_4CH{=}CHCH_2CH{=}CH(CH_2)_7CO{-}CoA$$

↓

↓ 3 cycles of β-oxidation

↓

$$CH_3(CH_2)_4CH{=}CHCH_2CH{=}CHCH_2CO{-}CoA \quad (cis \text{ double bond at C-3})$$

↓ *Enoyl-CoA isomerase*

$$CH_3(CH_2)_4CH{=}CHCH_2CH_2CH{=}CHCO{-}CoA$$

↓ *β-oxidation*

$$CH_3(CH_2)_4CH{=}CHCH_2CH_2CO{-}CoA \quad (cis \text{ double bond at C-4})$$

↓ *Acyl-CoA dehydrogenase*

$$CH_3(CH_2)_4CH{=}CHCH{=}CHCO{-}CoA \quad (2\text{-}trans\text{-}4\text{-}cis\text{-dienoyl-CoA})$$

NADPH + H$^+$ ⤸
              2,4-*Dienoyl-CoA reductase*
NADP ⤹

$$CH_3(CH_2)_4CH_2CH{=}CHCH_2CO{-}CoA \quad (3\text{-}trans \text{ double bond})$$

↓ *Enoyl-CoA isomerase*

$$CH_3(CH_2)_4CH_2CH_2CH{=}CHCO{-}CoA$$

↓

↓ 4 cycles of β-oxidation

↓

Figure 3.22 The oxidation of linoleic acid.

reductase by Kunau and Dommes showed that this enzyme could use NADPH to yield 3-*trans*-decenoyl-CoA as its product. The enoyl-CoA isomerase then moves the double bond to the 2-position and β-oxidation can proceed again as normal (Figure 3.22).

(f) Regulation of mitochondrial β-oxidation

There are two major products of β-oxidation. Complete operation of the cycle yields acetyl-CoA which can be fed into the TCA cycle as depicted in Figure 3.20. In some tissues, however, notably liver and the rumen epithelial cells of ruminant animals, acetoacetate accumulates. This compound, with its reduction product, β-hydroxybutyrate, and its decarboxylation product, acetone, make up a group of metabolites known as the ketone bodies. Free acetoacetic acid may accumulate in liver in two ways. The CoA derivative may be enzymatically hydrolysed to the free acid and CoA and the liver tissue lacks the thiokinase to reconvert the acid into its thiolester. Alternatively acetoacetyl-CoA may be converted into hydroxymethyl-glutaryl-CoA (HMG-CoA) which is subsequently cleaved to free acetoacetic acid:

$$CH_3COCH_2COSCoA + CH_3COSCoA \rightleftharpoons CH_3\underset{\underset{OH}{|}}{\overset{\overset{CH_2COSCoA}{|}}{C}}CH_2{-}COOH + CoA$$

$$CH_3COCH_2COOH + CH_3COSCoA$$

HMG is an important intermediate in cholesterol biosynthesis and this pathway provides a link between fatty acid and cholesterol metabolism. Ketone bodies are also excellent fuels for the liver and even brain during starvation, even though brain cannot utilize long chain fatty acids as fuels.

Acetyl-CoA is therefore the substrate for two competing reactions: with oxaloacetate to form citrate or with acetoacetyl-CoA to form ketone bodies (ketogenesis). Which reaction predominates depends partly on the rate of β-oxidation itself and partly on the redox state of the mitochondrial matrix which controls the oxidation of malate to oxaloacetate, hence, the amount of oxaloacetate available to react with acetyl-CoA. The proportion of acetyl groups going into the TCA cycle relative to ketogenesis is often referred to as the 'acetyl ratio'. The overall rate of β-oxidation may be controlled by a number of well-known mechanisms:

1. the availability of free fatty acids;
2. the rate of utilization of β-oxidation products, which in turn can either lead to specific inhibition of particular enzymes or to 'feedback inhibition' of the whole sequence.

The concentration of free fatty acids in plasma is controlled by glucagon (which stimulates) and insulin (which inhibits) breakdown of triacylglycerols in adipose tissue stores (section 4.6.3). Once the fatty acids enter cells they can be degraded to acetyl-CoA or used for lipid synthesis. The relative rates of these two pathways depends on the nutritional state of the animal, particularly on the availability of carbohydrate.

In muscle, the rate of β-oxidation is usually dependent on both the free fatty acid concentration in the plasma as well as the energy demand of the tissue. A reduction in energy demand by muscle will lead to a build-up of NADH and acetyl-CoA. Increased $NADH/NAD^+$ ratios lead to inhibition of the mitochondrial TCA cycle and increase further the acetyl-CoA/CoASH ratio. Kinetic studies with purified enzymes show that the major sites of β-oxidation inhibition are 3-hydroxyacyl-CoA dehydrogenase which is inhibited by NADH and 3-ketoacyl-CoA thiolase which is inhibited by acetyl-CoA (Figure 3.23).

In the liver, because of the interaction of lipid, carbohydrate and ketone body metabolism, the situation is more complex. McGarry and Foster

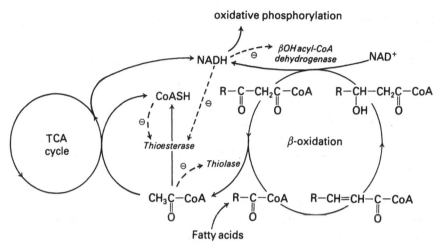

Figure 3.23 Regulation of β-oxidation in muscle. ⊖ = Inhibition of enzymes indicated.

proposed in 1980 that the concentration of malonyl-CoA was particularly important. In the fed state, when carbohydrate (glucose) is being converted to fatty acids, the level of malonyl-CoA is raised. Malonyl-CoA is a reversible inhibitor of carnitine palmitoyltransferase I (CPT I) and, therefore, reduces entry of acyl groups into mitochondria for oxidation. In a fasting state, lowered malonyl-CoA allows CPT I to function at high rates and this stimulates β-oxidation and ketogenesis (Figure 3.24).

Malonyl-CoA itself is generated by acetyl-CoA carboxylase which is hormonally regulated (section 3.2.7(a)). In the fasted condition a high glucagon/insulin ratio elevates cellular cAMP, thus allowing short-term inhibition of acetyl-CoA carboxylase by phosphorylation. Reduction of the glucagon/insulin ratio on feeding, reverses the effect. Thus, fatty acid synthesis and degradation are co-regulated by liver hormonal levels.

(g) Fatty acid oxidation in *E. coli*

When *E. coli* is grown on a medium containing fatty acids rather than glucose, a 200-fold induction of the enzymes of β-oxidation is seen. By the use of deficient mutants, it has been possible to identify a number of the genes responsible for coding the relevant enzymes. Interestingly, a single gene (*fad* AB) coded for enoyl-CoA hydratase, 3-hydroxyacyl-CoA dehydrogenase, 3-ketoacyl-CoA thiolase, *cis*-3-*trans*-2-enoyl-CoA isomerase and 3-hydroxyacyl-CoA epimerase. In fact these five enzyme activities are contained on a single mutlifunctional protein which has been

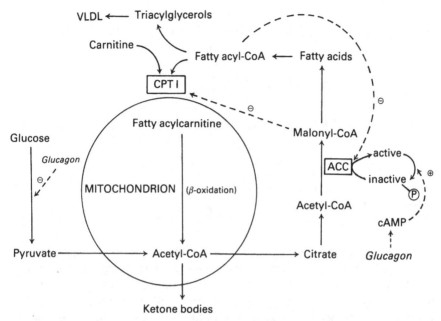

Figure 3.24 Regulation of fatty acid metabolism in liver. CPTI = carnitine palmitoyltransferase I, ACC = acetyl-CoA carboxylase, VLDL = very low density lipoproteins.

purified to homogeneity. The acyl-CoA dehydrogenase has also been isolated and is coded by the *fad* D gene. Three other genes are also involved, one of which apparently acts as a permease for long chain fatty acids.

Induction of the β-oxidation system is thought to occur by binding of long chain acyl-CoAs to a repressor protein. This results in the co-ordinated induction of all enzymes needed for β-oxidation except for electron transport flavoprotein ETF which is constitutive. Synthesis of the enzymes of β-oxidation is repressed strongly when glucose is present in the growth medium, even in the presence of fatty acids. This catabolite repression is caused by a lowering of the cellular concentration of cAMP. When the concentration of cAMP is low, the catabolite gene activator protein does not bind to the *fad* promoters thus preventing the efficient transcription of the *fad* regulon.

(h) β-Oxidation in microbodies

It was mentioned in section 3.3.1(a) that microbodies are an important (in plants exclusive) site of fatty acid oxidation. However, the details of the process show some differences from that in mitochondria (Figure 3.25). The

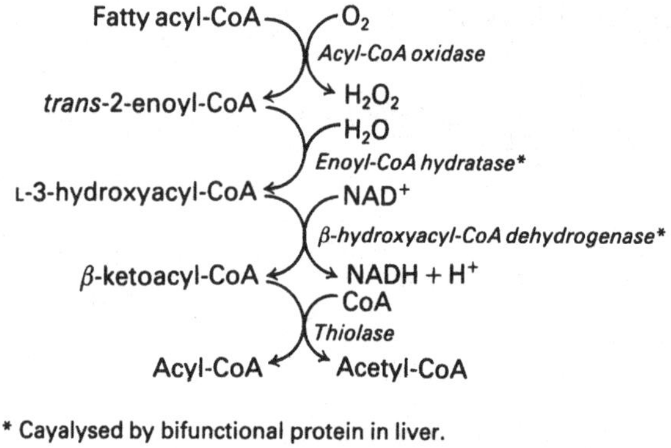

Fatty acyl-CoA $\longrightarrow$ O$_2$

Acyl-CoA oxidase

trans-2-enoyl-CoA $\longleftarrow$ H$_2$O$_2$

H$_2$O

Enoyl-CoA hydratase*

L-3-hydroxyacyl-CoA $\longleftarrow$ NAD$^+$

β-hydroxyacyl-CoA dehydrogenase*

β-ketoacyl-CoA $\longleftarrow$ NADH + H$^+$

CoA

Thiolase

Acyl-CoA $\longleftarrow$ Acetyl-CoA

* Cayalysed by bifunctional protein in liver.

Figure 3.25 Peroxisomal β-oxidation of fatty acids.

first step is catalysed by acyl-CoA oxidase which, in contrast to the acyl-CoA dehydrogenase of mitochondria, transfers two hydrogens from the substrate to oxygen to produce H_2O_2. The next two reactions (enoyl-CoA hydratase, 3-hydroxyacyl-CoA dehydrogenase) are catalysed by a bifunctional protein in rat liver. The thiolase enzyme shows high activity towards long chain substrates but little towards acetoacetyl-CoA.

Carnitine is not involved in the movement of fatty acids into microbodies. Either free fatty acids or acyl-CoAs appear to be able to cross the organelle's membrane. Moreover, because of the absence of an electron transport chain in microbodies there is no internal means of regenerating NAD^+. The reoxidation of NADH is thought to occur either by a glycerol phosphate shuttle or by movement of NADH to the cytosol and NAD^+ back.

In animal tissues, such as liver, the microbody system seems incapable of oxidizing long chain acyl-CoAs completely, due to the limited substrate specificity of the acyl-CoA oxidase. The medium chain products are transferred to carnitine by a peroxisomal medium chain carnitine acyltransferase. The acetate is transferred likewise using a carnitine acetyltransferase. The acylcarnitines can then be moved to the mitochondria for further oxidation. Studies on this system and on peroxisomal β-oxidation in general have been facilitated by the fact that hypolipidemic drugs like clofibrate, high fat diets, starvation or diabetes all increase peroxisomal oxidation considerably. Under average conditions peroxisomes are thought to contribute up to 50% of the total fatty acid oxidative activity of liver. A disease known as adrenoleukodystrophy is characterized by the accumulation of very long chain fatty acids in the adrenals. It may be due to a defect in peroxisomal β-oxidation.

In contrast to the chain length limitation of β-oxidation in liver peroxisomes, the glyoxysomes of fatty seedlings (and peroxisomes of plant leaves) are capable of completely oxidizing fatty acids to acetate. As mentioned before, no co-operation with mitochondria is necessary (apart from regeneration of NAD^+) and, indeed, plant mitochondria have not been found to contain the β-oxidation enzymes.

3.3.2 α-Oxidation of fatty acids is important when structural features prevent β-oxidation

α-Oxidation is important in animals for the formation of α-hydroxy fatty acids and for chain shortening, particularly in regard to the catabolism of molecules that cannot be metabolized directly by β-oxidation. Brain cerebrosides and other sphingolipids (section 6.2.4) contain large amounts of α-hydroxy fatty acids and a mixed function oxidase has been identified in microsomal fractions which requires O_2 and NADPH. Breakdown of α-hydroxy fatty acids also takes place in such fractions, needs NAD and O_2 and results in an acid one carbon shorter with the liberation of CO_2.

An α-oxidation system in liver and kidney is important for the breakdown of branched chain fatty acids – in particular, phytanic acid (3,7,11,15-tetramethyl palmitic acid) which is formed in animals from phytol, the side-chain of chlorophyll. In phytanic acid the methyl group at C_3 prevents β-oxidation. The methyl fatty acid is then shortened by 1 carbon by α-oxidation after which β-oxidation can continue, with the release of propionyl-CoA. Thus:

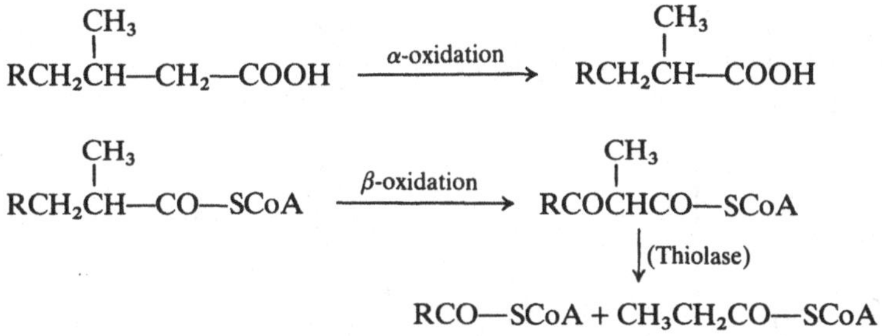

$$\underset{\displaystyle \overset{|}{\underset{}{RCH_2CH-CH_2-COOH}}}{\overset{\displaystyle CH_3}{}} \xrightarrow{\;\alpha\text{-oxidation}\;} \underset{\displaystyle \overset{|}{\underset{}{RCH_2CH-COOH}}}{\overset{\displaystyle CH_3}{}}$$

$$\underset{\displaystyle \overset{|}{\underset{}{RCH_2CH-CO-SCoA}}}{\overset{\displaystyle CH_3}{}} \xrightarrow{\;\beta\text{-oxidation}\;} \underset{\displaystyle \overset{|}{\underset{}{RCOCHCO-SCoA}}}{\overset{\displaystyle CH_3}{}}$$

$$\downarrow \text{(Thiolase)}$$

$$RCO-SCoA + CH_3CH_2CO-SCoA$$

In brain tissue it is believed that unesterified fatty acids are the substrates for α-oxidation but in liver it is not known whether acyl-CoA substrates are used. If they are, then the pathway shown above could use CoA esters throughout and there would be no need for an acyl-CoA synthetase step.

An inborn error of metabolism (Refsum's disease) due to a failure of α-oxidation is described in section 8.14.

Plants have an active α-oxidation system which is concerned, amongst other functions, in the turnover of the phytol moiety of chlorophyll as described above. The mechanism has been studied in several tissues. Although there is still some controversy as to whether O_2 or H_2O_2 is the substrate and which cofactors are involved, it seems probable that O_2 is activated by a reduced flavoprotein and a hydroperoxide intermediate is the active species which reacts with the fatty acid.

Until 1973 it seemed that there must be at least two different pathways for α-oxidation depending on the source of the enzymes. The pathways that had been studied in pea leaves and in germinating peanut cotyledons apparently had different cofactor requirements and different intermediates seemed to be involved. The discrepancies have now been resolved and a unified pathway has been proposed by Stumpf and his team in California (Figure 3.26).

They showed that molecular oxygen was a requirement and, in the peanut system, that a hydrogen peroxide generating system was involved. Enzymes which catalysed the reduction of peroxides, such as glutathione peroxidase, reduced α-oxidation and increased the production of D-hydroxypalmitate. This pointed to the existence of a peroxide, 2-hydroperoxypalmitate,

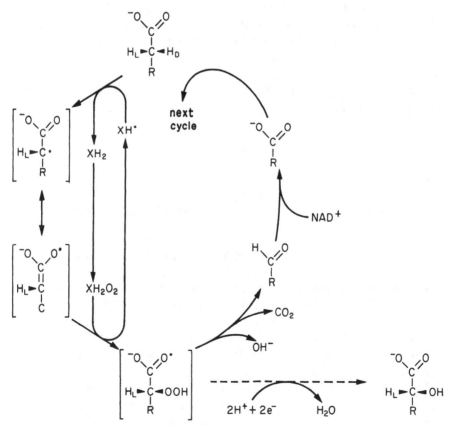

Figure 3.26 Proposed α-oxidation pathway.

intermediate. Accumulation of hydroxyacids in experiments was due to the intermediates being channelled into a dead-end pathway.

After the hydroperoxy intermediate, CO_2 is lost and a fatty aldehyde produced. This is oxidized by a NAD^+-requiring system to yield a fatty acid, one carbon less than the original (palmitic) substrate. α-Oxidation in plants is extra-mitochondrial and may be in the cytoplasm or associated with the endoplasmic reticulum (depending on the tissue).

3.3.3 ω-Oxidation uses mixed function oxidases

ω-Oxidation of straight-chain fatty acids yields dicarboxylic acid products. Normal acids are only slowly catabolized by such a system since β-oxidation is usually highly active. However, in substituted derivatives ω-oxidation is often an important first step to allow subsequent β-oxidation to take place. The oxidation has an ω-hydroxy fatty acid as an intermediate and the enzyme is a mixed function oxidase. The enzyme is very similar to, or identical with, the drug-hydroxylating enzyme system and involves cytochrome P_{450}. O_2 and NADPH are co-substrates.

$$H^+ + NADPH \quad \diagdown \quad \diagup \; Cyt. \; P_{450} \; ox. \diagdown \quad \diagup \; H_2O$$
$$\times \qquad \qquad \times \quad \rightarrow HOOC—R—CH_2OH$$
$$\qquad \qquad \qquad \qquad \qquad \qquad HOOC—R—CH_3$$
$$NADP^+ \quad \diagup \quad \diagdown \; Cyt. \; P_{450} \; red. \quad \diagdown \; O_2$$

An ω-hydroxylase has been studied in *Pseudomonas* by Coon and co-workers. They showed that the enzyme had non-haem iron as, apparently, the only prosthetic group involved directly in the hydroxylation reactions. The system was fractionated into three components: a non-haem iron protein (similar to rubredoxin), a flavoprotein and a final component needed for hydroxylase activity.

In plants, the ω-hydroxylase system is responsible for synthesis of the ω-hydroxy fatty acid components of cutin and suberin. Kolattukudy has studied the reactions in preparations from *Vicia faba*. NADPH and O_2 were cofactors and the enzyme showed typical properties of a mixed-function oxidase. However, the involvement of P_{450} in the plant system is unproven since, although the hydroxylation is inhibited by CO, the inhibition is not reversed by 420–460 nm light (a property typical for cytochrome P_{450} systems). Where dihydroxy fatty acids are being synthesized for cutin or suberin, the ω-hydroxy fatty acid is the substrate for the second hydroxylation. Like ω-oxidation, mid-chain hydroxylation also requires NADPH and O_2 and is located in the endoplasmic reticulum.

3.3.4 Chemical peroxidation is an important reaction of unsaturated fatty acids

One of the characteristic reactions of lipids which are exposed to oxygen is the formation of peroxides. Indeed, among non-enzymic chemical reactions which take place in the environment at ambient temperatures, the oxidation of unsaturated compounds is perhaps the most important both from an industrial and a medical point of view. In biological tissues, uncontrolled lipid peroxidation causes membrane destruction and is increasingly regarded as an important event in the control or development of diseases (section 8.11). In food, oxidation (either enzymically or chemically catalysed) can have desirable as well as adverse consequences (section 8.15).

In common with other radical chain reactions, lipid peroxidation can be divided into three separate processes – initiation, propagation and termination. During initiation a very small number of radicals (e.g. transition-metal ions or a radical generated by photolysis or high-energy irradiation) allow the production of $R\cdot$ from a substrate RH:

$$X\cdot + RH \rightarrow R\cdot + XH \qquad (3.13)$$

Propagation then allows a reaction with molecular oxygen:

$$R\cdot + O_2 \rightarrow ROO\cdot \qquad (3.14)$$

and this peroxide radical can then react with the original substrate:

$$ROO\cdot + RH \rightarrow ROOH + R\cdot \qquad (3.15)$$

Thus, reactions (3.14) and (3.15) form the basis of a chain reaction process.

Free radicals such as $ROO\cdot$ (and $RO\cdot$, $OH\cdot$ etc. which can be formed by extra side-reactions) can react at random by hydrogen abstraction and a variety of addition reactions to damage proteins, other lipids and vitamins (particularly vitamin A). Compounds which react rapidly with free radicals can be useful in slowing peroxidation damage. Thus, naturally-occurring compounds such as vitamin E are powerful anti-oxidants and tissues which are deficient in such compounds may be prone to peroxidation damage (section 8.11). Formation of lipid hydroperoxides can be readily detected by a number of methods of which the absorption of conjugated hydroperoxides at 235 nm is particularly useful.

Termination reactions may lead to the formation of both high and low molecular weight products of the peroxidation reactions. Depending on the lipid, some of the low molecular mass compounds may be important flavours (or aromas) of foods (section 5.2.3). For example, short to medium chain aldehydes formed from unsaturated fatty acids may give rise to rancidity and bitter flavours on the one hand or more pleasant attributes such as those associated with fresh green leaves, oranges or cucumbers on

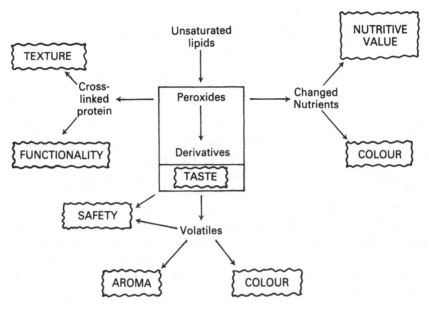

Figure 3.27 Reactions of polyunsaturated fatty acids leading to changes in food quality.

the other hand. Fish odours are attributed to a ketone. Some relevant changes in quality or nutritional value of food are indicated in Figure 3.27.

In general, it is considered desirable to reduce the initiation reaction as a means of controlling peroxidation. Apart from anti-oxidants (BHT; 3,5-di-*t*-butyl-4-hydrotoluene is often used as a food additive), metal binding compounds and phenolic compounds may be inhibitory as well as the superoxide dismutase and glucose oxidase-catalase enzyme systems.

3.3.5 Peroxidation catalysed by lipoxygenase enzymes

The second kind of peroxidation is catalysed by the enzyme lipoxidase (or lipoxygenase). The enzyme was originally thought to be present only in plants, but recently its presence in animal tissues has been demonstrated (section 3.4). The chief sources of the enzyme are in peas and beans (especially soybean), cereal grains and oil seeds. It was originally detected by its oxidation of carotene and has been used extensively in the baking industry for bleaching carotenoids in dough.

Lipoxidase catalyses the reaction:

$$R-CH=CH-CH_2-CH=CH-R_1 + O_2 \longrightarrow$$

$$\underset{cis}{R-CH=CH}-\underset{trans}{CH=CH}-CH-R_1$$

with OH, O above.

When Theorell and his colleagues in Sweden first purified and crystallized soybean lipoxygenase in 1947, they reported that it had no prosthetic group or heavy metal associated with it. Such a circumstance would make lipoxygenase unique among oxidation enzymes. Chan in England and Roza and Franke in The Netherlands demonstrated the presence of one atom of iron per mole of enzyme by atomic absorption spectroscopy. The product of the enzymic reaction – a hydroperoxide – is similar to the products of purely chemical catalysis but the lipoxidase reaction has a number of distinguishing features. The activation energy is smaller than that for chemical reactions, and the enzyme has very specific substrate requirements. In order to be a substrate, the fatty acid must contain at least two *cis* double bonds

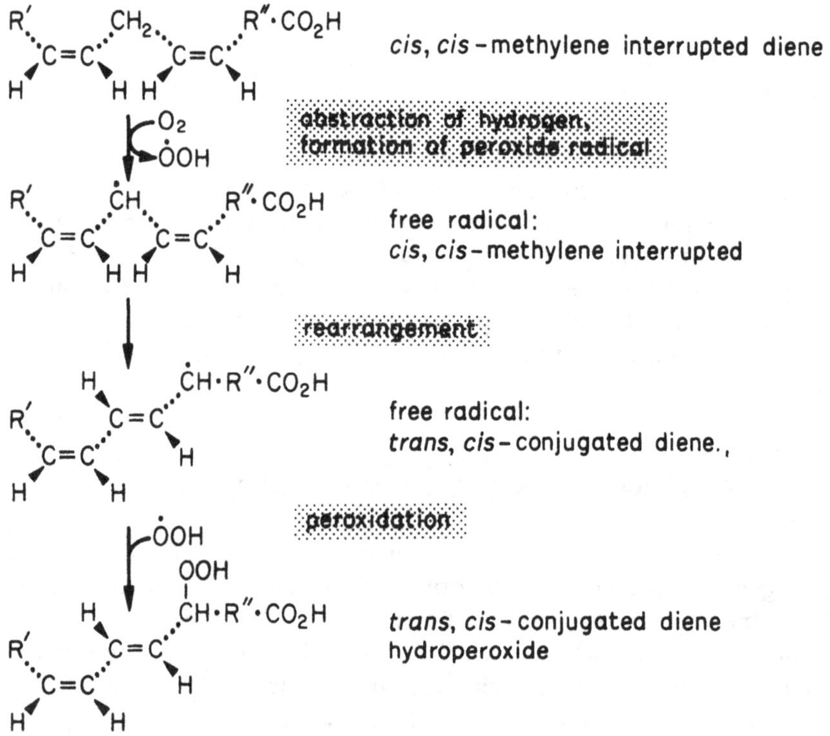

Figure 3.28 Possible mechanisms of lipoxygenase.

interrupted by a methylene group. Linoleic acid is the best known substrate. Most of the acids which have been tested and which show high reaction rates, have a double bond six carbon atoms from the methyl end of the chain. The distance of the carboxyl group from the diene system is less critical but it must not be stearically hindered, indicating that attachment to the enzyme surface may occur at the carboxyl end of the chain. Like chemically catalysed peroxidation, the lipoxidase reaction involves free radicals and can be inhibited by radical trapping reagents such as the tocopherols. The reaction sequence shown in Figure 3.28 has not been proved, but represents the most likely pathway.

Although we do not know the physiological function of lipoxygenase, the effect of the enzyme in plant tissues is to yield volatile products with characteristic flavours and aromas – either desirable or undesirable (section 3.3.4). The substrates for lipoxygenase are the free fatty acids released from storage lipids by lipases (section 4.5) or from membrane lipids (Chapter 7) by non-specific 'lipolytic acylhydrolases'.

3.4 ESSENTIAL FATTY ACIDS AND THE BIOSYNTHESIS OF EICOSANOIDS

In section 3.2.4 we described the biosynthesis of the polyunsaturated fatty acids and indicated that animals lack the $\Delta 12$ and $\Delta 15$-desaturases. They cannot, therefore, synthesize linoleic or linolenic acids. It turns out, however, that these polyunsaturated fatty acids are absolutely necessary for the maintenance of growth, reproduction and good health and must, therefore, be obtained in the diet from plant foods. They are called **essential fatty acids** and their role in nutrition and health will be described in detail in section 5.2.2. Chapter 5 will also describe the dual roles of the essential fatty acids and their longer-chain, more highly unsaturated metabolic products in membranes and as precursors of a variety of oxygenated fatty acids with potent biological activities now generally known collectively as the **eicosanoids**.

As early as 1930, two American gynaecologists, Kurzrok and Lieb, reported that the human uterus, on contact with fresh human semen, was provoked into either strong contraction or relaxation. Both von Euler in Sweden and Goldblatt in England subsequently discovered marked stimulation of smooth muscle by seminal plasma. Von Euler then showed that lipid extracts of ram vesicular glands contained the activity and this was associated with a fatty acid fraction. The active factor was named **prostaglandin** and was shown to possess a variety of physiological and pharmacological properties.

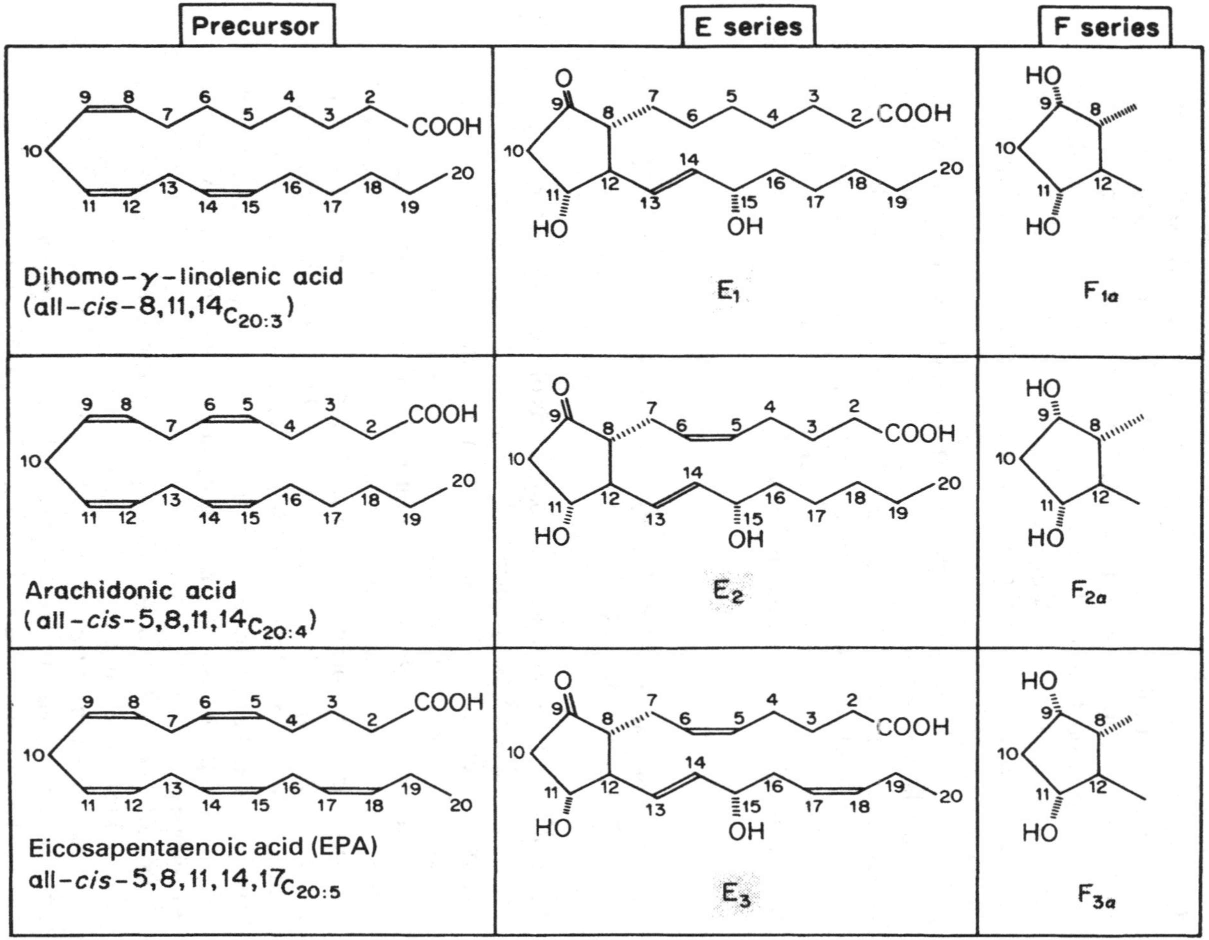

Figure 3.29 Structures of prostaglandins E and F and their precursors.

In 1947, the Swede, Bergstrom, started to purify these extracts and soon showed that the active principle was associated with a fraction containing unsaturated hydroxy acids. The work then lapsed until 1956, when with the help of an improved test system (smooth muscle stimulation in the rabbit duodenum) Bergstrom isolated two prostaglandins in crystalline form, (PGE_1 and $PGF_{1\alpha}$). Their structure, as well as that of a number of other prostaglandins, was elucidated by a combination of degradative, mass spectrometric, X-ray crystallographic and NMR studies (Figure 3.29). The nomenclature is based on the fully saturated 20C acid with C_8 to C_{12} closed to form a 5-membered ring; this is called prostanoic acid. Thus PGE_1 is designated 9-keto-11α,15α-dihydroxyprost-13-enoic acid. The 13,14 double bond has a *trans* configuration; all the other double bonds are *cis*. Figure 3.29 clearly brings out the difference between the 'E' and 'F' series, which have a keto and hydroxyl group at position 9 respectively; 'α' refers to the stereochemistry of the hydroxyl, and the suffix 1, 2 or 3 is related to the precursor fatty acid from which they are derived. That is, the 1, 2 or 3 refer to how many double bonds are contained in the prostaglandin structure. The name prostaglandin (and the related prostanoic acid structure) derives from the fact that early researchers believed that the prostate gland was the site of their synthesis.

3.4.1 The pathways for prostaglandin synthesis are discovered

Although prostaglandins were the first biologically-active eicosanoids to be identified, it is now known that the essential fatty acids are converted into a number of different types of eicosanoids. (Eicosanoid is a term meaning a C20 fatty acid derivatives.) The various eicosanoids are important examples of local hormones. That is, they are generated *in situ* and, because they are rapidly metabolized, only have activity in the immediate vicinity. A summary of the overall pathway for generation of eicosanoids is shown in Figure 3.30. Essential fatty acids can be attacked by lipoxygenases which give rise to leukotrienes or hydroxy and hydroperoxy fatty acids. Alternatively, metabolism by cyclooxygenase gives cyclic endoperoxides from which the classical prostaglandins or thromboxanes and prostacyclin can be synthesized. Because, historically, prostaglandin synthesis was elucidated before that of the other eicosanoids, we shall describe prostaglandin formation first.

After the structures of PGE and PGF had been defined, the subsequent rapid exploitation of this field, including the unravelling of the biosynthetic pathways, was done almost entirely by two research teams led by van Dorp in Holland and by Bergstrom and Samuelsson in Sweden. Both realized that the most likely precursor of PGE_2 and $PGF_{2\alpha}$ was arachidonic acid. This was

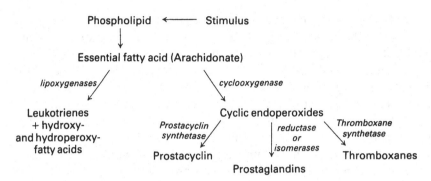

Figure 3.30 Overall pathway for conversion of essential fatty acids into eicosanoids.

then demonstrated by both groups simultaneously (with the same preparation of tritiated arachidonic acid) by incubation with whole homogenates of sheep vesicular glands. An important point that was noted, was that the unesterified fatty acid was the substrate and not an activated form (see later for discussion about release of free fatty acids in tissues). The ability of the arachidonic acid chain to fold allows the appropriate groups to come into juxtaposition for the ring closure to occur (Figure 3.29 and Figure 3.31). The reactions take place in the microsomal fraction but a soluble heat stable factor is required. This cofactor can be replaced by reduced glutathione. The reaction also requires molecular oxygen. Labelling with $^{18}O_2$ demonstrated that all three oxygen atoms in the final prostaglandin (Figure 3.31) are derived from the gas.

The reaction is catalysed by the enzyme prostaglandin endoperoxide synthetase. This enzyme has two distinct activities and is, thus, a multifunctional protein. It has been detergent-solubilized and purified to homogeneity. In spite of the complicated nature of the reactions it catalyses, the protein's subunit molecular mass is only 72 kDa. First, prostaglandin endoperoxide synthetase (PES) inserts two molecules of oxygen to yield a 15-hydroperoxy-9,11-endoperoxide with a substituted cyclopentane ring (PGG). This is the cyclooxygenase activity of the enzyme. A peroxidase activity then reduces PGG to its 15-hydroxy analogue (Figure 3.31) PGH. Immunocytochemical studies with cultured fibroblasts show that PES is localized on the endoplasmic reticulum and nuclear membranes.

PES requires an hydroperoxide activator which is used to remove a hydrogen atom from position 13 on the incoming fatty acid, and allow attack by oxygen. This is an unusual mechanism because, usually, oxygenases work by activating the oxygen substrate. There are other interesting features of the enzyme. For example, the free radical intermediates generated can also inactivate the enzyme. This self-deactivation of the cyclooxygenase occurs

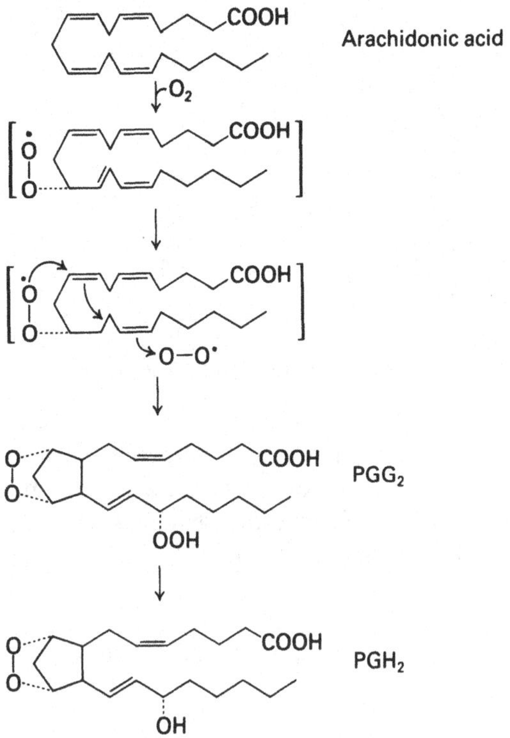

Figure 3.31 Mechanism of biosynthesis of the cyclic endoperoxide, PGH$_2$.

in vivo as well as for purified preparations. It may ensure that only a certain amount of endoperoxide is generated even when large quantities of precursor fatty acid are available. In addition, certain non-steroidal anti-inflammatory drugs, such as aspirin or indomethacin, interact with the cyclooxygenase. Vane found in 1971 that these drugs inhibited prostaglandin synthesis in a number of different tissues and cells. They compete with the fatty acid substrate for the cyclooxygenase active site. In addition, aspirin has been shown to acetylate a serine hydroxyl at or near the active site which inactivates the enzyme (Figure 3.32). This covalent modification cannot be repaired and, therefore, the cell can only restore PES activity by synthesizing fresh enzyme. Certain acetylenic fatty acids, such as eicosa-5,8,11,14-tetraynoic acid, also inactivate the cyclooxygenase, possibly by acting as suicide substrates (i.e. compounds which are converted to irreversible inhibitors by the enzyme's activity).

Prostaglandin H is the key intermediate for conversion to various active eicosanoids. The enzymes responsible for its further metabolism are present

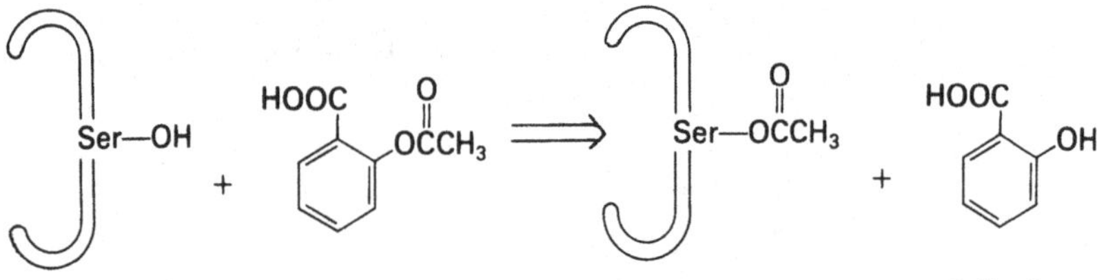

Prostaglandin endoperoxide synthetase (PES) –ACTIVE + Acetyl salicylic acid (Aspirin) ⟹ Acetylated enzyme –INACTIVE + Salicylic acid

Figure 3.32 Proposed mechanism for the inactivation of the cyclooxygenase component of prostaglandin endoperoxide synthetase by aspirin.

in catalytic excess to PES and, hence, are not regulatory except in the sense that the balance of their activities determines the pattern of prostaglandins and thromboxanes which are formed in a given tissue. A summary of the various possible reactions of a specific PGH (PGH$_2$) is shown in Figure 3.33.

3.4.2 Cyclic endoperoxides can be converted to different types of eicosanoids

The various eicosanoids produced from PGH$_2$ (Figure 3.33) have a remarkable range of biological activities. Moreover, the effect of a given eicosanoid varies from tissue to tissue. In the absence of specific inhibitors of their formation, and given the interacting effects of many eicosanoids, it has proved rather difficult to elucidate all their biological functions.

PGH to PGD isomerase activity is high in brain and the spinal cord and PGD$_2$ is also the main cyclooxygenase product of mast cells. It inhibits platelet aggregation, increases platelet cAMP content and has a membrane receptor distinct from that for PGI$_2$ (section 3.4.3). It can act as a peripheral vasoconstrictor, pulmonary vasoconstrictor and bronchoconstrictor. The latter activity can be demonstrated when PGD$_2$ is inhaled. It is thought to have various neuromodulatory actions (e.g. it can decrease noradrenaline release from adrenergic nerve terminals) and overproduction of PGD$_2$ may be involved in the hypotensive attacks of patients with mastocytosis.

PGE$_2$ is the main arachidonate metabolite in kidneys where it reduces ADH (antidiuretic hormone) – induced water reabsorption and may help to control renin release. It may also help to mediate the metabolism and interaction of macrophages with other cells.

PGF$_{2\alpha}$ seems to be the so-called luteolytic factor produced by mammalian uteri. When the compound is injected into cows it causes regression of the corpus luteum and induces ovulation and is used commercially as a regulator

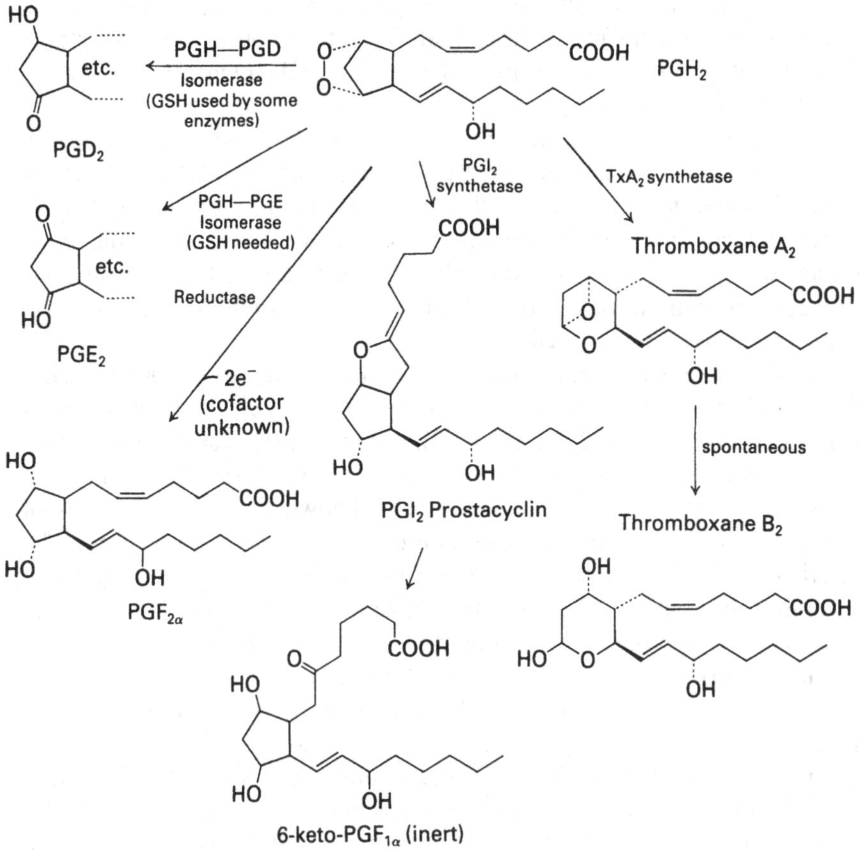

Figure 3.33 Conversion of PGH$_2$ into various biologically active eicosanoids.

of ovulation in dairy cows. PGF$_{2\alpha}$ is useful for the induction of abortions in women in midtrimester. In addition, if injected as a slow-releasable form it will act as a birth control agent probably by inhibiting implantation of the fertilized ovum in the uterine wall.

3.4.3 New eicosanoids are discovered

In 1973 the Swedes Hamberg and Samuelsson were studying the role of prostaglandins in platelet aggregation. They discovered two new cyclooxygenase products, one of which was highly active in stimulating aggregation. As little as 5 ng/ml caused platelets to aggregate and the aorta to contract. They called the substance **thromboxane** because of its discovery in thrombocytes. Thromboxane A_2 (Figure 3.33) is an extremely labile substance and is rearranged with a half-life of about 30 seconds to a stable and physiologically inert derivative, thromboxane B_2 (Figure 3.33). Even to this date, thromboxane A_2 has not been isolated and its structure has been deduced from the structure of its metabolites like thromboxane B_2. Because of the role of thromboxane A_2 in platelet aggregation considerable effort has been devoted to trying to find inhibitors of thromboxane synthase. Imidazole and pyridine or imidazole derivatives seem to be the most effective. In addition to causing platelet aggregation, thromboxane A_2 induces smooth muscle contraction and serotonin release (all at concentrations of less than 20 nM).

In 1976, Needleman and coworkers found an enol-ether (prostacyclin, PGI_2) which was produced in vascular endothelial cells and which would lead to coronary vasodilation. At the same time, Vane showed that prostacyclin had an antiaggregatory action on platelets. In fact, it is the most powerful inhibitor of platelet aggregation known and concentrations of less than 1 ng/ml prevent arachidonate-induced aggregation *in vitro*. The physiological effects of prostacyclin are essentially the opposite to those of thromboxane (Table 3.13). Like thromboxane A_2, it has a short half-life and rearranges to a stable and physiologically inert compound 6-keto-$PGF_{1\alpha}$ (Figure 3.33). PGI_2 synthetase is inactivated by a variety of lipid hydroperoxides and this explains why free radical scavengers (like Vitamin E) serve to protect the enzyme. Vasodepressor substances such as histamine or bradykinin along with thrombin all stimulate PGI_2 synthesis in cultured endothelial cells and may serve physiologically to limit the area of platelet deposition about a site of vascular injury. In general, PGI_2 elevates cAMP levels in responsive cells (platelets, vascular smooth muscle cells) and activation of adenylate cyclase through PGI_2 binding to its membrane receptor may represent its mechanism of action.

3.4.4 Prostaglandins and other eicosanoids are rapidly catabolized

We have already mentioned that thromboxane A_2 and prostacyclin have very short half-lives *in vivo*. In addition, it was shown by Vane and Piper in the late 1960s that prostaglandins like PGE_2 or $PGF_{2\alpha}$ were rapidly catabolized and did not survive a single pass through the circulation. The lung plays a major role in this inactivation process which is usually initiated by oxidation of the hydroxyl group at C_{15}. The $\Delta13$ double bond is next attacked and further degradation involves β- and ω-oxidation (sections 3.3.1 and 3.3.3). Concentrations of major active prostaglandin products in blood are less than 10^{-10} M and because of their rapid catabolism they can only act as local hormones or **autocoids** which modify biological events close to their sites of synthesis.

Table 3.13 Opposing effects of prostacyclin and thromboxanes on the cardiovascular system

Thromboxanes in platelets	*Physiological effect*	*Prostacyclin in arterial wall*
Stimulates	Platelet aggregation	Inhibits
Constricts	Arterial wall	Relaxes
Lowers	Platelet cAMP levels	Raises
Raises	Blood pressure	Lowers

3.4.5 Instead of cyclooxygenation, arachidonate can be lipoxygenated

For over 40 years it has been known that plant tissues contain **lipoxygenases** which catalyse the introduction of oxygen into polyunsaturated fatty acids. In 1974 Hamberg and Samuelsson found that platelets contained a 12-lipoxygenase and since that time 5- and 15-lipoxygenases have also been discovered. The immediate products are hydroperoxy fatty acids which for arachidonate substrate are hydroperoxy eicosatetraenoic acids (HPETEs).

The HPETEs can undergo three reactions (Figure 3.34). The hydroperoxy group can be reduced to an alcohol, thus forming an hydroxyeicosatetraenoic acid (HETE). Alternatively a second lipoxygenation elsewhere on the chain yields a dihydroxyeicosatetraenoic acid (diHETE) or a dehydration produces an epoxy fatty acid.

Epoxy fatty acids, such as leukotriene A_4 (Figure 3.34) can undergo non-enzymatic reactions to various diHETEs, can be specifically hydrated to a given diHETE or can undergo ring opening with GSH to yield peptide derivatives. Epoxy eicosatrienoic acids and their metabolic products are called **leukotrienes** – the name being derived from the cells (leukocytes) in which they were originally recognized.

As mentioned above, three lipoxygenases (5-, 12- and 15-) have been found in mammalian tissues. The 5-lipoxygenase is responsible for leukotriene production and is important in neutrophils, eosinophils, monocytes, mast cells and keratinocytes as well as lung, spleen, brain and heart. Products of 12-lipoxygenase activity also have biological activity, e.g. 12-HPETE inhibits collagen-induced platelet aggregation and 12-HETE can cause migration of smooth muscle cells *in vitro* at concentrations as low as one femtomolar (10^{-15} M). In contrast, few biological activities have been reported for 15-HETE although this compound has a potentially important action in inhibiting the 5- and 12-lipoxygenases of various tissues.

The opening of the epoxy group of leukotriene LTA_4 by the action of glutathione S-transferase attaches the glutathionyl residue to the 6-position

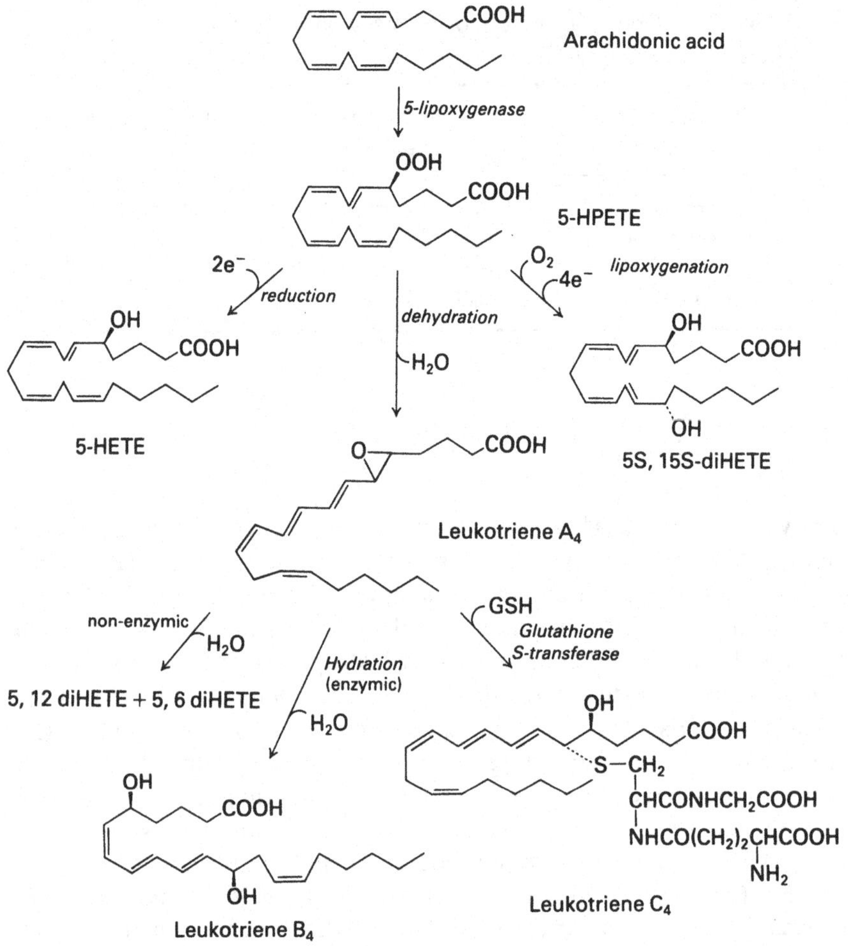

Figure 3.34 Formation of leukotrienes from arachidonic acid.

(Figure 3.34). This compound (LTC$_4$) can then lose a γ-glutamyl residue to give LTD$_4$ and the glycyl group is released in a further reaction to give LTE$_4$ (Figure 3.35). Prior to their structural elucidation, leukotrienes were recognized in perfusates of lungs as slow reacting substances (SRS) after stimulation with cobra venom or as slow reacting substances of anaphylaxis (SRS-A) after immunological challenge. LTC$_4$ and LTD$_4$ are now known to be major components of SRS-A.

3.4.6 Control of leukotriene formation

Unlike the cyclooxygenase products, the formation of leukotrienes is not solely determined by the availability of free arachidonic acid in cells (section

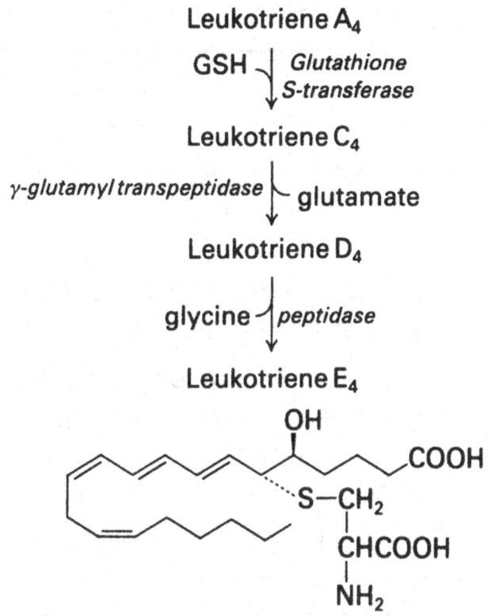

Figure 3.35 Formation of peptidoleukotrienes.

3.4.9). Thus, for example, 5-lipoxygenase requires activation in most tissues and various immunological or inflammatory stimuli are able to cause this.

In some cells both cyclooxygenase and lipoxygenase products are formed from arachidonic acid. It is possible that the arachidonic acid substrate comes from separate pools. For example, macrophages release prostaglandins but not leukotrienes when treated with soluble stimuli. In contrast, when insoluble phagocytic stimuli (such as bacteria) are used, both prostaglandin and leukotriene formation is increased.

We have already mentioned that 15-HETE, a 15-lipoxygenase product, can inhibit 5- and 12-lipoxygenases. Conversely, 12-HPETE has been shown to increase 5-HETE and LTB_4 formation. Thus, the various lipoxygenase pathways have interacting effects as well as the interdependent actions of cyclooxygenase and lipoxygenase products.

3.4.7 Physiological action of leukotrienes

Leukotrienes have potent biological activity. A summary of some of their more important actions is made in Table 3.14. The peptidoleukotrienes contract respiratory, vascular and intestinal smooth muscles. In general LTC_4 and LTD_4 are more potent than LTE_4. By contrast, LTB_4 is a chemotactic agent for neutrophils and eosinophils. Although it can cause plasma exudation by increasing vascular permeability, it is less potent than the peptidoleukotrienes. These actions of the leukotrienes have implications for asthma, immediate hypersensitivity reactions, inflammatory reactions and myocardial infarction.

Table 3.14 Biological effects of leukotrienes

Effect	Comments
Respiratory	Peptidoleukotrienes cause constriction of bronchi especially smaller airways and increase mucus secretion
Microvascular	Peptidoleukotrienes cause arteriolar constriction, venule dilation, plasma exudation
Leukocytes	LTB_4 chemotactic agent for neutrophils, eosinophils, e.g. increase degranulation of platelets, cell surface receptors and adherence of polymorphonucleocytes to receptor cells
Gastrointestinal	Peptidoleukotrienes cause contraction of smooth muscle (LTB_4 no effect)

The mechanism of action of the leukotrienes is unknown at present. Nevertheless, we do know that the compounds bind to receptors on the surface of target cells. These receptors show specificity for different types of leukotrienes, e.g. LTD_4 but not LTC_4.

The function of the 12- and 15-lipoxygenases is much less clear than that of the 5-lipoxygenase. Nevertheless, it is thought that the 15-lipoxygenase products (15-HETE) may play a role in endocrine (e.g. testosterone) secretion. A build-up of 12-HETE (46 × normal) has been reported in lesions of patients suffering from the skin disease psoriasis. This compound is known to be a chemoattractant and topical application of 12-HETE causes erythema (similar to sunburn). Interestingly, one of the most effective anti-erythema treatments is etretinate which inhibits the 12-lipoxygenase. Etretinate has been used for psoriatic patients but it is particularly effective for anti-sunburn treatment.

3.4.8 The cyclooxygenase products also exert a range of activities

The eicosanoids produced by the cyclooxygenase reaction exert a range of profound activities at concentrations down to 10^{-9} g per gram tissue. These are now known to include: effects on smooth muscle contraction, inhibition or stimulation of platelet stickiness, bronchoconstriction/dilation and vasoconstriction/dilation with a consequent influence on blood pressure. The effect on smooth muscle contraction was the earliest to be recognized and has been much used as a biological assay for eicosanoids. This use has been largely replaced by modern analytical methods such as gas chromatography-mass spectrometry or a combination of HPLC separations followed by radioimmuno assay.

The effects of prostaglandins, in particular on smooth muscle contraction, is utilized in both medicine and agriculture (section 3.4.2). PGE_2 is known to have a cytoprotective effect in the upper gut while PGI_2 (prostacyclin) is used in dialysis and the treatment of peripheral vascular disease. Continuous perfusion of very small amounts (e.g. 0.01 μg PGE_1 min^{-1}) into pregnant women causes uterine activity similar to that encountered in normal labour without any effect on blood pressure. Prostaglandins and synthetic analogues are now being used for induction of labour and for therapeutic abortion in both humans and domestic animals. The exact time of farrowing of piglets, for example, can be precisely controlled by prostaglandin treatment. For abortions in women, a solution of PGF_2 at a concentration of 5 μg ml^{-1} is administered by intravenous infusion at a rate of 5 μg min^{-1} until abortion is complete (about five hours). The method appears, so far, to be safe with few side effects and no further surgical intervention is required.

PGE is also an effective antagonizer of the effect of a number of hormones on free fatty acid release from adipose tissue (section 4.6.3). It is likely that its mode of action, and that of thromboxanes and prostacyclin, in smooth muscle and platelets involves interaction with the adenyl cyclase enzyme in the cell membrane, so that their physiological activities are largely expressed through regulation of cellular cAMP levels. The sequence of events is probably: 1. release of arachidonic acid from a membrane phospholipid by the activation of phospholipase A (section 3.4.9); 2. the formation of the active compound *via* the cyclic endoperoxide. (It still has to be discovered what controls the relative activity of the different pathways after the cyclic endoperoxide intermediate); 3. the reaction of the active molecules with specific receptor sites on the target cell membrane; 4. the positive or negative regulation of adenyl cyclase.

In fat cells, PGE acts to reduce cAMP levels; in platelets prostacyclins tend to raise cAMP levels, while thromboxanes have the reverse effect, thereby initiating aggregation (Table 3.13). Ca^{2+} is required for the aggregation process and there is some evidence that thromboxane A_2 or other cyclooxygenase products can act as Ca^{2+} ionophores. The interactive effects of thromboxane A_2 and prostacyclin with regard to platelet cAMP levels are depicted in Figure 3.36.

In healthy blood vessels, prostacyclin is produced which counteracts the thromboxane effect on cAMP levels and, therefore, reduces platelet adhesion. In artifical or damaged vessels the thromboxane effect is unopposed and adhesion takes place.

Another area of prostaglandin action which has been studied in detail is their role in water-retention by the kidney. The final event in urine production occurs in the terminal portion of the renal tubule where ADH

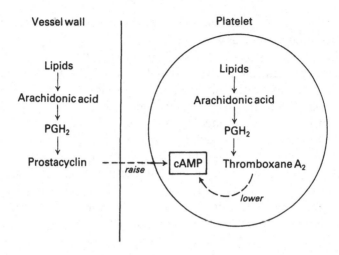

Figure 3.36 Interaction of thromboxanes and prostacyclins in regulating platelet cAMP levels.

(antidiuretic hormone) interacts with receptors on the tubule to cause movement of water back into the tissue. ADH does this by activating adenylate cyclase and increasing cellular cAMP levels. (cAMP itself has been shown to have a hydroosmotic effect also.) It has been known for some time that low concentrations of PGE_2 inhibit the ADH-induced cAMP formation and, hence, the hydroosmotic effect. The mechanism for this antagonism is unclear but it has been suggested that PGE_2 may prevent ADH activating the adenylate cyclase by modifying the properties of one of the GTP binding proteins at the receptor.

3.4.9 For eicosanoid synthesis, an unesterified fatty acid is needed

A necessary prerequisite for eicosanoid formation is the availability of an appropriate unesterified fatty acid. In mammals this is usually arachidonic acid. The concentration of unesterified arachidonic acid in cells is well below the K_m for prostaglandin H synthetase. Thus, the first stage for eicosanoid formation will normally be an activation of the release of arachidonate from the *sn*-2 position of phosphoglycerides which contain it. Two classes of phospholipids are thought to play major roles as sources of arachidonate in cells – phosphatidylcholine (the major membrane constituent) and phosphoinositides (by virtue of the high enrichment of arachidonate at their *sn*-2 position). Thus, hydrolysis of phosphoinositides not only produces two second messengers directly (section 8.8) but may also initiate an arachidonate cascade. Other potential sources of arachidonic acid are the plasmalogens (section 6.2.1) which also have a high enrichment at the *sn*-2 (acyl) position. Because plasmalogens are poor substrates for phospholipase A_2, they are hydrolysed by a plasmalogenase first. Thus, release of arachidonic acid from plasmalogens could be controlled independently to that from diacyl-phosphoglycerides.

Two main types of stimuli increase arachidonate release. These can be called physiological (specific) and pathological (non-specific). Physiological stimuli, such as adrenaline, angiotensin II and certain antibody–antigen complexes, cause the specific release of arachidonic acid. In contrast, pathological stimuli, such as mechanical damage, ischaemia, membrane-active venoms such as mellitin or tumour promoters like phorbol esters, have generalized effects on cellular membranes and promote release of all fatty acids from the *sn*-2 position.

It is also of interest that phospholipase A_2, whose activity is needed to initiate eicosanoid production, is also needed to produce another type of biologically active lipid : platelet activating factor (section 7.1.10).

3.4.10 Essential fatty acid activity is closely related to double bond structure and to the ability of such acids to be converted into a physiologically active eicosanoid

Work from the Dutch school and from Holman's laboratory at the Hormel Institute, Minnesota, originally showed that only those fatty acids (including new synthetic odd-numbered acids) that act as precursors for biologically-active eicosanoids have EFA activity. Moreover, the 'n-6, n-9' hypothesis has had to be dropped since some of the new synthetic acids do not contain this structure (Table 3.15). Van Dorp and his colleagues at the Unilever Laboratories at Vlaardingen have made the postulate that only those fatty acids capable of being converted into the $\Delta 5,8,11,14$-tetraenoic fatty acids of chain lengths 19C, 20C and 22C will show essential fatty acid activity because only these tetraenoic acids can give rise to physiologically active eicosanoids. However, it is now known that columbinic acid (*trans*-5, cis-9, cis-12–18 : 3), when given to EFA-deficient rats, normalizes growth and cures the dermatitis but is unable to form an eicosanoid.

Moreover, although there is a correlation between EFA activity and the potential to be converted to eicosanoids, one cannot cure EFA deficiency by infusion of eicosanoids because they are rapidly destroyed and because different cells produce their own special pattern of eicosanoids. However, one of the first organs to show EFA deficiency is the skin, whose water permeability is very much increased in this condition. Topical application of EFA to skin can reverse the deficiency symptoms and there is now evidence

Table 3.15 Relationship between fatty acid structure and EFA activity

Fatty acid chain length	Position of double bonds		EFA Potency (units/g)
	From carboxyl end (Δ)	From methyl end (ω)	
18 : 2	9, 12	6, 9	100
18 : 3	6, 9, 12	6, 9, 12	115
18 : 3	8, 11, 14	4, 7, 10	0*
18 : 4	6, 9, 12, 15	3, 6, 9, 12	34
18 : 4	5, 8, 11, 14	4, 7, 10, 13	0*
19 : 2	10, 13	6, 9	9
20 : 2	11, 14	6, 9	46
20 : 3	8, 11, 14	6, 9, 12	100
20 : 3	7, 10, 13	7, 10, 13	0*
20 : 4	5, 8, 11, 14	6, 9, 12, 15	139
21 : 3	8, 11, 14	7, 10, 13	56
22 : 3	8, 11, 14	8, 11, 14	0*
22 : 5	4, 7, 10, 13, 16	6, 9, 12, 15, 18	139

* Cannot give rise to any prostaglandins.

that topical application of prostaglandins can also be very effective. (See section 8.12 for other comments about lipids and skin diseases.)

We are still a long way from being able to account for the fate of all the EFA that enter the body and the fate of the eicosanoids that may be formed from them. Eicosanoids are metabolized very rapidly and their metabolites excreted in the urine or bile. The detection, isolation and analysis of such metabolites is, therefore, one approach to studying the daily eicosanoid production. In this way it has been estimated that 1 mg of prostaglandin metabolites is formed in 24 hours in man – considerably less than the 10 g of EFA which are thought to be necessary daily. However, the demonstration that EFAs like linoleic acid may have additional functions (section 8.12) apart from being eicosanoid precursors may account for this apparent discrepancy and raises the possibility that new roles for EFAs may be discovered in the future.

3.5 SUMMARY

Many hundreds of different fatty acid structures have been found in Nature but only a few such compounds occur commonly. These are usually even chain fatty acids of 16–22 carbons in length. The most common saturated fatty acids are palmitic and stearic with oleic acid the most prevalent monounsaturated molecule. When polyunsaturated fatty acids occur, their double bonds are almost invariably separated by 3 carbons (i.e. methylene-interrupted). Linoleic and α-linolenic acids are found in plants, arachidonic acid is common in animals while marine organisms contain significant amounts of 20 or 22 carbon polyunsaturates.

Fatty acids are made initially by a combination of the action of acetyl-CoA carboxylase and fatty acid synthetase. The acetyl-CoA carboxylase enzymes from different organisms vary in their structure with multifunctional proteins being found in mammals and higher plants while *E. coli* uses a multienzyme complex. The mechanism of acetyl-CoA carboxylase is, however, broadly similar and takes place in two main stages. First biotin is carboxylated by an ATP-dependent reaction. Second, the carboxyl group is transferred to acetyl-CoA thus producing malonyl-CoA.

Acetyl-CoA carboxylase can be regarded as the first committed step of fatty acid synthesis. It is not surprising, therefore, that the enzyme is under careful regulation. The control mechanisms have been studied best in mammals where allosteric regulation is provided by tricarboxylic acids or acyl-CoAs while long term regulation by diet involves changes in the amount of carboxylase protein. Different regulatory mechanisms are used by *E. coli* and by higher plants.

Fatty acid synthetases are divided into Type I, Type II and Type III. Type I enzymes are multifunctional proteins which contain covalently bound acyl carrier protein (ACP) and release products as unesterified fatty acids (mammals) or transfer them to CoA (yeast). In animals the fatty acid synthetase is a homo dimer with each chain containing all of the partial reactions. By contrast, yeast fatty acid synthetase contains two dissimilar peptides and operates as an $\alpha_6\beta_6$ complex.

Type II fatty acid synthetases are found in bacteria such as *E. coli*, cyanobacteria and the chloroplasts of algae, higher plants and *Euglena*. They are multienzyme complexes which comprise a number of dissociable proteins which are responsible for acyl transfer, condensation, dehydration and reduction reactions. ACP can also be isolated as a small molecular mass acidic protein. In *E. coli* the fatty acid synthetase has the additional feature that it produces monounsaturated fatty acids (mainly *cis*-vaccenate) in addition to long chain saturated products.

Type III synthetases are responsible for the chain lengthening of pre-formed fatty acids and are also called elongase enzymes. Most fatty acid elongases are membrane-bound enzyme complexes located on the endoplasmic reticulum. Like the Type I and Type II fatty acid synthetases they use malonyl-CoA as the source of 2C addition units. However, some elongation activity occurs in mitochondria where acetyl-CoA is used for additionally a reversal of β-oxidation.

Once fatty acids have been made *de novo* they can either be modified by elongation (above), by desaturation or by other reactions. Desaturation usually takes place by an aerobic mechanism – an exception being in the *E. coli* fatty acid synthetase. Aerobic desaturases differ from each other by the nature of the acyl substrate they use, the type of reduced cofactor and the position at which the double bond is introduced into the acyl chain. Particularly notable are the desaturases which produce the polyunsaturated linoleic and α-linolenic acids. These enzymes use complex lipid substrates rather than acyl-thioesters.

Like acetyl-CoA carboxylase, the different fatty acid synthetases and desaturases are under careful metabolic regulation. Various mechanisms are used to control their activity, depending on the organism.

Fatty acids are oxidized by chemical or enzymatic mechanisms. Chemical oxidation is particularly important for polyunsaturated fatty acids and gives rise to the formation of free radicals which can, in turn, damage tissues. Similar peroxidation reactions can also be catalysed by lipoxygenase enzymes.

Oxidation of saturated and unsaturated fatty acids can take place by α-, β- or ω-oxidations. α-Oxidation is common in plants and brain tissue and causes the release of CO_2 and formation of a new fatty acid, one carbon shorter. The more important β-oxidation mechanism causes the release of acetyl-CoA. The process requires a cycle of four reactions (oxidation, hydration, oxidation, thiolysis) and is localized in mitochondria and microbodies in eukaryotes; the proportional distribution varying with organism. β-Oxidation is the most important process by which energy is released from fat stores in animals and plants.

Eicosanoids are 20C biologically active compounds which derive from essential fatty acids. Arachidonic acid can be converted to a cyclic endoperoxide and this, in turn, can form prostaglandins, thromboxanes or prostacyclin. In contrast, if arachidonate is acted upon by a lipoxygenase then the leukotrienes or hydroxy fatty acids can be formed. All these eicosanoids have very potent, and often interacting, biological actions and affect almost every tissue in mammals.

REFERENCES

General

Applewhite, T.H. (ed.) (1988) *Proceedings of the World Conference on Biotechnology of the Fats and Oils Industry*. Am. Oil Chem. Soc., Champaign, Il., USA.

Gunstone, F.D., Harwood, J.L. and Padley, F.B. (eds) (1986) *The Lipid Handbook*, Chapman and Hall, London.

Harwood, J.L. and Russell, N.J. (1984) *Lipids in Plants and Microbes*, George Allen and Unwin, Hemel Hempstead.

Mead, J.F., Alfin-Slater, R.B., Howton, D.R. and Popjak, G. (1986) *Lipids: Chemistry, Biochemistry and Nutrition*, Plenum Press, New York.

Numa, S. (ed.) (1984) *Fatty Acid Metabolism and its Regulation*, Elsevier, Amsterdam.

Vance, D.E. and Vance, J.E. (eds) (1985) *Biochemistry of Lipids and Membranes*, Benjamin/Cummings, Menlo Park, CA, USA.

Fatty acid synthesis

Clark, D. and Cronan, J.E. (1981) Bacterial mutants for the study of lipid metabolism. *Method. Enzymol.*, **72**, 293–307.

Cook, H.W. (1985) Fatty acid desaturation and chain elongation in eucaryotes, in *Biochemistry of Lipids and Membranes* (eds D.E. Vance and J.E. Vance), Benjamin/Cummings, Menlo Park, CA, USA, pp. 181–212.

Brenner, R.R. (1981) Nutritional and hormonal factors influencing desaturation of essential fatty acids. *Prog. Lipid Res.*, **20**, 41–47.

Fulco, A.J. (1983) Fatty acid metabolism in bacteria. *Prog. Lipid Res.*, **22**, 133–160.

Hardie, D.G., Carling, D. and Slim, A.T.R. (1989) The AMP – activated protein kinase: a multisubstrate regulator of lipid metabolism. *Trends Biochem. Sci.*, **14**, 20–23.

Harwood, J.L. (1988) Fatty acid metabolism. *Annual Rev. Plant Physiol.*, **39**, 101–138.

Jackowski, S. and Rock, C.O. (1987) Acetoacetyl-ACP synthase, a potential regulator of fatty acid biosynthesis in bacteria. *J. Biol. Chem.*, **262**, 7927–7931.

Jaworski, J.G. (1987) Biosynthesis of monoenoic and polyenoic fatty acids, in *The Biochemistry of Plants* (eds P.K. Stumpf and E.E. Conn), *Vol. 9*, Academic Press, New York, pp. 159–174.

Jeffcoat, R. (1979) The biosynthesis of unsaturated fatty acids and its control in mammalian liver. *Essays Biochem.*, **15**, 1–36.

Mangold, H.K. and Spener, F. (1980) Biosynthesis of cyclic fatty acids, in *The Biochemistry of Plants* (eds P.K. Stumpf and E.E. Conn), *Vol. 4*, Academic Press, New York, pp. 647–663.

McCarthy, A.D. and Hardie, D.G. (1984) Fatty acid synthase – an example of protein evolution by gene fusion. *Trends Biochem. Sci.*, **10**, 60–63.

de Renobales, M., Cripps, C., Stanley-Samuelson, D.W., Jurenka, R.A. and Blomqvist, G.J. (1987) Biosynthesis of linoleic acid in insects. *Trends Biochem. Sci.*, **12**, 403–406.

Rock, C.O. and Cronan, J.E. (1985) Lipid metabolism in procaryotes, in *Biochemistry of Lipids and Membranes* (eds D.E. Vance and J.E. Vance), Benjamin/Cummings, Menlo Park, CA, USA, pp. 73–115.

Stumpf, P.K. (1987) The biosynthesis of saturated fatty acids, in *The Biochemistry of Plants* (eds P.K. Stumpf and E.E. Conn), *Vol. 9*, Academic Press, New York, pp. 121–136.

Wakil, S.J., Stoops, J.K. and Joshi, V.C. (1983) Fatty acid synthesis and its regulation. *Annual Rev. Biochem.*, **52**, 537–579.

Fatty acid breakdown

Esterbauer, H. and Cheeseman, K.H. (eds) (1987) Lipid peroxidation: biochemical and biophysical aspects. Special issue of *Chemistry and Physics of Lipids, Vol. 44*.

Galliard, T. (1980) Degradation of acyl lipids: hydrolytic and oxidative enzymes, in *Biochemistry of Plants* (eds P.K. Stumpf and E.E. Conn), *Vol. 4*, Academic Press, New York, pp. 85–116.

Galliard, T. and Chan, H.W.S. (1980) Lipoxygenases, in *Biochemistry of Plants* (eds P.K. Stumpf and E.E. Conn), *Vol. 4*, Academic Press, New York, pp. 131–162.

Kindl, H. (1987) β-Oxidation of fatty acids by specific organelles, in *Biochemistry of Plants* (eds P.K. Stumpf and E.E. Conn), *Vol. 9*, Academic Press, New York, pp. 31–53.

McGarry, J.D. and Foster, D.W. (1980) Regulation of hepatic fatty acid oxidation and ketone body production, *Annual Rev. Biochem.*, **49**, 395–420.

Schulz, H. (1985) Oxidation of fatty acids, in *Biochemistry of Lipids and Membranes* (eds D.E. Vance and J.E. Vance), Benjamin/Cummings, Menlo Park, CA, USA, pp. 116–142.

Eicosanoid metabolism

Hammarstrom, S. (1983) Leukotrienes. *Annual Rev. Biochem.*, **52**, 355–377.

Holman, R.T. (ed) (1981) Essential fatty acids and prostaglandins. *Progress in Lipid Research, Vol. 20*, Pergamon Press, New York.

Needleman, P., Turk, J., Jakschik, B.A., Morrison, A.R. and Lefkowith, J.B. (1986) Arachidonic acid metabolism. *Annual Rev. Biochem.*, **55**, 69–102.

Sammuelsson, B. *et al.* (1978) Prostaglandins and thromboxanes. *Annual Rev. Biochem.*, **47**, 997–1029.

Smith, W.L. and Borgeat, P. (1985) The eicosanoids: prostaglandins, thromboxanes, leukotrienes and hydroxy-eicosaenoic acids, in *Biochemistry of Lipids and Membranes* (eds D.E. Vance and J.E. Vance), Benjamin/Cummings, Menlo Park, CA, USA, pp. 325–360.

4 *Lipids as energy stores*

4.1 INTRODUCTION: MANY PLANTS AND ANIMALS NEED TO STORE ENERGY FOR USE AT A LATER TIME. LIPID FUELS ARE MAINLY TRIACYLGLYCEROLS OR WAX ESTERS

All living organisms require a form of energy to sustain life. Whereas the basic mechanisms for powering the life-sustaining anabolic chemical reactions through the high energy bonds of ATP and similar molecules are common to animals and plants, the primary sources of energy are very different. Plants use sunlight as the primary fuel source to enable them to synthesize carbohydrates. They are then able to synthesize fatty acids from the degradation products of carbohydrates as described in section 3.2. Animals do not have the facility directly to use sunlight but must take in their fuel in the diet as lipid or carbohydrate from plants or from other animals that have themselves synthesized their body tissues from plant materials.

Energy needs are of two main types. There is a requirement to maintain, throughout life, on a minute to minute basis, the organism's essential biochemistry. Either carbohydrates or fatty acids can fulfil this function, although tissues do vary in their ability to utilize different fuels: the mammalian brain, for example, is reliant on glucose as a fuel source, since it cannot utilize fatty acids directly. More intermittent and specialized needs for fuels arise, however, which demand that the organism maintains a store of fuel that can be mobilized when required. Thus, animals may need to use such reserves after a period of starvation or hibernation or for the energy demanding processes of pregnancy and lactation, while plants need abundant supplies of energy for seed germination. Animals and plants each make use of different types of fuels that may be stored in different tissues. Thus practically all animals use lipids as a long term form of stored energy and almost all use triacylglycerols as the preferred lipid. Some marine animals, however, make use of wax esters as their energy store. Most animals store their energy in a specialized tissue, the adipose tissue, but some fish use their flesh or the liver as a lipid store. In general, plants store the fuel required for germination in their seeds either as lipid or starch. Oil bearing seeds usually contain triacylglycerols but some desert plants, like jojoba, use wax esters as a fuel.

This chapter is about storage lipids in general, but because of their predominance as a fuel and their importance in biochemistry, the discussion will largely be concerned with the triacylglycerols.

4.2 THE NAMING AND STRUCTURE OF THE ACYLGLYCEROLS (GLYCERIDES)

4.2.1 Types of natural acylglycerols: acylglycerols or fatty acid esters of glycerol are the major components of natural fat and oils

This chapter is mainly about the esters of the trihydric alcohol, glycerol, with fatty acids. The preferred term is now acylglycerols although you will certainly come across the older synonymous term glycerides in other literature. Likewise triacylglycerol, diacylglycerol and monoacylglycerol will be used for the specific compounds rather than the older triglyceride, diglyceride and monoglyceride.

In triacylglycerols, all three glycerol hydroxy groups are esterified with a fatty acid. Partial acylglycerols have only one or two positions esterified and are called mono- and di-acylglycerols respectively (Figure 4.1). Acylglycerols are the chief constituents of natural fats (solids) and oils (liquids), names that are often used synonymously for them, although it is important to remember that natural fats and oils also contain minor proportions of other lipids, for example steroids and carotenoids. The most abundant fatty acids in natural acylglycerols are palmitic, stearic, oleic and linoleic; plant acylglycerols usually have a relatively higher proportion of the more unsaturated fatty acids. Coconut and palm oils are exceptions (section 5.1.2). Milk fats have a much higher proportion of short chain fatty acids (C2–C10) than other animal fats (section 5.1.2). Odd chain or branched chain fatty acids are only minor constituents of acylglycerols; seed oils, especially, contain a variety of unusual fatty acids with oxygen containing groups and ring systems (section 3.1). Some acylglycerols that contain ricinoleic acid (section 3.1.5 and 3.2.3) may have further fatty acids esterified with the hydroxy group of the ricinoleic acid. Thus, tetra-, penta-, and hexa-acid acylglycerols occur in some plant oils.

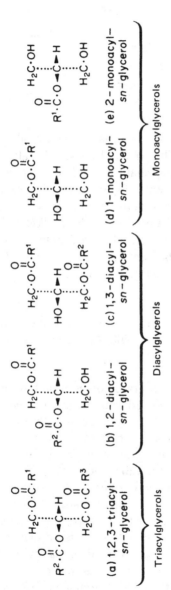

Figure 4.1 Some common acylglycerol structures.

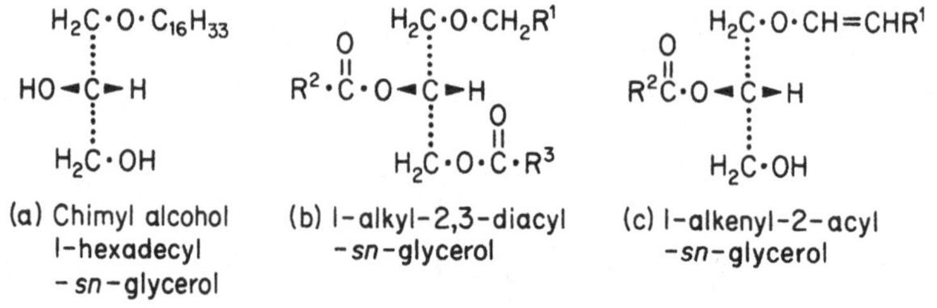

(a) Chimyl alcohol
1-hexadecyl
– sn – glycerol

(b) 1-alkyl-2,3-diacyl
– sn-glycerol

(c) 1-alkenyl-2-acyl
– sn-glycerol

Figure 4.2 Structures of some glyceryl ethers.

Alkylglycerols contain hydrocarbon chains linked to glycerol by an ether rather than an ester linkage. Chimyl, batyl and selachyl alcohols are glyceryl monoethers derived from long chain alcohols corresponding to C16:0, C18:0 and C18:1 respectively (Figure 4.2a). These are the major naturally occurring glyceryl ethers and may be esterified in addition with one or two fatty acids (Figure 4.2b,c). Usually both alkyl and alkenyl ethers occur together in varying proportions. They are abundant in the liver oils of fish such as sharks and in the central nervous system. Only very small quantities of ether glycerides are found in plants.

4.2.2 Distribution of fatty acids: many different combinations of fatty acids are possible. All natural oils are complex mixtures of molecular species

Natural acylglycerols contain a vast range of different fatty acids. Even if the number of fatty acids were restricted to three, it would be possible to have 18 different arrangements of these fatty acids within the triacylglycerol molecule. There are, however, many hundreds of fatty acids and therefore the potential number of possible molecular species of triacylglycerols is enormous. This assumes, however, that each fatty acid may occupy any of the three positions relative to any other acid; in other words that the distribution is random. This would mean that during their biosynthesis there would be no selectivity at any step in the pathway. Many people, notably the great pioneering lipid chemist Hilditch, considered that this was the case, until some years ago studies with purified preparations of the enzyme, pancreatic lipase (which rapidly hydrolyses the outer ester bonds of a triacylglycerol but not the central one, sections 4.5 and 5.3.1), revealed that the proportion of unsaturated to saturated fatty acids in the resulting 2-monoacylglycerol was considerably greater than on the outer positions of the original triacylglycerol. It was argued that as the outer positions were equivalent and could not be distinguished from each other, then the remaining fatty acids must be randomly distributed between these two positions. This was known as the **1,3-random**, **2-random** or **restricted random** hypothesis. By analysing the composition of total triacylglycerol fatty acids and either the acids released by lipase hydrolysis or those present in the resulting monoacylglycerols and by making the assumption set forth in this hypothesis, then one can calculate the range of molecular species to be expected in any given fat.

The most direct way to find out what species are present in a particular fat is to separate them from each other and analyse each one directly. This is now possible by thin layer chromatography on silica gel impregnated with silver nitrate and other techniques described in Chapter 2. Biochemists have also realized that, owing to the prochiral nature of glycerol (Chapter 1), positions 1 and 3 are not equivalent. Techniques are available for distinguishing between positions 1 and 3 and it has been revealed that in many cases fatty acids are not randomly distributed between the outer positions. Complete generalization about acylglycerol composition is impossible because while many acylglycerols are known to have a stereospecific distribution of fatty acids, others seem to conform to the restricted random distribution, while yet others have a completely random distribution. Lard and human milk fat are unusual in having a preponderance of palmitic acid (a saturated fatty acid) at position 2.

During the last century, most chemists (with the notable exception of Berthelot) assumed that natural fats were mixtures of simple, single acid acylglycerol. The lack of techniques for separating fat components prolonged this state of ignorance. For a complete analysis, the acylglycerols themselves must be separated into components. The chief methods are, in order of historical importance: fractional crystallization, low pressure fractional distillation, countercurrent distribution, argentation TLC, gas liquid chromatography and high performance liquid chromatography (HPLC). The epic work that has provided the solid basis for our knowledge was that of Hilditch and his school beginning in 1927. They refined the technique of fractional crystallization and also introduced an oxidative method for the determination of trisaturated acylglycerol species. This depended on the fact that all acylglycerols containing unsaturated fatty acids could be oxidized by permanganate in acetone to yield oxidized fats whose physical properties were very different from the remaining saturated acylglycerols and which could therefore be easily separated from them. It was these techniques that led Hilditch to develop his theories of even or random distribution. The next step forward came with the introduction of countercurrent distribution, a technique that separates mixtures according to their partition between immiscible solvents. The invention of an automatic apparatus that could perform a large number of transfers enabled a much higher degree of resolution of natural fats, especially highly unsaturated ones, than that achieved by crystallization. Nevertheless, the technique is quite laborious, consumes large amounts of solvents and is not now widely used for small scale analytical work, although it can be useful for preparative scale work.

4.2.3 Structural analysis: by combining analytical techniques, the structures of molecular species in a mixture can be determined

Two methods in combination have been responsible for most of the present day analyses of natural fats – lipase hydrolysis and thin layer

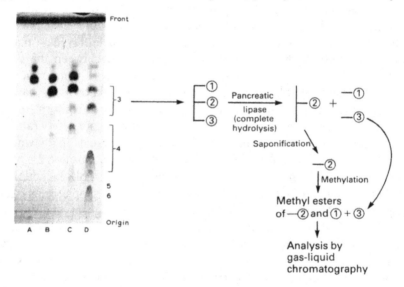

Figure 4.3 Analysis of acylglycerols by combination of argentation-thin layer chromatography and gas–liquid chromatography.

First, triacylglycerols are separated by argentation-TLC into different 'species' according to the total number of double bonds in the molecule. The left-hand side of the figure shows thin layer separations of natural triacylglycerol mixtures on silica gel G impregnated with silver nitrate. Solvent: isopropanol–chloroform, 1.5 : 98.5 (v/v). Spots were located by spraying with 50% sulphuric acid and charring. A = palm oil; B = olive oil; C = groundnut oil; D = cottonseed oil. The numbers represent the total number of double bonds in each triacylglycerol molecule.

The right-hand side shows how compounds from a lane on the chromatogram *that has not been sprayed with sulphuric acid* can be removed for analysis of the fatty acids on positions 1/3 and 2. The area of silica gel corresponding to each species is removed from the plate, the compound eluted with ether and subjected to lipase hydrolysis which releases fatty acids 1 and 3. These are converted into methyl esters for analysis by GLC. The fatty acid remaining in the monoacylglycerol after lipase hydrolysis representing position 2 can be released by saponification and also analysed by GLC.

chromatography – each combined with gas-liquid chromatography. The method capable of the best resolution is argentation TLC (section 2.4.4); in fact it is the only method whereby an unique separation of a glyceride mixture has been achieved. Even so, it is dependent on the total number of double bonds in the molecule, so that in a complex mixture, one must expect there to be several components that cannot be resolved. Identification of the components depends on being able to synthesize model acylglycerols; this may impose a practical limitation on the method. However, a complete analysis of the mixture can usually be made as follows: 1. fractionation of the acylglycerols into groups by argentation TLC; 2. lipase hydrolysis of each triacylglycerol component; 3. normal silicia gel TLC to separate the fatty acids specifically released from positions 1 and 3 from the monoacylglycerols that contain the fatty acid originally in position 2; 4. GLC of the released fatty acids and the monoacylglycerol fatty acids (Figure 4.3). Methods have been introduced for the fractionation of whole acylglycerols (as opposed to their fatty acid constituents) by GLC and more recently by HPLC, but it is doubtful whether the resolution will ever be as great as that achieved by argentation TLC.

4.2.4 Stereospecific analysis: the different positions on the glycerol molecule can be distinguished by techniques making use of enzyme specificity

The realization that positions 1 and 3 of glycerol are not equivalent led several biochemists to try to find methods for distinguishing between them and hence to discover whether there were differences in the fatty acids occupying these positions. This is known as stereospecific analysis. One of the first methods for stereospecifically analysing all three positions of a triacylglycerol was devised by the Canadian biochemist, Hans Brockerhof, and involved a partial hydrolysis of the triacylglycerols with lipase, followed by a fairly complicated chemical conversion of the resulting diacylglycerols into artificial phospholipids. A more convenient method was invented by the American biochemist, Lands. This is described in Figure 4.4 and beautifully illustrates the asymmetry of glycerol by the use of the enzyme diacylglycerol kinase, which phosphorylates the 3-hydroxyl of a 1,2-diacylglycerol and not the 1-hydroxyl of a 2,3-diacylglycerol. Again, use is made of the specificity of phospholipase A (section 7.2.2) to identify the fatty acid in position 2, the fatty acid in position 1 being derived from analysis of the lysophospholipid. A slight disadvantage of methods involving lipase hydrolysis, which restricts their use somewhat, is that they do not yield a random sample of fatty acids when the triacylglycerol contains (a) long chain polyunsaturated fatty acids such as 20:5 and 22:6 (characteristic of

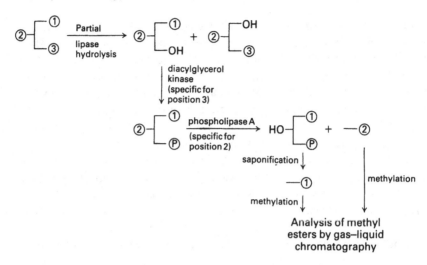

Figure 4.4 Stereospecific analysis of triacylglycerols.

fish oils) which are hardly hydrolysed at all; or (b) very short chain fatty acids, which are hydrolysed much more rapidly than normal ones.

As these methods have increased in use, so more natural acylglycerols have been shown to possess a stereospecific distribution of fatty acids rather than a random or 1,3-random one. Examples are the milk fats, in which the characteristic short chain fatty acids accumulate in position 3. Animal depot fats (with the notable exceptions of lard and human milk fat) have saturated fatty acids in position 1, short chain and unsaturated fatty acids at position 2; position 3 seems to have a more random population, although polyenoic acids tend to concentrate at position 3 in mammals but in position 2 in fish and invertebrates. Certain seed oils contain acetate residues which occur only at position 3. This inherent asymmetry may lead to optical activity, which although extremely small, is measurable, especially in the extreme cases of acylglycerols containing acetate.

4.3 THE STORAGE OF TRIACYLGLYCEROLS IN ANIMALS AND PLANTS

4.3.1 Adipose tissue depots: animals store their reserve fuel in a specialized tissue

A triacylglycerol, when burned chemically in a bomb calorimeter, yields 39 kilojoules of energy per gram (gross energy, GE), and is therefore a more concentrated form of fuel than either protein or carbohydrate, which supply about 16 and 22 kJ g^{-1} respectively. According to Hess's law, this holds true in the body, too, although the pathways between the initial and final states in the oxidation of the fuel are much more complex. The net energy available to the body (metabolizable energy, ME, 38 kJ g^{-1}) is rather less than the gross energy because of excretory losses. When the energy supply from dietary acylglycerols exceeds the energy demands of the body, they are deposited in adipose tissue. As a storage fuel, they have the advantage that they can be stored in anhydrous form and represent more energy for less bulk that the glycogen stored in the liver or muscle which is heavily hydrated. The glycogen store, too, has little capacity to expand, whereas the adipose tissue is seemingly capable of enormous expansion.

There are two types of adipose tissue in the body, known as brown and white adipose tissue. White adipose tissue is the more abundant and is the tissue involved in the storage of body fat. Brown adipose tissue has a more specialized function in energy metabolism (5.5.2). White adipose tissue is widely distributed throughout the body. A large proportion in humans is located just beneath the skin (subcutaneous adipose tissue) and is the tissue that influences the contours of the body. It also provides an insulating and protective layer. Fat contributes a larger proportion of the body weight in women than in men and their subcutaneous adipose tissue is correspondingly more abundant. The contribution of subcutaneous adipose tissue to body mass is particularly noticeable in overweight or obese individuals (section 5.5.2). The tissue is also located internally, for example surrounding the kidneys (perirenal adipose tissue), along the intestinal tract (mesenteric adipose tissue) and in the omentum.

Although adipose tissue contains many types of cells, the ones responsible for fat storage are the adipocytes which are bound together with connective tissue and supplied by an extensive network of blood vessels.

Figure 4.5 presents a scanning electron microscope picture of white adipose tissue and its associated connective tissue. There is still uncertainty about the origins of adipose tissue, some authors believing that adipocytes orginate from fibroblasts that accumulate lipid and subsequently develop all the enzymic functions of fat cells, others regarding the adipocytes as having a more specialized origin. The characteristic feature of the adipocyte is a large single lipid globule surrounded by a narrow ring of cytoplasm. An early stage in the development of the fat cell, the **adipoblast**, is characterized by a large number of lipid inclusions, about 1–2 μm in diameter composed largely of triacylglycerols. The appearance of lipid is thought to correspond with the loss of capacity of cells to divide. In the next stage, the **preadipocyte**, the droplets begin to fuse and surround the nucleus and the droplets themselves are surrounded by large numbers of mitochondria. These cells are similar in appearance to brown fat cells (section 5.5.2). In the final stage, the lipid

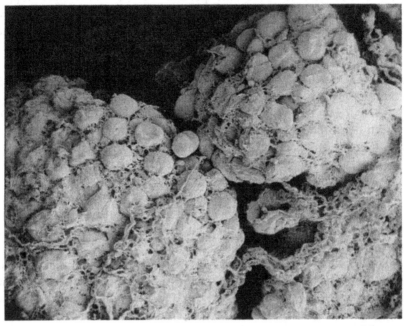

50 μm

Figure 4.5 White adipose tissue as seen by scanning electron microscopy. The bar indicates the scale: 1 cm represents 50 μm. Reproduced with kind permission of The Society of Chemical Industry, from *Chemistry and Industry* (1986), 18 September, p. 768, Figure 1.

droplets fuse to a single lipid globule that pushes the nucleus and cytoplasm to the periphery of the cell.

Adipocytes are unusual in being able to expand to many times their original size by increasing their content of stored fat. Thin sections through adipose tissue indicate not only the big differences in average size of fat cells between lean and obese individuals (Figure 4.6a, b) but also the wide range of cell size within a single piece of tissue. The adipocytes can take in fat from circulating lipoproteins. This involves hydrolytic breakdown of the triacylglycerols in the lipoproteins and release of fatty acids catalysed by the enzyme lipoprotein lipase (section 5.3.5(g)). The fatty acids are transported into the cell and incorporated back into triacylglycerols (section 4.4).

Figure 4.6 Sections through adipose tissue of (a) lean and (b) obese animals. Tissue fixed in formalin, stained with haematoxylin/eosin, embedded in wax and cut in sections a few μm thick. The bars indicate the scale: 6 mm represents 50 μm.

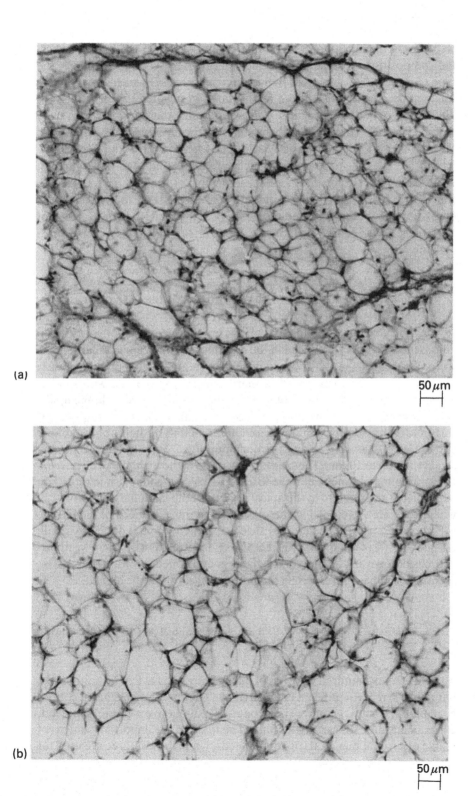

(a)

50 μm

(b)

50 μm

Because circulating lipoproteins are influenced by the fat content of the diet, the composition of the adipose tissue can provide a good indication of the type of dietary fat eaten. This is illustrated in Table 4.1 which shows figures for the composition of adipose tissue (lard) in pigs fed different types of dietary fat. Alternatively, fat cells can synthesize their lipid *de novo* from glucose which is transported into the cell from the bloodstream (section 4.4). When there is a demand for fatty acids elsewhere in the body, they can be mobilized by breakdown of the triacylglycerols in the lipid globule, catalysed by the enzyme hormone sensitive lipase (section 4.5), followed by transport out of the cell. The integration of the systems for synthesis and breakdown, storage and mobilization is orchestrated by a variety of hormones and is geared to the needs and nutritional status of the animal (section 4.6).

4.3.2 Milk as an energy store for the needs of the newborn

It would be a mistake to think that all energy storage in mammals resides in adipose tissue. Milk, produced in the mammary gland, can be regarded as a store of energy for the newborn animal. The cream fraction of milk consists of milk fat globules composed primarily of triacylglycerols, with smaller amounts of cholesterol, fat soluble vitamins and other hydrophobic lipids, some of which may be environmental contaminants. The milk fat globule is surrounded by a unit membrane containing mainly protein, phospholipid and cholesterol and has an average diameter of 1–2 μm, although some may attain sizes up to 10 μm. The fat droplets found in the mammary cell have no discernible membrane. They are formed within the endoplasmic reticulum membranes and migrate to the apical regions of the mammary secretory cell where they are enveloped in the plasma membrane. A neck of membrane is pinched off and the resulting vesicle is expelled into the lumen as a milk fat globule. Like the triacylglycerols in adipose tissue, their synthesis and breakdown in the mammary gland is controlled by hormones whose secretion is in part governed by the nutritional status of the animal.

4.3.3 Storage of triacylglycerols in oil seeds: some plants use lipids as a fuel, which is stored as minute globules in the seed

There are many hundred varieties of plants known to have oil bearing seeds, but only a few are significant commercially. They are the soya bean, cotton, groundnut (peanut), sunflower, coconut, oil palm, olive, sesame, rapeseed, cocoa bean, linseed, castor and tung. The first ten are important sources of edible oils for human foods or animal feeds, while the last three are used for

Table 4.1 Fatty acid composition of some animal storage fats important in foods

Food fat	4:0–12:0	14:0	16:0	16:1	18:0	18:1	18:2	20:1 22:1	LC PUFA	Others
Lard (1)	0	1	29	3	15	43	9	0	0	0
Lard (2)	0	1	21	3	12	46	16	0	0	1
Poultry	0	1	27	9	7	45	11	0	0	1
Beef suet	0	3	26	9	8	45	2	0	0	0
Lamb	0	3	21	4	20	41	5	0	0	6
Milk – cow	10	12	26	3	11	29	2	0	0	4
Milk – goat	21	11	27	3	10	26	2	0	0	0
Egg yolk	0	0	29	4	0	43	11	0	0	4
Cod liver oil	0	6	13	13	3	20	2	18	20	5

Lard (1) represents fat from pigs given a normal cereal-based diet whereas Lard (2) represents fat from pigs given diets to which linoleic acid-rich oil-seeds have been added. Contrast the linoleic acid content of storage fats from simple-stomached animals (pigs, poultry) with those from ruminants in which much of the polyunsaturated fatty acid content is reduced by hydrogenation. In fish oils, long-chain polyunsaturated fatty acids (LC PUFA) predominate. These are largely $C_{22:5}$ and $C_{22:6}$.

other industrial purposes such as paints, varnishes and lubricants. Although the seed is the most important organ for the storage of triacylglycerols, some species store large quantities of oil in the mesocarp or pericarp of the fruit surrounding the seed kernel. The avocado is a familiar example of a plant with an oily mesocarp and has been the subject of extensive studies of fatty acid biosynthesis. The mesocarp of the oil palm is a commercially important source of palm oil, used in soap and margarine making, which is quite distinct in chemical and physical properties from the seed oil of the same plant: palm kernel oil (Figure 4.7 and Table 4.2). Droplets of triacylglycerols are also present in the seeds of plants that store predominantly carbohydrates (starch) as the primary source of fuel for seed germination. These include legumes such as peas or beans and cereals like wheat or barley.

Microscopic examination of a mature oil seed or one in the active phase of oil accumulation reveals a cytoplasm packed with spherical organelles that consist mainly of triacylglycerols, called oil bodies (Figure 4.8). You may

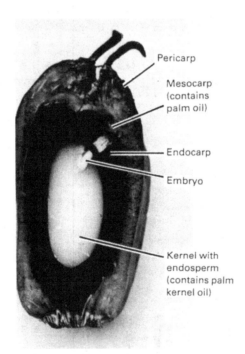

Figure 4.7 Section through an oil palm fruit. Reproduced with kind permission of Dr L.H. Jones.

Table 4.2 The fatty acid composition of some vegetable oils important in foods

Oil	8:0	10:0	12:0	14:0	16:0	18:0	18:1	18:2	18:3	20:1 + 22:1	Others
						g/100 g total fatty acids					
Avocado	0	0	0	0	20	1	60	18	0	0	1
Coconut	8	7	48	16	9	2	7	2	0	0	1
Corn	0	0	0	1	14	2	30	50	2	0	1
Olive	0	0	0	trace	12	2	72	11	1	0	2
Palm	0	0	trace	1	42	4	43	8	trace	0	2
Palm kernel	4	4	45	18	9	3	15	2	0	0	0
Peanut	0	0	trace	1	11	3	49	29	1	0	6
Rape – high erucic	0	0	0	trace	4	1	24	16	11	43	1
Rape – low erucic	0	0	0	trace	4	1	54	23	10	trace	8
Safflower – high oleic	0	0	0	0	5	2	73	17	1	0	2
Safflower – high linoleic	0	0	0	0	6	3	15	73	1	0	2

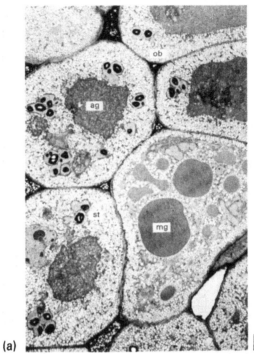

(a)

10 μm

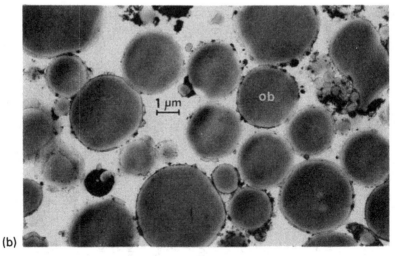

(b)

also find reference to terms such as oleosomes, lipid-containing vesicles and reserve oil droplets which are synonymous. Some authors have used the term spherosomes to refer to the same organelle, but it should strictly be used to describe another particle in plant cells that has a high content of phospholipids. The core of the oil body is composed mainly of triacylglycerols, with minor quantities of other hydrophobic lipids such as sterols, hydrocarbons and carotenoids. The oil body is surrounded by a membrane of protein and phospholipids. There is controversy about the origin of the oil body but a current hypothesis is that newly formed triacylglycerols accumulate between the two layers of the endoplasmic reticulum double membrane (section 6.5). When the oil-filled membrane vesicle reaches a certain critical size, it separates from the reticulum, becoming an independent particle with a half unit membrane. Several authors have described the presence of some enzymes of lipid biosynthesis in the membrane of the oil body. If the seeds are crushed gently, the oil bodies can be separated by flotation in the centrifuge, like the cream from milk (Figure 4.8). Further disruption of the oil bodies themselves yields membranes that can be separated by centrifugation to separate them from the floating fat. Unlike the fat globule of the fat cell, the oil bodies in the seed cells always stay as discrete particles with an average diameter of about 1 μm without ever coalescing; fat cell oil globules may reach more than 200 μm in diameter. Recent research reveals the presence of specific proteins in the oil body membrane resembling the apoprotein of plasma LDL (section 5.3.5).

A closer parallel to the seed oil bodies in the animal kingdom is found in the cells of the brown adipose tissue which are packed with small discrete oil droplets, 1–3 μm in diameter. These cells contain a large number of specialized mitochondria, adapted for oxidizing the fatty acids from the oil droplets which they surround. Similarly, seed oil bodies are surrounded by glyoxysomes which, during seed germination, receive fatty acids from the oil bodies for oxidation prior to the synthesis of carbohydrates, which occurs actively at this time.

Figure 4.8 Oil bodies (a) in an intact seed and (b) after isolation: ob = oil bodies; mg = myrosin grains; ag = aleurone grains; st = starch grains. Bars indicate scale in μm. In (a) 1 cm represents 10 μm; (b) 5 mm represents 1 μm. Reproduced with kind permission from C.G. Smith.

4.4 THE BIOSYNTHESIS OF TRIACYLGLYCEROLS

There is an important difference between animals and plants with respect to acylglycerol composition and metabolism. Plants must of necessity synthesize their acylglycerols from simple starting materials according to their requirements, since they have no dietary source of preformed lipids. Unlike animals, they have the ability to synthesize the linoleic acid that they normally possess in abundance: animals rely on plants for this essential nutrient (section 5.2.2). The fatty acid composition of animal acylglycerols is greatly influenced by their diet and, therefore, ultimately by the vegetables they eat. The way in which dietary glycerides are modified by animals may differ between species and from organ to organ within a species. Such modifications are not necessarily always effected by the animal's own cells. For example, in ruminants such as cows, sheep or goats, the microorganisms present in the rumen hydrogenate the double bonds of dietary poly-unsaturated fatty acids like linoleic and α-linolenic acids to form a mixture of mainly saturated and *cis* and *trans* monoenoic acids (section 3.2.5). An outstanding and significant feature of acylglycerol fatty acid composition is that it is quite distinctly different from that of the phospholipids or the non-esterified fatty acid (NEFA) pool. An understanding of this cannot be obtained simply from analyses of fatty acid distributions but must depend upon the study of different metabolic pathways, and of the individual enzymes in those pathways, involved in the synthesis of each lipid class. We shall discuss these in the following sections.

4.4.1 Pathways for complete (*de nove*) synthesis: these pathways build up the molecules from small basic constituents

(a) The glycerol phosphate pathway: a product of the glycolysis pathway provides the glycerol backbone of the triacylglycerols. By this means phospholipid and triacylglycerol metabolism interlink

We inferred in the last section that glycerides may arise in different ways depending on whether the tissue needs to start at the beginning from component parts or whether it requires to modify existing, possible dietary acylglycerols. Historically, the *de novo* pathway, now usually known as the glycerol phosphate pathway (Figure 4.9) was worked out first. This pathway was first proposed by the American biochemist, Kennedy, based on earlier work of Kornberg and Pricer, who first studied reactions 1 and 4, the formation of phosphatidic acid by acylation of glycerol phosphate.

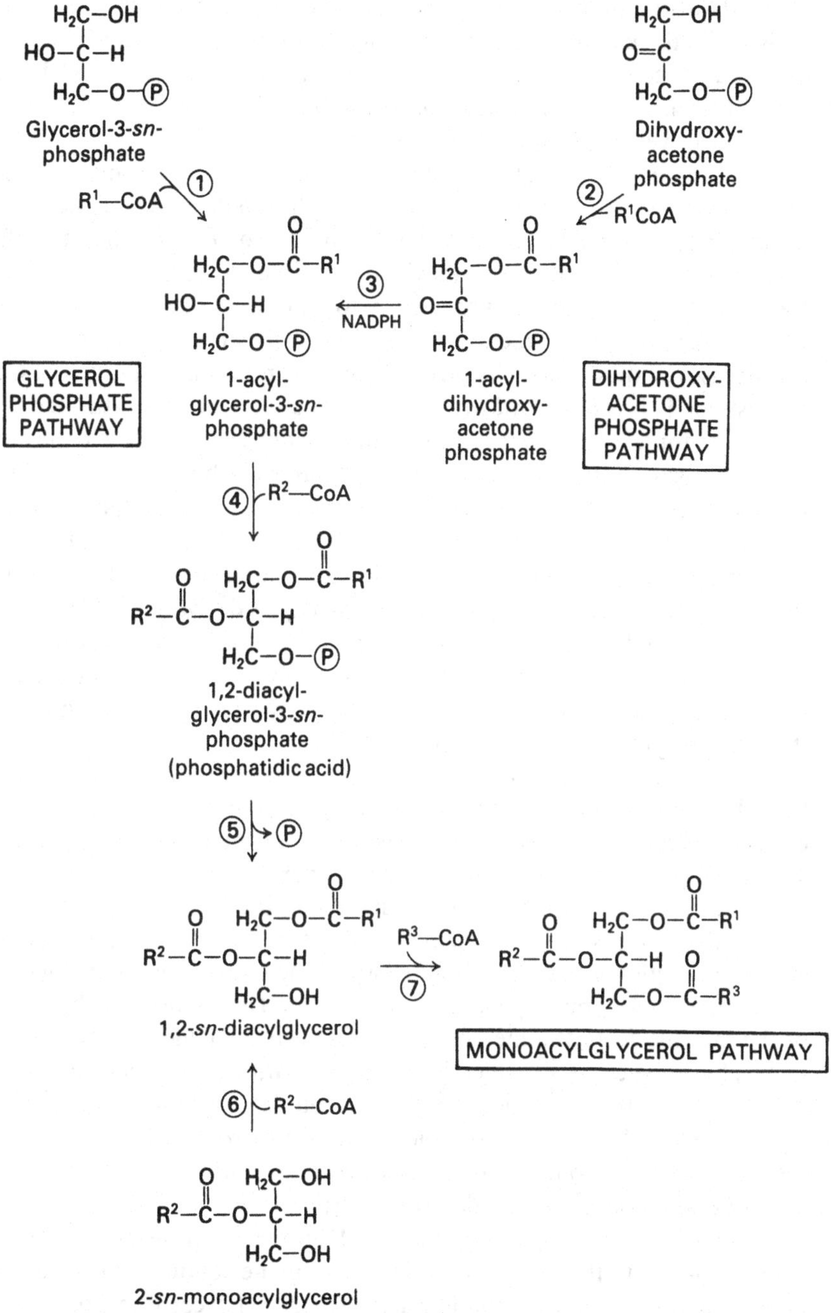

Figure 4.9 The biosynthesis of triacylglycerols.

Kennedy also demonstrated the central role of phosphatidic acid in both phospholipid (section 7.1.2) and triacylglycerol biosynthesis and one of his outstanding contributions has been to point out that the diacylglycerol derived from phosphatidic acid forms the basic building block for triacylglycerols as well as phosphoglycerides.

It is now known that steps 1 and 4 in Figure 4.10, the transfer of acyl groups from acyl-coenzyme A to glycerol-3-phosphate are catalysed by two distinct enzymes specific for positions 1 and 2. The enzyme that transfers acyl groups to position 1 (acyl-CoA: glycerol phosphate 1-O-acyl transferase) exhibits marked specificity for saturated acyl-CoA thiolesters whereas the second enzyme (acyl-CoA: 1-acyl glycerol phosphate 2-O-acyl transferase) shows specificity towards mono and dienoic fatty acyl-CoA thiolesters. This is in accord with the observed tendency for saturated fatty acids to be found in position 1 and unsaturated ones in position 2 in the lipids of living tissue. Although these acyltransferases are bound to intracellular membranes, the rate of the esterification reaction is stimulated *in vitro* by addition of the soluble fraction derived by cell fractionation. The factor responsible for this stimulation is now thought to be a small protein of 12 000 molecular mass that very tightly binds to acyl-CoA derivatives. This protein is known as the fatty acid binding protein or sometimes, the Z-protein. It is thought to desorb fatty acyl-CoAs that have bound to intracellular membranes remote from the sites of triacylglycerol synthesis and direct them to the synthetic site (section 5.3.3).

The next step (5 in Figure 4.10), catalysed by phosphatidate phosphohydrolase is thought to be the rate-limiting step that controls the activity of the pathway and the overall rate at which cells synthesize triacylglycerols. The activity of this enzyme generally runs parallel to the potential of the tissue to synthesize triacylglycerols and is modified by a large number of factors that suggest its importance in the control of triacylglycerol biosynthesis (section 4.6.2). The major part of the enzyme in the living cell is membrane bound, but a proportion is soluble. This could arise by release of the enzyme from the membranes during cell breakage but may have a more physiological basis which is concerned with the metabolic control of acylglycerol synthesis (section 4.6.2). The stimulation of microsomal triacylglycerol biosynthesis by the soluble fraction *in vitro* may be partly due to the presence of the solubilized phosphohydrolase and partly to the presence of fatty acid binding proteins described earlier.

The final step in the pathway (step 7, Figure 4.10) transfers a fatty acid from acyl-CoA to the diacylglycerol formed by the action of phosphatidate phosphohydrolase on phosphatidic acid. The diacylglycerol acyl transferase usually seems to have a wide fatty acid specificity.

(b) The dihydroxyacetone phosphate pathway: a slight variant in which the glycerol backbone derives from dihydroxyacetone phosphate rather than glycerol phosphate

A slight variation of the glycerol phosphate pathway was discovered in the late 1960s. The microsomal fractions from hamster intestinal mucosa and from rat liver are able to catalyse the formation of triacylglycerols from dihydroxyacetone phosphate (DHAP) and glyceraldehyde phosphate (GAP) as well as from glycerol phosphate (GP). The microsomal fraction does not contain glycerol phosphate dehydrogenase and therefore the formation of acylglycerols from DHAP or GAP cannot proceed via their conversion into GP. Moreover, an inhibitor of triose phosphate isomerase completely inhibits acylglycerol synthesis from GAP but not from DHAP, indicating that DHAP is the immediate precursor of the glycerol moiety of the acylglycerol (Figure 4.10). Some researchers have argued that the acyl transferases that transfer the fatty acid from fatty acyl-CoA to position 1 in GP and in DHAP are identical enzymes. There is, however, some evidence that the DHAP acyl transferase in the mitochondria of rat liver and adipose tissue is quite distinct from GP acyl transferase because its activity is affected by diet in a different way. An important difference between the two pathways is that the conversion of glucose into triacylglycerol via the DHAP pathway requires NADPH, whereas the conversion via the GP pathway involves NADH. NADPH is normally associated with reactions of reductive synthesis such as fatty acid biosynthesis and there is some evidence to support the view that the activity of the DHAP pathway is enhanced under conditions of increased fatty acid synthesis and relatively reduced in conditions of starvation or when the animal is fed a relatively high fat diet (particularly unsaturated fat). The relative importance of this pathway is still quite obscure.

(c) Formation of triacylglycerols in plants: formation of triacylglycerols in plants involves the cooperation of different subcellular compartments

As in animals, triacylglycerols constitute the major lipid store in plants, sometimes representing as much as 80% of the dry weight of the seeds. However, our knowledge of the details of triacylglycerol formation in plants is less extensive than in mammals, partly because of the necessity of working with developing seeds of a precise age. As seeds or fruits mature, storage material is not deposited at a constant rate. Indeed, three phases are usually seen (Figure 4.10). Thus, in phase 1, cell division is rapid but there is little

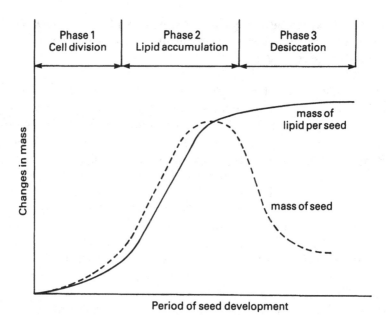

Figure 4.10 The accumulation of lipid during seed development.

deposition of storage material, whether it be protein, lipid or carbohydrate. In phase 2, there is a fast accumulation of storage material. Moreover, if the lipid stores have an unusual chemical composition, the special enzymes needed for synthesis are only active at this stage. Finally, in phase 3, desiccation takes place with little further metabolism.

Fatty acid biosynthesis *de novo* is concentrated in the plastid of the plant cell. By a combination of acetyl-CoA carboxylase and fatty acid synthetase, palmitoyl-ACP is produced. This is chain-lengthened by a specific condensing enzyme to stearoyl-ACP, which is desaturated to oleoyl-ACP by a $\Delta 9$-desaturase (section 3.2.4(b)). Under most conditions, palmitate and oleate are the main products of biosynthesis in plastids (chloroplasts in leaves). These acyl groups can be transferred to glycerol-3-phosphate within the organelle or hydrolysed, converted into CoA-esters and exported outside the plastid. If fatty acids are esterified to glycerol-3-phosphate within the plastid, then 16C acids (mainly palmitate) are attached at the *sn*-2 position while 18C acids (mainly oleate) are attached at the *sn*-1 position. This combination of acids is characteristic of plastids and, because it takes place within the organelle (and also resembles the situation in cyanobacteria, from which chloroplasts may derive) has been termed a **prokaryotic pattern**. The German biochemist, Ernst Heinz, together with the New Zealander, Gratton Roughan, have done much to establish the existence of prokaryotic and eukaryotic pathways in plants (Figure 4.11).

Since triacylglycerol biosynthesis takes place on the plant endoplasmic reticulum, an essential feature involves the hydrolysis of acyl groups from acyl-ACPs to encourage their export. Thus, in developing seeds, acyl-ACP thioesterase activity must be much higher than plastid glycerol-3-phosphate acyltransferase. Furthermore, phosphatidate phosphohydrolase, which is often thought to be rate-limiting for triacylglycerol biosynthesis, will be low in plastids since there is no need for a rapid supply of diacylglycerols to these organelles when the carbon flux is mainly to the endoplasmic reticulum.

After non-esterified fatty acids are attached to CoA in the plastid envelope they can, apparently, move rapidly to the endoplasmic reticulum. It is not known how this transport takes place, although lipid-binding

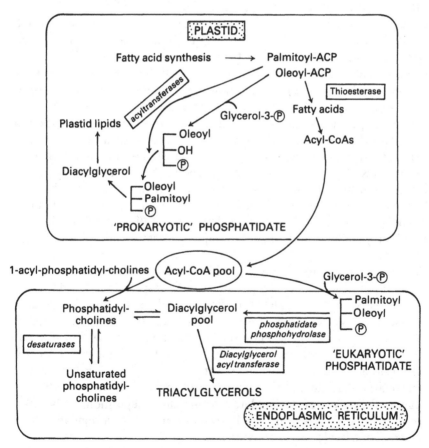

Figure 4.11 The 'plant pathway' for triacylglycerol biosynthesis.

transport proteins have been purified from a number of plant tissues. It is clear, however, that acyl-CoAs (rather than acyl-ACPs as in the plastid) serve as substrates for the endoplasmic reticulum glycerol-3-phosphate acyltransferases. These two enzymes show a substrate specificity that causes saturated fatty acids to be preferentially esterified in the *sn*-1 position while oleate is very much preferred to palmitate at the *sn*-2 position. Thus, the phosphatidate produced in the endoplasmic reticulum has the opposite positional distribution of fatty acids to that formed in the plastid, and is termed **eukaryotic**.

A phosphatidate phosphohydrolase located in the endoplasmic reticulum then acts to generate diacylglycerols. The diacylglycerols have three main fates. In all plants they form the backbone of the membrane phospholipids and are incorporated into molecules like phosphatidylcholine (section 6.2.1). However, during oil accumulation in seeds, net phospholipid synthesis appears to be low, so that carbon is preferentially channelled to lipid stores. Nevertheless, in many plants, the CDP-choline: diacylglycerol choline phosphotransferase plays another role. The reaction for this enzyme has been shown to be approximately in equilibrium. Thus, it can allow the rapid exchange of diacylglycerol between its pool and that of the newly synthesized phosphatidylcholine. Since phosphatidylcholine is the substrate for oleate (and linoleate) desaturation in seeds, the reversible nature of the cholinephosphotransferase allows the diacyglycerol pool to become enriched in polyunsaturated fatty acids. This process has been studied in particular detail in safflower (which has an oil rich in polyunsaturated fatty acids, comprising 75% linoleate) by the Swede, Sten Stymne and the British biochemist Keith Stobart. They have also shown that the acyl-CoA pool can also be utilized through the activity of an acyl-CoA: lysophosphatidyl-choline acyltransferase. In contrast, for seeds where less unsaturated oils accumulate (e.g. avocado) the subsidiary flux of diacylglycerols through phosphatidylcholines and their consequent desaturation, is much less important.

The third fate for diacylglycerols, which occurs in all fat-storing seeds, is the acylation of its *sn*-3 position. The enzyme responsible, diacylglycerol acyltransferase, usually has less specificity than the acyltransferases adding the first two fatty acids. Thus, in many plants, the fatty acids that accumulate at the *sn*-3 position depend upon the composition of the acyl-CoA pool. Nevertheless, in a few tissues, the substrate specificity of the diacylglycerol acyltransferases may also have an important role in determining the nature of the final stored lipid. From measurements of enzyme activities *in vitro* and because diacylglycerol accumulates during lipid deposition, it is often considered that diacylglycerol acyltransferase may have a regulatory role in the rate of triacylglycerol biosynthesis.

From the above (Figure 4.11) it will be obvious that the plastid and endoplasmic reticulum compartments of the plant cell have to integrate their metabolism during lipid storage. In addition to its role in triacylglycerol biosynthesis, the endoplasmic reticulum is also involved in the modification of fatty acids in those seeds where unusual acids are found in the storage lipid. Thus in the older oil-seed rape varieties, erucic acid is a major constituent (sections 5.1.2 and 8.15.2). This 22C unsaturated fatty acid is formed by the elongation of oleoyl-CoA in the endoplasmic reticulum. As mentioned at the start of this section, the biosynthesis of unusual acids is confined to phase 2 of seed development (Figure 4.11) and is, therefore, coordinated with triacylglycerol accumulation. It is noteworthy that unusual fatty acids do not appear in membrane lipids, where their presence would presumably be detrimental to normal cellular function.

4.4.2 Pathways for restructuring – the monoacylglycerol pathway: organisms often find it convenient to retailor existing molecules rather than start from basics. For example, the monoacylglycerol pathway is important mainly in the restructuring of absorbed dietary fat

The main function of the monoacylglycerol pathway (Figure 4.10) is to resynthesize triacylglycerols from the monoacylglycerols formed during the digestion of fats in the small intestine. It involves a stepwise acylation first of monoacylglycerols, then diacylglycerols and was first discovered by Georg Hübscher's research team in Birmingham, UK in 1960. The reactions are catalysed by enzymes in the endoplasmic reticulum of the enterocytes of many species of animals. The 2-monoacylglycerols are the preferred substrates compared with 1-monoacylglycerols and the rate of the first esterification is influenced by the nature of the fatty acid esterified in position 2. Monoacylglycerols with short chain saturated or longer chain unsaturated fatty acids are the best substrates. Diacylglycerol acyltransferase is specific for 1,2-*sn*-diacylglycerols and will not acylate the 2,3- or 1,3-isomers. Diunsaturated or mixed acid diacylglycerols are better substrates than disaturated compounds, but we have to be cautious when we interpret results of this kind. Lipids containing unsaturated fatty acids are more easily emulsified than saturated ones, so that we may not be observing differences in specificity of the enzyme for fatty acid composition but differences in solubility of the substrates when the enzyme assays are performed *in vitro*. In the living animal, the substrates for this pathway, the monoacylglycerols, arise mainly from the hydrolysis of dietary triacylglycerols by the enzyme lipase in the intestinal lumen (section 5.3.1). Therefore, this mechanism is one by which existing acylglycerols are modified rather than one by which new fat is formed (sections 5.3.2 and 5.3.3).

The monoacylglycerol pathway has also been demonstrated in the adipose tissue of the hamster and the rat. Under certain conditions it appears to compete with the glycerol phosphate pathway for acyl groups and may serve to regulate the activity of the latter pathway. The origin of the monoacylglycerol substrate in adipose tissue is not known.

4.4.3 The biosynthesis of ether bonds

We mentioned the occurrence of glyceride ethers in section 4.2. The mechanism of formation of this ether bond is precisely the same as in the plasmalogen phospholipids (section 7.1.9).

4.5 THE CATABOLISM OF ACYLGLYCEROLS

The word catabolism is used to describe metabolic reactions involved in the breakdown of complex biological molecules. One of the main themes running through this book is that, with very few exceptions, lipids in biological tissues are in a dynamic state: they are continually being synthesized and broken down. This is known as turnover. Enzymes that catalyse the hydrolysis of the ester bonds in acylglycerols are called lipases and they belong to a whole family, whose members differ in respect to their substrates and the positions in substrates of the bonds that they hydrolyse. There are also many enzymes called esterases that hydrolyse ester bonds in general, but lipases form a distinct class and the distinction lies in the physical state of their substrates. The milieu in which a lipase acts is heterogeneous: the lipid substrate is dispersed as an emulsion in the aqueous medium and the enzyme acts at the interface. If, by some means, a one-phase system is obtained, for example when the triacylglycerol contains short chain fatty acids (as in triacetin) or when a powerful detergent is present, then the lipid may be hydrolysed by an esterase, not a lipase.

4.5.1 The nature and distribution of lipases: before the constituent fatty acids can be used for energy or metabolism in other pathways, they must be released by hydrolytic enzymes termed lipases which have a range of specificities

Lipases are widespread in nature and are found in animals, plants and microorganisms. The initial step in the hydrolysis is generally the splitting of the fatty acids esterified to the primary hydroxyl groups of glycerol (Figure

4.3). Even though these are now known to be distinguishable, this reaction is not'stereospecific and the fatty acids in positions 1 and 3 are initially removed at equal rates. Once one fatty acid has been removed, however, the resulting diacylglycerols are more slowly hydrolysed than the original triacylglycerols and the monoacylglycerols only if the fatty acids are racemized to position 1. The preference for the removal of fatty acids in the primary positions together with the diminished rate of hydrolysis of partial glycerides results in an accumulation of monoacylglycerols and non-esterified fatty acids as the major products of lipase hydrolysis. This is especially true *in vivo* where controls such as the resynthesis of di- and tri-acylglycerols regulate the degree of hydrolysis. *In vitro*, the hydrolysis is more likely to go to completion, especially if bile salts are added as detergents to effect better emulsification of the substrate. The racemization of 2-monoacylglycerols to 1-monoacylglycerols can also allow further hydrolysis to occur.

Some lipases not only hydrolyse fatty acids on the primary positions of acylglycerols but will also liberate the fatty acid esterified in position 1 of phosphoglycerides. In most cases, the rate of hydrolysis is independent of the nature of the fatty acids released. There are, however, a number of exceptions to this rule: fatty acids with chain lengths less than 12 carbon atoms, especially the very short chain lengths of milk fats, are cleaved more rapidly than the normal chain length fatty acids (14 to 18 carbons) while the very long chain polyenoic acids (20:5 and 22:6) found in the oils of fish and marine mammals are very slowly hydrolysed. The lipases in some microorganisms have fatty acid specificity. For example, the fungus *Geotrichum candidum* possesses a lipase that seems to be specific for oleic acid in whichever position it is esterified.

4.5.2 Animal lipases play a key role in the digestion of food and in the uptake and release of fats by tissues

Much of our information about lipases comes from studies on the enzyme isolated from pancreatic juice; this is the enzyme primarily responsible for the digestion of dietary lipids in the animal digestive tract, a process that is described in detail in section 5.3.1. Pancreatic lipase is a glycoprotein of molecular mass 50 kDa. The first step in the catalytic process is the adsorption of the enzyme on to the hydrophobic interface of the substrate micelles. Detergent molecules, like bile salts, tend to compete with the lipase for binding sites and the adsorption of the enzyme is assisted by a helper molecule called co-lipase, which is a small protein. *In vitro*, it can be shown that the activity of lipase is inhibited by high concentrations of bile salts and that this inhibition can be overcome by the addition of co-lipase.

In vivo, the co-lipase probably forms a stoichiometric complex with lipase, anchoring the enzyme to the interface and enhancing its catalytic activity.

Different lipases are present in other tissues and differ mainly in their substrate specificities. One of the most important of these in animals is lipoprotein lipase, sometimes referred to in some of the older literature as clearing factor lipase. This enzyme is clearly distinguished by the fact that it hydrolyses triacylglycerols associated with proteins and its main function is in the catabolism of serum lipoproteins (section 5.3.5(g)). The enzyme will hydrolyse emulsions of simple triacylglycerols much more rapidly if a protein is added to the incubation mixture, in particular apoprotein C (section 5.3.5). Fielding's purification of lipoprotein lipase illustrates beautifully the way in which the specific property of an enzyme to form complexes with its substrate can be used in enzyme purification. Addition of lipoproteins to the partially purified enzyme created a complex with a density lower than that of the original enzyme and much of the contaminating protein. When centrifuged in a medium of suitable density, the complex floated to the top of the centrifuge tube and could be separated from the more dense material.

Another lipase, present in adipose tissue, is activated by hormones such as the catecholamines (section 4.6.3). Adipose tissue also contains a lipase that is much more active on monoacylglycerols than are other lipases. A lingual lipase is thought to be important in the digestion of fat by new-born animals (section 5.3.1). Lipases are still being discovered and it will be a long time before the whole complex jigsaw puzzle of glyceride breakdown can be pieced together.

4.5.3 Plant lipases: breakdown of stored lipids in seeds is aided by the formation of a specialized organelle, the glyoxysome

Seeds that contain lipid may have as much as 80% of their dry weight represented by triacylglycerols. Plants such as soya beans face two particular problems in using such energy reserves. First, these plants have to mobilize the lipid rapidly and break it down to useful products. This overall process involves the synthesis of degradative enzymes as well as the production of the necessary membranes and organelles that are the sites of such catabolism. Secondly, plants with lipid-rich seeds must be able to form water-soluble carbohydrates (mainly sucrose) from the lipid as a supply of carbon to the rapidly elongating stems and roots. Animals are unable to convert lipid into carbohydrate (Figure 4.12) because of the decarboxylation reactions of the Krebs (tricarboxylic acid) cycle (isocitrate dehydrogenase and α-ketoglutarate dehydrogenase). Thus, for every two carbons entering the Krebs cycle from lipid as acetyl-CoA, two carbons are lost as CO_2. In plants, these decarboxylations are avoided by a modification of the Krebs cycle which is called the **glyoxylate cycle**. This allows lipid carbon to contribute to the synthesis of oxaloacetate, which is an effective precursor of glucose and hence, sucrose.

When water is imbibed into a dry seed, there is a sudden activation of metabolism once the total water content has reached a certain critical proportion. So far as lipid-storing seeds are concerned, triacylglycerol lipase activity is induced and studies by Huang in California and others suggest that this lipase interacts with the outside of the oil droplet through a specific protein which aids its binding. This interaction has some similarities to the binding of lipoprotein lipase to very low density lipoproteins in animals (section 5.3.5). The lipases in seeds hydrolyse the *sn*-1 and 3 positions of triacylglycerols and because the acyl group of the 2-monoacylglycerol products can migrate rapidly to the *sn*-1 position, the lipases can completely degrade the lipid stores.

The fatty acids that are liberated are activated to CoA-esters and broken down by a modification of β-oxidation which takes place in specialized microbodies. Because these microbodies also contain the enzymes of the glyoxylate cycle (see above), they have been termed glyoxysomes by the Californian biochemist, Harry Beevers. Beevers and his group worked out methods for the isolation of glyoxysomes from germinating castor bean seeds and showed by careful study that fatty acid β-oxidation was confined to this organelle. The glyoxysomes have only a temporary existence. They are formed during the first two days of germination and, once the lipid stores have disappeared (after about 6 days), they gradually break down. Nevertheless, in leaf tissues, the role of glyoxysomes in the β-oxidation of fatty acids is replaced by other microbodies (section 3.3.1).

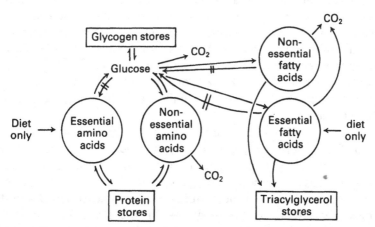

Figure 4.12 Permitted and forbidden interconversions of fats, carbohydrates and proteins in animals.

4.6 THE INTEGRATION AND CONTROL OF ANIMAL ACYLGLYCEROL METABOLISM

The major function of triacylglycerols in animals is as a source of fatty acids to be used as metabolic fuel. Any discussion about the integration of their metabolism is therefore a description of the regulation of fuel economy in the body and the transport of fatty acids between tissues involved in the synthesis, storage and oxidation of fatty acids and their glycerol esters.

4.6.1 Fuel economy: because of the essentiallity of maintaining constant blood glucose concentrations, the interconversion of different types of fuels has to be hormonally regulated

The maintenance of fuel reserves within fairly narrow limits is referred to as fuel homeostasis. There are a number of rules that must be obeyed by the body. Rule one is to maintain blood glucose concentrations within very narrow limits and to return to normal levels as quickly as possible, if that level is perturbed. Rule two is that glycogen is stored as an emergency fuel (mainly in the liver and muscle) and when the stores have been filled, excess carbohydrate is converted into fat which is stored in adipose tissue. Rule three is to maintain the optimum supply of protein needed for growth, tissue repair and enzyme synthesis. Excess protein can also be converted into fat. Protein reserves are carefully conserved and when energy is needed, first glycogen and then fat are used preferentially, although during a period of fast, even overnight, some proteins are used.

Carbohydrate and protein are interconvertible and themselves convertible into fat, but once fat has been formed, the biochemical evidence suggests that it is only stored (section 4.3.1) or oxidized (section 3.3.1), not converted into proteins or carbohydrates (Figure 4.13). When an animal has an external supply of energy (i.e. when it is eating or has just eaten food), its current energy requirements are first satisfied by oxidation of glucose or fatty acids and protein is synthesized or replaced. Unused fuel goes into filling glycogen stores and, as these are limited, any remaining is stored as fat in adipose tissue.

Animals have a continual demand for energy, particularly when they have not eaten from some time, during growth, when undergoing vigorous physical activity or during pregnancy or lactation. In these circumstances, first glucose levels are topped up by glycogen breakdown and some release of muscle protein. Next, additional fuel comes from adipose tissue in the form of glycerol (converted into glucose in the liver) and fatty acids (oxidized in various tissues, particularly muscle, Figure 4.13.) It is important to remember that one of the main reasons why maintenance of blood glucose

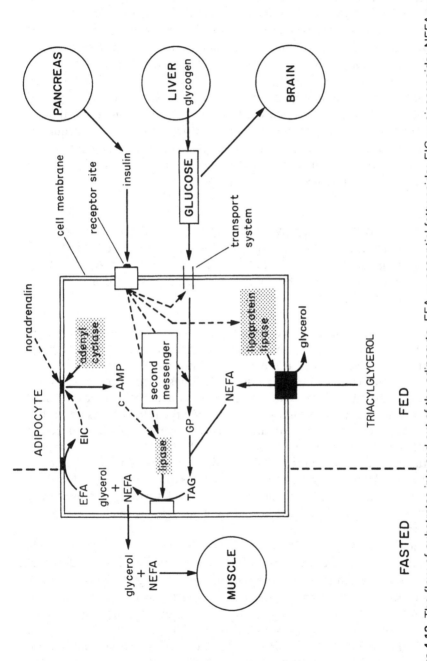

Figure 4.13 The flow of substrates into and out of the adipocyte. EFA = essential fatty acids; EIC = eicosanoids; NEFA = non-esterified fatty acids; TAG = triacylglycerols; GP = glycerol phosphate

concentration is the first priority is that the brain can only use glucose as a fuel, not free fatty acids (although it is able to utilize ketone bodies).

The control of these complex processes is mediated by the amount of energy in the diet, the nature of the dietary constituents and the types of circulating hormones and their concentrations. It is likely, however, that the effects of diet in controlling glyceride metabolism are mediated through the secretion of the appropriate hormones. The dominant role is played by insulin. Its concentration in the blood helps to coordinate the flow of fuel either into storage or from the stores into various tissues as required. High concentrations of insulin characterize the fed state when ample fuel is available from the diet; low levels signal the starved state when the animal's own reserves need to be called upon.

4.6.2 The control of acylglycerol biosynthesis: although the need for metabolic fuels is important, other requirements for lipids, such as membrane formation, may take precedence, requiring close integration of storage and structural lipid formation

An important concept of metabolism is that of turnover which envisages the continual renewal, involving biosynthesis and breakdown, of body constituents. The rates of turnover may differ widely between different biological molecules in different tissues. The rates of forward and backward reactions may also differ. Thus, when there is net synthesis of the tissue component, the rate of synthesis is faster than the rate of breakdown, but both may be proceeding simultaneously. When metabolic control is excercised, by whatever means, it may affect the rate of synthesis, the rate of degradation, or both. Turnover allows a fine degree of control of metabolic pathways. Control may be exercised on the synthesis or degradation of enzymes catalysing the metabolic reactions at the level of gene transcription or translation or on the allosteric control of enzymes by small molecules or cofactors.

In animals, net triacylglycerol synthesis occurs when energy demands exceed immediate requirements. Most diets contain both fat and carbo-hydrates, but when there is a preponderance of energy as carbohydrates, the tissues convert them into fatty acids which are then esterified into acylglycerols. When there is a preponderance of fat in the diet, fat synthesis from carbohydrate is depressed in the tissues. The products of fat digestion are then converted into lipoproteins (section 5.3) which circulate in the bloodstream. When the lipoproteins reach the tissues, fatty acids are released from the acylglycerols at the endothelial surfaces of cells, a process catalysed by the enzyme lipoprotein lipase, and are taken up into the cells

(Figure 4.14). Once inside, they are esterified into acylglycerols. In discussing the control of acylglycerol synthesis, we shall be discussing the esterification of fatty acids synthesized *de novo* or released from circulating lipoproteins. The control of fatty acid synthesis itself is discussed in section 3.2.7.

Although acylglycerols may be synthesized in many animal tissues, the most important are the small intestine, which resynthesizes triacylglycerols from components absorbed after digestion of dietary fats, the liver, which is concerned mainly with synthesis from carbohydrates and redistribution, the adipose tissue, which is concerned with longer term storage and the mammary gland, which synthesizes milk fat during lactation. It is assumed that the load of dietary fatty acids is the main factor controlling the rate of synthesis in the gut via the monoacylglycerol pathway, whereas in other tissues, the glycerol phosphate pathway is predominant. Triacylglycerols may also be synthesized in some tissues, not only in response to a surfeit of dietary energy but as a response to metabolic stress. Good examples are in liver, heart and skeletal muscle. Under stress conditions, fatty acids are mobilized from adipose tissue as a result of stimulation by stress hormones such as the catecholamines (section 4.6.3) and when their supply exceeds the need or the capacity of cells to oxidize them, the net result is a synthesis and deposition of triacylglycerols. The formation of triacylglycerols relieves the potentially toxic effects of free fatty acids and CoA esters.

The cellular concentrations of the substrates for the acyltransferases that catalyse acylglycerol esterification, acyl-CoA and glycerol phosphate are influenced by nutritional status. The main factors influencing the amounts of glycerol phosphate available for acylglycerol synthesis are those that regulate the levels and activities of the enzymes of glycolysis and gluco-neogenesis. Starvation reduces intracellular concentrations of glycerol phosphate severely and the original concentration is rapidly restored by refeeding the animal. Intracellular concentrations of acyl-CoA increase during starvation. There is little evidence, however, that intracellular concentrations of glycerol phosphate, acyl-CoA or the activities of the transferases are important in regulating acylglycerol synthesis. Since the main metabolic competition to acylglycerol synthesis is β-oxidation, however, the regulation of the relative activities of glycerolphosphate acyltransferase and carnitine palmitoyltransferase (Figure 4.14) may be expected to control the branch point between these two pathways. Carnitine palmitoyltransferase can be acutely regulated by inhibition with malonyl-CoA (section 3.3.1) but similar metabolic controls on the acyltransferases involved in acylglycerol synthesis are still obscure. These remarks apply mainly to the liver, on which most research has been done. Aspects of control of acylglycerol metabolism may be rather different in other tissues (consult references).

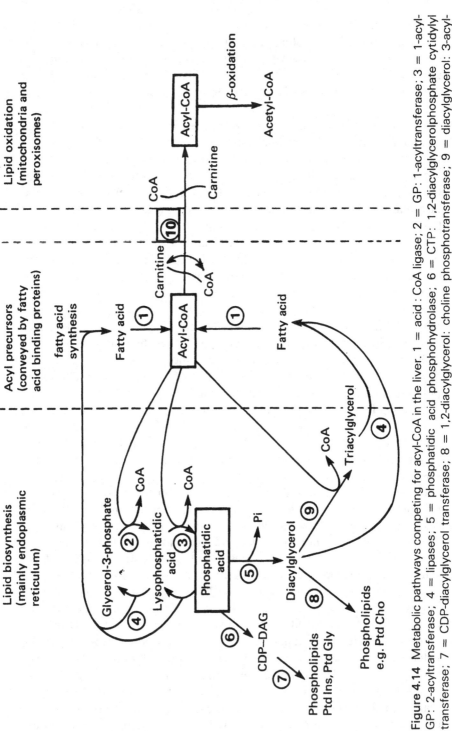

Figure 4.14 Metabolic pathways competing for acyl-CoA in the liver. 1 = acid : CoA ligase; 2 = GP: 1-acyl-GP: 2-acyltransferase; 3 = 1-acyltransferase; 4 = lipases; 5 = phosphatidic acid phosphohydrolase; 6 = CTP: 1,2-diacylglycerolphosphate cytidylyl transferase; 7 = CDP-diacylglycerol transferase; 8 = 1,2-diacylglycerol: choline phosphotransferase; 9 = diacylglycerol: 3-acyl-transferase; 10 = carnitine acyltransferase.

The rate limiting enzyme in mammalian acylglycerol biosynthesis is phosphatidate phosphohydrolase (PAP) and its activity in a tissue generally runs parallel to the potential for overall acylglycerol synthesis in that tissue. Its activity in liver is increased by high levels of dietary sucrose and fat, by ethanol and by conditions, such as starvation, that result in high concentrations of plasma non-esterified fatty acids. It is increased during the regeneration of liver after partial hepatectomy and in obese animals. It is decreased in diabetes and by administration of drugs that result in a reduction of circulating lipid concentrations. The factors that tend to increase the activity of PAP are also those that result in an increased supply of saturated and monounsaturated fatty acids to the liver, namely those fatty acids normally esterified in simple acylglycerols. If PAP activity is low, the substrate for the enzyme – phosphatidic acid, the central intermediate in lipid metabolism – does not accumulate but becomes a substrate for the biosynthesis of acidic membrane phospholipids such as phosphatidylinositol (sections 7.8 and 8.8). Phospholipid metabolism makes more extensive demands on a supply of unsaturated fatty acids than does simple acylglycerol metabolism and the activities of enzymes that divert phosphatidic acid into phospholipid rather than triacylglycerol metabolism tend to be elevated in conditions where unsaturated fatty acids predominate (Figure 4.14).

Recent research throws light on how the activity of PAP may be controlled (Figure 4.15). PAP exists in the cytosol and in the endoplasmic reticulum. It was originally assumed that the soluble enzyme was an artefact arising from damage to intracellular membranes during cell breakage and fractionation. Recently, evidence has been obtained that the cytosolic enzyme is physiologically inactive until it translocates on to the membranes on which phosphatidate is being synthesized. The translocation process seems to be regulated both by hormones and substrates. Cyclic-AMP (cAMP), for example, displaces the enzyme from membranes, whereas increasing concentrations of non-esterified fatty acids and their CoA esters promote its attachment. Insulin, which has the effect of decreasing intracellular concentrations of cAMP, ensures that the translocation is more effective at lower fatty acid concentrations. The mechanisms that cause these changes are not yet established but may involve the reversible phosphorylation of PAP. It seems, therefore, that the membrane-associated PAP is the physiologically active form of the enzyme and the cytosolic form represents a reservoir of potential activity.

This phenomenon is a recently discovered form of metabolic control and enzymes that exist in different locations in the cell and can regulate metabolism by moving from one location to another are called ambiquitous enzymes. Another example in lipid metabolism is the CTP: phosphorylcholine cytidylyl transferase which regulates the biosynthesis of phosphatidylcholine (7.1.4). The diacylglycerols formed by the action of PAP

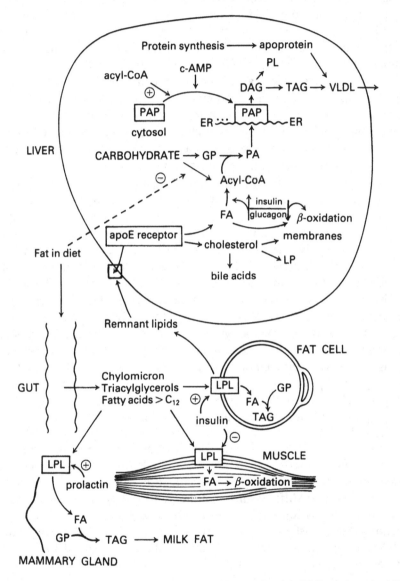

Figure 4.15 The control of triacylglycerol biosynthesis. PAP = phosphatidate phosphohydrolase; LPL = lipoprotein lipase; DAG = diacylglycerol; TAG = triacylglycerol; PL = phospholipid; GP = glycerol-3-phosphate; PA = phosphatidic acid; FA = fatty acids; ER = endoplasmic reticulum; LP = lipoproteins; VLDL = very low density lipoprotein.

can either be diverted into triacylglycerol synthesis (Figure 4.9) or into phosphoglyceride synthesis (Figure 7.1). The biosynthesis of phospholipids takes precedence over that of triacylglycerols when the rate of synthesis of diacylglycerols is relatively low. This ensures the maintenance of membrane turnover and bile secretion which are more essential processes in physiological terms than the accumulation of triacylglycerols. The precise mechanisms for the preferred synthesis of phosphoglycerides is not certain, however. Among the factors involved are probably a relatively low K_m of choline phosphotransferase for diacylglycerols. A factor that may eventually limit the biosynthesis of phosphatidylcholine is a limitation in the supply of choline and it is well recognized that a major effect of choline deficiency is fatty liver indicating a diversion away from phospholipid biosynthesis into triacylglycerol biosynthesis.

The role of diet in the control of glyceride metabolism is almost certainly mediated by changes in the concentrations of circulating hormones which induce or repress the biosynthesis of enzymes such as PAP. This is a poorly understood area of lipid biochemistry because of the difficulties of isolating and purifying enzymes that are predominantly membrane-bound in order to study their induction or allosteric control (section 3.2.7).

The fate of fatty acids entering the liver is markedly influenced by the ratio of insulin to glucagon in the circulation. High ratios favour esterification of fatty acids into acylglycerols and a low rate of β-oxidation. Phospholipid biosynthesis is far less affected than that of triacylglycerols. The activity of PAP, however, is probably not directly dependent on insulin since it is increased in diabetes and starvation, both conditions where circulating insulin concentrations are low.

Insulin has several important metabolic roles relating to acylglycerol biosynthesis. One effect is to increase the activity of the glucose transporter in the fat cell membrane allowing rapid passage of glucose into the cell. Another effect is to induce the synthesis of the enzyme lipoprotein lipase which catalyses the splitting of circulating triacylglycerols into fatty acids and glycerol. Inside the cell, insulin stimulates the synthesis of lipids by increasing the activities of lipogenic enzymes in general and also inhibits the breakdown of triacylglycerols from the fat globule by hormone-sensitive triacylglycerol lipase by causing a decrease in production of cAMP (Figure 4.16).

Thyroid hormones stimulate the rate of triacylglycerol biosynthesis in liver but have the opposite effect in adipose tissue. This may be due to an increased rate of general metabolic turnover rather than a specific effect on PAP. Recent research indicates that the hormones most immediately implicated in the control of PAP activity are the glucocorticoids. High serum cortisol concentrations are associated with most of the conditions described above that lead to an increased activity of PAP and an elevated rate of acylglycerol synthesis. The response of the liver cell to this increased activity is to export newly synthesized triacylglycerols in the form of lipoproteins, resulting in elevated blood concentrations of lipoproteins. However, a factor that may limit the export of lipoproteins is the capacity of the liver cell to synthesize the protein moiety of the lipoproteins. When protein synthesis cannot keep pace with lipid synthesis, triacylglycerols accumulate in the liver resulting in the toxic condition, **fatty liver**.

Further research into the hormonal control of these processes will give insights into how to control the common diseases of lipid metabolism (section 5.5).

4.6.3 The control of glyceride breakdown: mobilization from the fat stores is regulated by hormonal balance which in turn is responsive to nutritional and physiological states

In physiological states demanding the consumption of fuel reserves, the resulting low concentrations of insulin turn off the biosynthetic pathways and release the inhibition of triacylglycerol lipase within the cell. The activity of this enzyme is controlled by a complex cascade mechanism illustrated in Figure 4.16.

The lipase must be converted into an active form. This conversion is regulated by cyclic-AMP produced from ATP by the enzyme adenylate cyclase. The activity of the cyclase is under the control of the catecholamine hormones such as noradrenalin which stimulate its activity, or prostaglandins (formed from dietary essential fatty acids, section 3.4) which inhibit its activity. Another feature of the control is the destruction of cyclic-AMP by a specific phosphodiesterase which is activated by xanthines, an example of which is caffeine, found in coffee.

The demand for the consumption of fuel reserves can be regarded as a form of metabolic stress. This is characterized by a low activity of insulin relative to stress hormones: catecholamines, corticotropin, glucocorticoids and glucagon. Such an hormonal balance can occur in starvation, diabetes, trauma and under the influence of certain toxins. As we have seen earlier, it can also be a response to the ingestion of large amounts of ethanol, fructose or fat.

Fatty acids released by the lipase are transported out of the cell, bound to plasma albumin (section 5.3.5) and transported to those tissues, such as muscle, that utilize fatty acids as major sources of fuel. Whether fatty acids are directed into β-oxidation or acylglycerol synthesis may be governed by the competition for available acyl-CoA molecules by the acyltransferases involved in the esterification of acylglycerols (section 4.6.2) and the carnitine palmitoyl transferase of the mitochondrial membrane (Figure 4.14

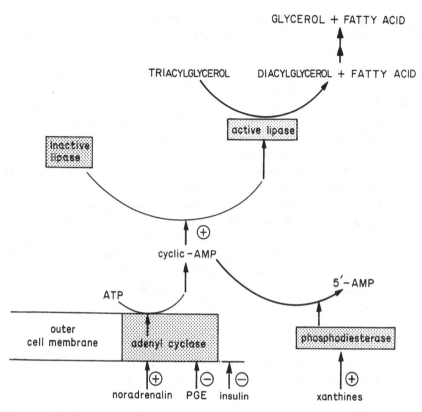

Figure 4.16 The control of triacylglycerol catabolism.

and section 3.3.1). The latter enzyme is increased during starvation when the animal most needs to oxidize fatty acids as a reserve fuel.

In plants, triacylglycerols are mainly synthesized and stored in the cotyledons of the seed (e.g. sunflower, soya, rape), the mesocarp of the fruit (avocado, oil palm) or the endosperm tissue (castor bean, coconut). Although the main pathways for the biosynthesis of triacylglycerols in plants have now been elucidated, the control of lipid biosynthesis in plants is little understood.

4.7 WAX ESTERS

4.7.1 Occurrence and characteristics

Wax esters are esters of long chain fatty acids with long chain fatty alcohols having the general formula: RCOOR. They occur in some species of bacteria, notably the mycobacteria and corynebacteria but are mainly important for providing a form of energy reserve that is an alternative to triacylglycerols in certain seed oils (e.g. jojoba), the oil of the sperm whale, the flesh oils of several deep sea fishes (e.g. orange roughy) and in zooplankton. Thus a major part of jojoba oil consists of wax esters, 70% of sperm whale oil and as much as 95% of orange roughy oil. In global terms, however, the zooplankton of the oceans may be of greatest importance. At certain stages of their life cycles, the zooplankton may synthesize and store massive amounts of wax esters, as much as 70% of their dry weight.

The component fatty acids and alcohols of wax esters have an even number of between 10 and 30 carbon atoms. Normally, they are straight chain saturated or monounsaturated chains of 16–24C. Branched and odd numbered chains are rare in both constituents except for the bacterial waxes. Wax esters from cold-water organisms exist as oils down to about 0°C and their physico-chemical properties differ from those of the triacylglycerols in that they are more hydrophobic and less dense, which may help to provide greater buoyancy for marine animals.

4.7.2 Biosynthesis of wax esters involves the condensation of a long chain fatty alcohol with fatty acyl-CoA

It is concluded that zooplankton synthesize their own wax esters *de novo* since their primary food, the phytoplankton, are devoid of these lipids. Indeed, all marine organisms so far studied, mammals and fish as well as crustaceans, can synthesize wax esters from simple carbohydrate and amino acid precursors. The fatty acids are synthesized by the malonyl-CoA pathway (section 3.2.2), although they can also be derived from the dietary lipids. A membrane-bound enzyme system catalysing the reduction of long chain fatty acids to fatty alcohols seems to be present in high activity in organisms elaborating wax esters. It is specific for reduced pyridine nucleotides and the CoA thiolesters of long chain fatty acids.

$$RCOSCoA + 2NAD(P)H + H^+ \rightarrow RCH_2OH + 2NAD(P)^+ + CoASH$$

$$RCHOH + RCOSCoA \rightarrow RCOOCHR$$

The condensation of the fatty alcohol and fatty acyl-CoA is catalysed by non-specific ester synthetases. In animals, the predominant source of the fatty acid moiety is thought to be dietary lipids while the alcohol moiety arises by reduction of fatty acids synthesized *de novo* from dietary carbohydrate.

4.7.3 Digestion and utilization of wax esters

The wax esters of fish and of jojoba seed oils are poorly hydrolysed by the pancreatic lipases of the human digestive system so that these lipids have poor nutritive value for Man. Fish, such as salmon and herring grow rapidly when feeding on zooplankton rich in wax esters, yet do not contain these lipids themselves. Their digestive systems are adapted to the efficient hydrolysis of wax esters, most of the products being absorbed and resynthesized into triacylglycerols. Tissue breakdown of wax esters so that the products can be utilized as metabolic fuels presumably involves lipases or esterases analogous to the hormone sensitive lipases in adipase tissue of organisms that store triacylglycerols, but these reactions have been little studied.

Some authorities now believe that wax esters have to be considered as key compounds in the transmission of carbon through the marine food chain. Wax esters seem to be found in greatest amounts in regions where animals experience short periods of food plenty followed by long periods of food shortage, for example in polar regions, where the short summer limits the period of phytoplankton growth, or in deep waters with a low biomass.

4.7.4 Surface lipids include not only wax esters but a wide variety of lipid molecules

Although true waxes are esters of long chain fatty acids and fatty alcohols, the term wax is often used to refer to the whole mixture of lipids found on the surfaces of leaves, or on the skin or the fur of animals. These surface lipids comprise a complex mixture of wax esters, long chain hydrocarbons, non-esterified long chain fatty acids, alcohols and sterols. The surface lipids are responsible for the water-repellant character of the surface and are important in conserving the organism's water balance and providing a barrier against the environment. Some components, notably the long chain non-esterified fatty acids, have antimicrobial properties and this also contributes to their protective role in a physiological as well as a physical way.

The hydrocarbon components of the surface lipids (sections 6.6. and 8.12) have very long chain lengths (around 30C) and are formed from long chain fatty acids (section 3.2.2) by two sorts of mechanisms. Odd chain hydrocarbons can be formed by decarboxylation of an aldehyde. The first product is an α-hydroxyacid formed by α-oxidation of the long chain fatty acid (section 3.3.2). This compound first loses its carboxyl carbon to form an aldehyde which then loses its carbonyl carbon to yield the alkane.

4.8 SUMMARY

Animals and plants require long term reserves of metabolic fuels that can, at times of need, be mobilized and oxidized to drive vital metabolic processes. Thus, in times of starvation or strenuous exercise, animals can utilize fuel reserves to maintain basal metabolism or sustain muscular activity, respectively. Plants may use their reserves to fuel the energy demanding process of seed germination. Triacylglycerols are the most common forms of energy reserve on account of the high amount of energy stored per gram of fuel compared with carbohydrates, and because they can be stored in a compact, unhydrated form. Wax esters, however, have some advantages as reserve fuels for certain desert plants and marine animals that live in relatively hostile conditions.

Animals store triacylglycerols in adipose tissue, although some marine species utilize the flesh or liver as storage organs. Plants store triacylglycerols (or wax esters) in the seed or sometimes in the fleshy fruit mesocarp. In both animals and plants, the predominant pathway for the biosynthesis of triacylglycerols is the glycerol phosphate pathway. Glycerol phosphate is acylated to phosphatidic acid which is then dephosphorylated to a diacylglycerol by phosphatidate phosphohydrolase, the rate-limiting and controlling step in the pathway. This step provides a branch point in metabolism since the diacylglycerols can be either channelled into phospholipid biosynthesis or acylated to form triacylglycerols. In the intestine, the components of digested and absorbed dietary fat are resynthesized into triacylglycerols via the stepwise acylation of monoacylglycerols. Breakdown of triacylglycerols in the animal gut and in the tissues of animals and plants is catalysed by a range of hydrolytic enzymes termed lipases.

Fuel homeostasis in animals is under complex dietary and hormonal control. Conditions favouring high circulating concentrations of insulin favour fuel storage, while utilization is stimulated by the stress hormones. The enzymology of triacylglycerol turnover in plants has become well established but our understanding of its regulation is more rudimentary. In preparation for germination, lipases are induced, that are capable of complete degradation of the stored fat. Specialized organelles, the glyoxysomes, are then formed to utilize the fatty acids and convert them into simple carbohydrates needed for stem and root formation. The conversion of fatty acids into carbohydrates is not possible in animals.

REFERENCES

Glycerides

Brindley, D.N. (1985) Metabolism of triacylglycerols, in *Biochemistry of Lipids and Membranes* (eds D.E. Vance and J.E. Vance), Benjamin/Cummings, Menlo Park, CA, USA.

Brindley, D.N. (1988) *Phosphatidate Phosphohydrolase*, CRC Press, Florida.

Cryer, A. and Van, R.L.R. (eds) (1985) *New Perspectives in Adipose Tissue Structure, Function and Development*, Butterworths, London.

Stymne, S. and Stobart, A.K. (1987) Triacylglycerol biosynthesis, in *Biochemistry of Plants, Vol. 9* (eds P.K. Stumpf and E.E. Conn), Academic Press, New York, pp. 175–214.

Wax esters

Bauermeister, A. and Sargent, J.R. (1979) Wax esters: major metabolites in the marine environment, *Trends in Biochemical Sciences*, September 1979, pp. 209–211.

Kolattukudy, P.E. (1980) Cutin, suberin and waxes, in *The Biochemistry of Plants*, Vol. 4, (eds P.K. Stumpf and E.E. Conn), Academic Press, New York, pp. 571–645.

Kolattukudy, P.E. (1987) Lipid-derived defensive polymers and waxes and their role in plant-microbe interaction, in *Biochemistry of Plants* (eds P.K. Stumpf and E.E. Conn), Academic Press, New York, pp. 291–314.

Sargent, J., Henderson, R.J. and Tocher, D.R. (1989) The lipids, in *Fish Nutrition*, 2nd edn (ed. J.E. Halver), Academic Press, New York.

5 *Dietary lipids: implications for health and disease*

The main emphasis of this chapter will be on lipids in the human diet, their metabolism in the healthy body and their relation to disease processes. Lipid metabolism in other animals will be touched on, however, for two main reasons. First, a great deal of the food we eat is supplied by farm animals and the types of lipids we consume from this source are determined by the animals' own metabolism, their diets and by farming practices. Secondly, rigorously controlled studies of human metabolism are difficult and expensive to achieve and scientists are, therefore, frequently obliged to make inferences about the metabolic effects of lipids in man from research with experimental animals, generally laboratory rodents.

5.1 LIPIDS IN FOOD

5.1.1 Types of dietary lipids: the fats in foods are derived from the structural and storage fats of animals and plants

With few exceptions, all food originates from living matter; therefore, the lipids in the diet of man are the structural and storage lipids of animals and plants. Table 5.1 lists the sources of fat in the British diet. Qualitatively, we can expect similar types in most developed countries, although the proportions may vary widely.

Animal storage fats (mainly triacylglycerols with some dissolved cholesterol and fat soluble vitamins) will be present in meat that retains parts of the adipose tissue. Lard is a cooking fat derived from the adipose tissue of the pig. Milk fat occurs in full fat milk, cheese, cream, full fat yoghurt, butter

Table 5.1 Contribution of fat to the diet by different foods

	1981	1983	1985	1987
Fat in g/person/day				
All dairy products	32	30	27	24
All meat	28	26	25	25
Margarine and other fats	26	26	24	26
Biscuits, cakes, pastries	6	7	6	7
All other foods	12	12	14	14
All foods	104	101	96	96
Fat as % dietary energy				
Total fat	42	43	43	42
Saturated fatty acids	19	19	18	17
Ratio of polyunsaturated to saturated fatty acids (P/S)	0.25	0.29	0.32	0.37

Source: National Food Survey (1987) Ministry of Agriculture, Fisheries and Food.

and low fat dairy spreads. Plant storage fats (mainly triacylglycerols with some fat soluble vitamins) are present in their original form in nuts (e.g. peanuts), cereal grains and some fruits such as the avocado. They are also familiar in cooking oils, salad oils and mayonnaises. Both animal and plant storage fats are used in the manufacture of spreading fats: margarines and low fat spreads. Eggs are a rich source of fat and because it is present as a lipoprotein (section 5.3.5), it contains significant quantities of triacyl-glycerols, cholesterol and phospholipids. The predominance of fat in foods like cooking oils, margarines, butter, and meat fat is obvious: we call these the visible fats. Where fat is incorporated into the structure of the food, either naturally, as in peanuts or in cheese, or because it is added in the cooking or manufacturing process, as in cakes and pastries or prepared meats, it is not obvious to the consumer and we call these hidden fats.

Because all food derives from biological tissues and because all tissues contain biological membranes, all foods contain, to a greater or lesser extent, structural fats. These are predominantly phospholipids and glycolipids (Chapter 6) with, in animal tissues, cholesterol and in plant tissues, the plant sterols. Thus, the lean part of meat contains the structural lipids, phospholipids and cholesterol of the muscle membranes and minor quantities of glycolipids. Consumption of tissues like brain would introduce larger quantities of animal sphingolipids into the diet. Dairy products contribute small amounts of structural lipids because of the presence of the fat globule membrane. The structural lipids of plants that are important in the diet are those in green leafy vegetables, which are predominantly galactolipids (section 6.2.3) and the membrane lipids of cereal grains, vegetables and fruits.

5.1.2 Fatty acids in the diet: the fatty acid composition of dietary lipids depends on the relative contributions of animal and plant structural or storage lipids, which vary widely in their fatty acid contents

Perhaps the most significant aspect of lipids in the human diet is their content of different types of fatty acids: saturated, monounsaturated and polyunsaturated (section 3.1). It is important to point out at the outset that all natural fats contain complex mixtures of all three types of fatty acids. It is incorrect to describe a fat as saturated or polyunsaturated: only the constituent fatty acids can be so described. Yet fats in which the saturated acids form the largest single fraction are frequently categorized as saturated fats to contrast them with fats in which polyunsaturated fatty acids predominate. As an example of where this can be misleading to a layman, lard is frequently categorized as a saturated fat, yet over half (about 55%) of the fatty acids are unsaturated (Table 4.1).

In general, the animal storage fats are characterized by a predominance of saturated and monounsaturated fatty acids (Table 4.1). Storage fats from ruminant animals tend to have a greater proportion of saturated fatty acids because of the extensive hydrogenation of the animal's dietary fats in the rumen (section 3.2.5). Nevertheless, ruminant fats contain a substantial proportion of monounsaturated fatty acids because of the presence in mammary gland and in adipose tissue of desaturases (section 3.2.4). Milk fat is also characterized by the high proportion of its saturated fatty acids which have chain lengths of twelve carbon atoms or less. These are synthesized in the mammary gland itself (section 3.2.2). The fatty acid composition of simple stomached animals is more dependent on their diet since they do not extensively hydrogenate the dietary unsaturated fatty acids but can incorporate them directly into the adipose tissue. (Studies comparing germ-free animals with those containing a conventional gut microflora indicate that non-ruminants do contain microorganisms that can hydrogenate fatty acids to a limited extent in the colon.) Thus pig adipose tissue can be significantly enriched in linoleic acid, simply by feeding the animals rations supplemented with soya bean oil. Even ruminant fats can be enriched with polyunsaturated fatty acids but in order to do so the dietary polyunsaturated fatty acids have to be protected from hydrogenation in the rumen. This can be achieved by treating the oilseeds used to supplement the feed with formaldehyde, which cross-links the protein and renders the seed oil unavailable to the rumen microorganisms. When the food reaches the acid environment of the true stomach, the cross-links are digested and the polyunsaturated acids are available for absorption just as in simple stomached animals.

Such feeding methods are theoretically useful for modifying the fatty acid composition of food products, if higher contents of polyunsaturated fatty acids are required. Polyunsaturated ruminant products are not generally available, however: they are expensive and subject to oxidative deterioration (section 3.3.4) which may result in poor taste and appearance. Exceptions to the general rule that animal storage fats are predominantly saturated and monounsaturated are the oils of fish and marine mammals, which are rich in polyunsaturated fatty acids of the n-3 family (sections 3.1 and 5.2.2).

Seed oils contain a wide variety of fatty acids, the composition of which is characteristic of the family to which the plant belongs. Generally one fatty acid predominates (Table 4.2). It may either be one of the normal fatty acids, palmitic, oleic, or linoleic, as exemplified by palm oil, olive oil and sunflower seed oil or it may be an unusual acid, for example erucic acid in older varieties of rape seed oil. Coconut oil and palm kernel oil are unusual among seed oils in having a preponderance of saturated fatty acids in which the acids of medium chain length predominate. It is therefore an unjustified generalization to characterize all vegetable oils as unsaturated.

The fatty acid composition of the structural lipids of the animal tissues used as food is remarkably uniform irrespective of the animal's diet or whether it is simple stomached or a ruminant (Table 5.2), although there are clearly some species differences and diet can influence membrane composition in subtle ways (sections 5.2.2 and 8.9). A predominant fatty acid is arachidonic acid, a member of the n-6 family, and meat provides most of our dietary supply of this fatty acid. Plant structural fats are similarly dominated by polyunsaturated acids and supply mainly α-linolenic acid, a member of the n-3 family.

5.1.3 The effects of industrial processing: modification of the fatty acid composition of foods by industrial processing results in a general reduction in unsaturation as well as positional and geometric isomerization of double bonds

We have learned that some degree of influence on the fatty acid composition of animal food lipids can be exerted by animal feeding practices. Likewise, the fatty acid composition of plant fats can be modified by breeding. The best example of this is the change that was brought about by breeding out the erucic acid in rapeseed in exchange for its metabolic precursor, oleic

Table 5.2 Fatty acid composition of some structural lipids important in foods

Food	16:0	16:1	18:0	18:1	18:2	18:3	20:4	LC* PUFA	Others
Beef – muscle	16	2	11	20	26	1	13	0	11
Lamb – muscle	22	2	13	30	18	4	7	0	4
Lamb – brain	22	1	18	28	1	0	4	14	12
Chicken – muscle	23	6	12	33	18	1	6	0	1
Chicken – liver	25	3	17	26	15	1	6	6	1
Pork – muscle	19	2	12	19	26	0	8	0	14
Cod – flesh	22	2	4	11	1	trace	4	52	4
Green leaves	13	3	trace	7	16	56	0	0	5

* Longer chain polyunsaturated fatty acids.

acid (section 8.15.2). By far the greatest influence on food fatty acid composition, however, is by industrial processing, especially catalytic hydrogenation. The objective of industrial hydrogenation is to reduce the degree of unsaturation and there are two main reasons for this. The first is to improve oxidative stability. The more highly unsaturated a fat, the greater is its susceptibility to oxidation (section 3.3.4), resulting in poor flavour and colour and even generating toxic compounds (section 8.15.4). The second is to improve the physical properties of the fat. Highly unsaturated oils have low melting points and as such are unsuitable for many food uses where a solid texture is desired. Industrial hydrogenation (hardening) raises the melting point, thereby giving better textural properties to many food fats. By careful choice of catalyst and temperature, the oil can be selectively hydrogenated so as to achieve the desired characteristics. Indeed, the process is seldom taken to completion, since fully saturated fats, especially those that would be derived from the very long chain fatty acids of fish oils, would have melting points that were too high.

Hydrogenation is carried out in an enclosed tank in the presence of 0.05–0.20% of a finely powdered catalyst (usually nickel) at temperatures up to 180°C after which all traces of the catalyst are removed by filtration. Chemically, there are three main results of hydrogenation:

1. the total number of double bonds in the fatty acid molecule is reduced;

2. a proportion of the double bonds that are present in the original oil in the *cis* geometric configuration are isomerized to the *trans* form; and

3. the double bonds are shifted along the hydrocarbon chain from their original positions in the natural fat (Table 5.3; Figure 5.1).

These changes may have nutritional significance that will be discussed in section 8.15.3. It is important to note that *trans* double bonds do occur in natural fats but generally much less abundantly than *cis* bonds. Thus some seed oils have a significant content of fatty acids with *trans* unsaturation (Table 3.3) although these are not important sources of fat in the human diet. All green plants contain small quantities of *trans*-3-hexadecenoic acid (Table 3.2). The most important naturally occurring isomeric fatty acids in human foods are those in ruminant fats. In principle, the chemical outcome of rumen biohydrogenation (section 3.2.5) is similar to that of industrial hydrogenation listed above. Whether of natural or of industrial origin, fatty acids with *trans* bonds may be monounsaturated or polyunsaturated; the latter may have *cis* and *trans* bonds within the same molecule. Fatty acids containing *trans*-unsaturation in ruminant products make roughly the same contribution to the UK diet as those produced by industrial hydrogenation. Together, these foods provide about 6–7 g per day of *trans*-unsaturated fatty acids.

Industrial processing or domestic cooking which involves heating in the presence of oxygen can result in oxidative degradation of the unsaturated fatty acids in food fats (section 3.3.4 and 8.15.4). This can lead to a reduction in the polyunsaturated fatty acid content of the fat and a concomitant reduction in the antioxidants in the food, especially tocopherols (Table 5.4). Oxidative rancidity can occur upon storage even under ambient or

Table 5.3 Changes in the fatty acid composition of a vegetable oil during hydrogenation

	(g/100 g total fatty acids)	
Fatty acid	*Soya bean oil*	*Hydrogenated soya bean oil*
16:0	11	11
18:0	4	7
18:1 *cis*	22	33
trans	0	12
18:2 *cis*-9, *cis*-12	54	22
cis-9, *trans*-12 } *trans*-9, *cis*-12 }	0	8
cis-9, *trans*-12 } *trans*-8, *cis*-12 }	0	4
18:3 all isomers	8	2

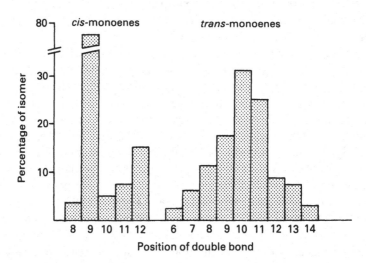

Figure 5.1 Formation of positional and geometrical isomers during industrial hydrogenation of a vegetable oil. N.B. The concentration of *cis*-isomers is about twice that of *trans*-isomers. Reproduced with kind permission of Professor G.J. Brisson and MTP Press Ltd, from Brisson, G.J.P. (1982), *Lipids in Human Nutrition*, p. 38, Figure 19.

Table 5.4 Changes in the composition of a vegetable oil during frying

	Frying time (hours)				
Characteristics of the oil	*0*	*10*	*32*	*88*	*231*
Carbonyl value (mEq/kg)	6	44	96	159	126
Peroxide value (mEq/kg)	1.1	5.0	8.9	7.6	5.3
Oxidized fatty acids (g/100 g)	0	0.5	1.7	5.1	4.7
Dimers (g/100 g)	0	0.6	1.0	2.8	2.3
cis-9, *cis*-12–18 : 2 (g/100 g)	20.7	19.4	17.5	15.2	14.2
α-Tocopherol (μg/g)	136	0	0	0	0

Source: Yizaki *et al.* (1987) *Lipids*, **19**, 326.

refrigeration temperatures and antioxidants such as butylated hydroxy-toluene, butylated hydroxyanisole and gallate are used in the food industry to prevent this deterioration.

5.2 ROLES OF DIETARY LIPIDS

5.2.1 As a source of metabolic energy

(See sections 4.3.1, 4.4 and 4.6.3).

5.2.2 As a source of essential nutrients

Dietary lipids contain two types of essential nutrients for animals and man: the essential fatty acids and the fat soluble vitamins.

(a) The essential fatty acids: discovered in the late 1920s, certain polyunsaturated fatty acids play a vital role in animal physiology. They have to be obtained from the diet

Discovery of the essential fatty acids
The student is recommended to read this section in conjunction with section 3.2.4 which give the structural and metabolic relationships between the different polyunsaturated fatty acids.

The great period of isolation and identification of vitamins lay between 1840 and the late 1920s. During this time the deficiency effects of all the water soluble and lipid soluble compounds were demonstrated, though not surprisingly, elucidation of their structures and total synthesis took another 35 years. By the end of the 1920s, it was thought that all the major accessory food factors had been discovered and that carbohydrate and fat were important only in so far as their energy contribution was concerned.

This whole concept was overturned in 1929 when the Americans Burr and Burr described how acute deficiency states could be produced in rats by feeding fat-free diets and that these deficiencies could be eliminated by adding only certain specific fatty acids to the diet. It was shown that linoleic and later arachidonic acid were responsible for this effect and the term vitamin F was coined for them. This name, however, has now been reserved for some factors of the vitamin B complex, so that the term essential fatty acid (EFA) is now *de rigeur*.

Essential fatty acid deficiency can be produced in a variety of animals, including man, but data for the rat are the best documented (Table 5.5). The disease is characterized by skin symptoms, such as dermatosis and the skin becomes more 'leaky' to water. Growth is retarded, reproduction is impaired and there is degeneration or impairment of function in many organs of the body. Biochemically, essential fatty acid deficiency is characterized by changes in the fatty acid composition of many tissues,

Table 5.5 Major effects of essential fatty acid deficiency in rats

Skin symptoms	Dermatosis; water permeability ↑, sebum secretion ↓, epithelial hyperplasia
Weight	↓
Circulation	Heart enlargement
	Capillary resistance ↓
Kidney	Enlargement; intertubular haemorrhage
Lung	Cholesterol accumulation
Endocrine glands	Adrenals: weight: ↓ ♀, ↑ ♂
	Thyroid: weight ↓
Reproduction	Females: irregular oestrus; impaired lactation and reproduction
	Males: degeneration of the semeniferous tubules
Metabolism	Changes in tissue fatty acid composition
	Cholesterol in plasma ↓
	Cholesterol in liver; adrenals; skin ↑
	Mitochondria: swelling and uncoupling of oxidative phosphorylation
	Liver: triacylglycerol output ↑

↑ indicates increase; ↓ indicates decrease.

especially their biological membranes, whose function is impaired and, in the mitochondria, the efficiency of oxidative phosphorylation is much reduced. Well documented EFA deficiency in man is rare, but was first seen in children fed virtually fat-free diets. Four hundred infants were fed milk formulas containing different amounts of linoleic acid. When the formulas contained less than 0.1% of the dietary energy as linoleic acid, clinical and chemical signs of EFA deficiency ensued. The skin abnormalities were very similar to those described in rats and these and other signs of EFA deficiency disappeared when linoleic acid was added to the diet.

Which fatty acids are essential?
The effects of different acids on the relief of deficiency symptoms were first compared by measurement of growth response, but later Thomasson at the Unilever Laboratories in the Netherlands used a test based on disturbances in water metabolism, since an important result of EFA deficiency is an altered permeability of biological membranes. On this basis, Thomasson was able to catalogue the relative effects of the natural di-, tri- and tetra-unsaturated acids, showing that linoleic (9,12–18:2), γ-linolenic (6,9,12–18:3) and arachidonic (5,8,11,14–20:4) acids (all in the n-6 family) had a similar order of activity whereas that of α-linolenic acid (9,12,15–18:3, n-3) was much lower. As research continued, classification on the basis of double bond position measured from the methyl terminal group seemed to show that all the active acids possessed *cis* double bonds in the n-6, n-9 positions. Thomasson's colleague van Dorp, however, made a wide range of synthetic polyunsaturated fatty acids, with different chain lengths, some with an odd carbon number and with double bonds in different positions and these were tested for essential fatty acid activity. Not all those with biological activity had double bonds in the n-6, n-9 positions, so this hypothesis had to be abandoned. All fatty acids with essential fatty acid activity, as measured by restoration of growth or water permeability or curing of dermal symptoms in rats, have a *cis, cis* methylene-interrupted double bond system. The conversion of *cis* into *trans* double bonds eliminates biological activity, although the presence of a *trans* bond in conjunction with a *cis, cis*-methylene interrupted sequence does not necessarily destroy activity (for example, columbinic acid, *trans*-5,*cis*-9,*cis*-12–18:3 from *Aquilegia vulgaris* is able to restore growth and correct dermal symptoms although it is not a direct eicosanoid precursor).

After nearly 60 years of intensive research, we still do not have complete answers to the questions: what, in structural terms, determines whether a fatty acid will have essential fatty acid activity; why, exactly, are they essential; and, how much do we need in the diet? In the following sections we shall try to give you a summary of current knowledge.

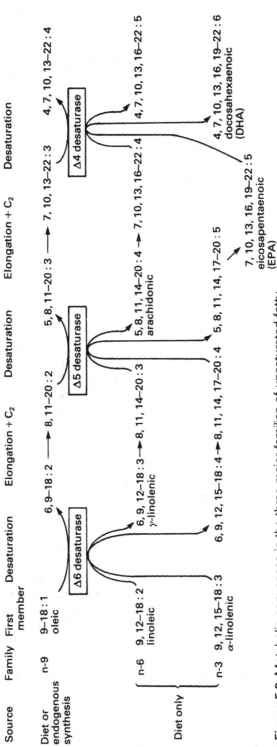

Figure 5.2 Metabolic sequences in the three major families of unsaturated fatty acids leading to the long chain polyunsaturated fatty acids. A minor family, n-7, has been omitted for simplicity.

With hindsight, it is clear that the reasons why some polyunsaturated fatty acids have EFA activity and some do not can only be understood by studying the biochemical pathways by which they are interconverted and metabolized to more complex end-products.

In section 3.2.4 we described the pathways for the biosynthesis of the polyunsaturated fatty acids, and they are illustrated in Figure 5.2.

This sequence of desaturations and elongations enables tissues to produce a variety of polyunsaturated fatty acids tailored to their needs. Because, during the course of evolution, animals have lost the ability (retained by plants) to insert double bonds in positions 12 and 15, the members of these four families (n-3, n-6, n-7, n-9) cannot be interconverted in animal tissues. Linoleic acid and its relatives are termed essential because without them animals will die. Therefore, the first member of the series has to be supplied in the diet from plant sources. Arachidonic acid, the main product of the elongation and desaturation of linoleic acid, has essential fatty acid activity in that it can cure the signs of EFA deficiency described earlier but it is not essential in the human diet as long as linoleic acid is supplied, i.e. it is an essential metabolite but not an essential nutrient for man.

The fact that the $\Delta6$-desaturase can introduce its double bond at position 6 into the first member of each fatty acid family (n-3, n-6, n-7 and n-9) has extremely important consequences. All four precursors can compete for desaturation by the same enzyme as illustrated in Figure 5.2. The affinity of the substrates for the $\Delta6$-desaturase is in the order 18:3 > 18:2 > 18:1. In man and many other mammals, the most important metabolic pathway is the one in which linoleic acid is converted into arachidonic acid. Normally, the diet contains sufficient linoleic acid for this pathway to be able continually to supply the quantity of arachidonic acid needed by body tissues. There are circumstances when this is not so, however:

1. When the absolute amount of linoleic acid in the diet is low. This can happen, for example, when hospital patients are on enteral or parenteral feeds low in fat, or when babies are fed artificial formulas that contain little or no linoleic acid, or in malnutrition. Thus, attention has often focused on protein-energy malnutrition in third world countries, but essential fatty acid deficiency should also be considered when chronic malnutrition results in a low fat intake. As illustrated in Figure 5.3, the serum lipids of malnourished children are often deficient in linoleic and related acids.
2. When the dietary intake of linoleic acid is normally adequate but a person is unable to absorb dietary fat (section 5.3.4). This may occur as a temporary result of a generalized illness or as a direct and permanent consequence of a specific disease involving impairment of absorption, for example, cystic fibrosis, achrodermatitis enteropathica and Crohn's disease.

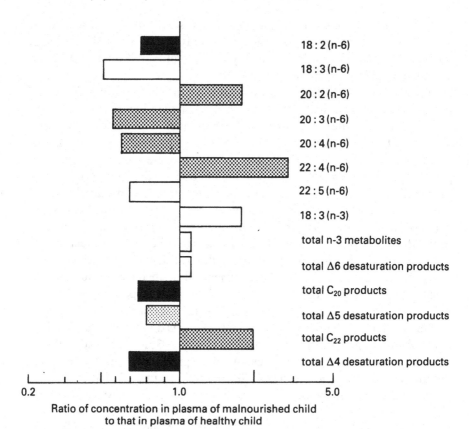

Figure 5.3 Polyunsaturated fatty acid profiles in plasma phospholipids of children with chronic malnutrition compared with well-nourished children; □ = difference not significant; ▦ = $p < 0.05$; ■ = $p < 0.01$. Reproduced with kind permission of Professor R.T. Holman and the American Oil Chemists' Society, from *Dietary Fats and Health*, (eds E.G. Perkins and W.J. Visek), p. 354, Figure 4, 1983.

3. As a consequence of genetic disorders resulting in the lack of specific desaturase enzymes. The American biochemist Ralph Holman who was a pioneer in our understanding of the essential fatty acids has used the technique of serum fatty acid profiling to pinpoint aberrations in the pathways of polyunsaturated fatty acid metabolism, which might be due to defects in desaturases. Examples are in diseases such as Sjorgren–Larsson syndrome, Reyes syndrome and Prader–Willi syndrome (Figure 5.4). It is also possible that some people may have low Δ6-desaturase activity, not as a result of a specific gene defect but as a result of normal biological variation. It has been suggested that such people may benefit

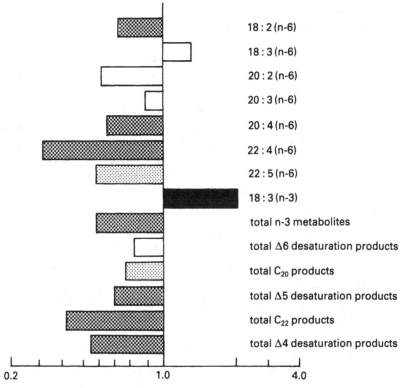

Figure 5.4 Polyunsaturated fatty acid profiles in plasma phospholipids of children with cystic fibrosis compared with those in normal children; □ = difference not significant; ▦ = $p < 0.05$; ▦ = $p < 0.01$; ■ = $p < 0.001$. Reproduced with kind permission of Professor R.T. Holman and the American Oil Chemists' Society, from *Dietary Fats and Health*, (eds E.G. Perkins and W.J. Visek), p. 255, Figure 5, 1983.

from the presence in the diet of γ-linolenic acid (present in evening primrose, blackcurrant and borage). The rate limiting 6-desaturase step is thus by-passed.

4. When the diet contains a small, but normally adequate, amount of linoleic acid which is swamped by enormous amounts of other fatty acids in the diet (e.g. oleic acid or isomeric fatty acids, section 8.15.3). In this situation, the fatty acids in excess compete successfully with linoleic acid for the Δ6-desaturase and generate a series of polyunsaturated fatty acids that have no essential fatty acid activity and cannot substitute for the linoleic acid family. The effect of this competition is illustrated in Figure

5.2. The most likely alternative pathway is the one beginning with oleic acid and this generates a 20-carbon acid with the structure all-*cis*-5,8,11-eicosatrienoic acid in place of arachidonic acid. This acid is not normally present in tissues in more than minute amounts and its accumulation provides a biochemical diagnosis of the occurrence and extent of EFA deficiency. Experiments with laboratory animals have shown that even very small quantities of linoleic acid are sufficient to protect the animal from EFA deficiency as long as the remainder of the diet is low in fat. Increasing the intake of non-essential fat however, while maintaining the same low level of linoleic acid, can induce overt EFA deficiency. This principle has been demonstrated by feeding rats diets rich in triolein or hydrogenated fats containing a high proportion of isomeric fatty acids in the presence of low concentrations of linoleic acid (section 8.15.3).

From this discussion it will be apparent that while all EFA are polyunsaturated, not all polyunsaturated fatty acids are EFA. Nutritionists and biochemists have used the ratio of 5,8,11–20:3/5,8,11,14–20:4 (often called the triene : tetraene ratio) as a biochemical index of essential fatty acid status (Figure 5.5). For many years, it was held that a ratio greater than 0.4 was diagnostic of EFA deficiency but recent work has suggested that it may be prudent to revise this figure downwards to nearer 0.2.

In the foregoing discussion we examined the consequences of competition between n-6 (essential) and n-9 (non-essential) fatty acids. It is clearly also possible to have competition between n-6 and n-3 fatty acids. Normally, the human diet contains very little n-3 fatty acids in comparison with n-6. The main sources are green leaves, some seed oils (Table 3.3) and fish oils. In the UK, the consumption of fatty fish, which has been hitherto rather low, is now beginning to increase and there is much discussion about the nutritional benefits of fish oils. Some consideration of the essentiality of n-3 fatty acids is therefore in order.

As far as we know, linoleic acid is essential for almost every animal species at a dietary level of about 1% of energy, although there is evidence that some insects and protozoa can produce this fatty acid. There is less certainty about α-linolenic acid. Fish, whose lipid metabolism is geared to processing a high dietary intake of n-3 fatty acids, seem to have a definite and high requirement (3% of energy) for fatty acids of this family. The brain and retinal rods of most species, including the rat, are characterized by a high proportion of the n-3 fatty acids and we can infer that, since these acids cannot be synthesized in the body, there is some dietary requirement for them, however small. The requirement may be limited to early life.

Capuchin monkeys maintained on a diet containing adequate linoleic acid but little or no α-linolenic acid suffered symptoms closely resembling those of classical EFA deficiency. These were cured by the addition of linseed oil

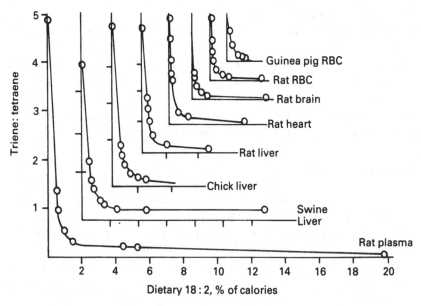

Figure 5.5 Influence of dietary linoleic acid on the ratio of 20 : 3(n-9) to 20 : 4(n-6) [the 'triene : tetraene ratio'] in different species. The relative scales are similar in all plots; the inset plots are to indicate the similarity in relationship over a wide range of tissues and species. Reproduced with kind permission of Professor R.T. Holman and Pergamon Press Ltd, from *Progress in the Chemistry of Fats and Other Lipids* (1970), **9**, 607–82.

(which contains appreciable amounts of linolenic acid) to the diet. In man, the essentiality of α-linolenic acid remained in doubt until Holman and his colleagues in 1982 described a case of a young girl who displayed neurological symptoms 4–5 months after being on total parenteral nutrition in which the fat component contained mainly linoleic acid and only a minute amount of linolenic acid. When safflower oil was replaced by soybean oil containing much more linolenic acid, the neurological symptoms disappeared. More recently a Norwegian group provided evidence for linolenic acid deficiency in elderly patients fed by gastric tube. It seems quite certain that only very small amounts of this nutrient are needed in human diets and you will find this topic succinctly reviewed by Zollner (see references).

Cats are unusual among mammals in that neither linoleic nor α-linolenic acid alone is sufficient to protect against the effects of fatty acid deficiency: these animals require arachidonic acid in their diets because they lack the desaturases that insert double bonds at positions 6, 12 and 15.

Why are some fatty acids essential?
Two distinct roles, to account for their essentiality have been discussed for the EFA. These are as membrane components and as precursors for biologically active metabolites, the eicosanoids. Each will be discussed in turn, followed by a discussion of the links between the two roles. You will need to read the membrane section in conjunction with Chapter 6 and the eicosanoid section in conjunction with Chapter 3. In addition, a specific role for linoleic acid in skin function is discussed in section 8.12.

Essential fatty acids in membranes
Among the results of EFA deficiency are changes in the properties of biological membranes of which polyunsaturated fatty acids are major constituents of the structural lipids. These changes in properties, for example, the permeability of the membrane to water and small molecules such as sugars and metal ions, can be correlated with changes in the fatty acid composition of the membrane. Membranes of liver mitochondria from EFA deficient animals have smaller proportions of linoleic and arachidonic acids and larger proportions of oleic acid and all-*cis*-5,8,11-eicosatrienoic acid than those of healthy animals (Table 5.6). β-Oxidation and oxidative phosphorylation (section 3.3.1) are less efficient. These changes at the molecular and cellular levels are reflected in the animal's poorer performance in converting food energy into metabolic energy for growth and in the maintenance of body function.

Organs and tissues performing functions such as storage (e.g. adipose tissue), chemical processing (e.g. liver), mechanical work (e.g. muscle) and excretion (e.g. kidney) tend to have membranes in which the n-6 fatty acids

Table 5.6 Changes in the fatty acid composition of mitochondrial phospholipids in essential fatty acid deficiency

| Fatty acid | (g/100 total fatty acids) | |
	EFA sufficiency	EFA deficiency
16 : 0	31	31
18 : 0	10	9
18 : 1 (n-9)	13	32
18 : 2 (n-6)	21	10
20 : 3 (n-9)	0.6	3.7
20 : 3 (n-6)	1.7	1.0
20 : 4 (n-6)	14	7
20 : 3 (n-9)/20 : 4 (n-6)	0.04	0.53

predominate, with arachidonic acid as the major component. In contrast, nervous tissue, reproductive organs and the retina of the eye have a greater proportion of the longer chain acids with 5 or 6 double bonds, predominantly of the ń-3 family. It has been assumed that fatty acids of the n-3 family have specific roles in vision, reproduction and nerve function and there is some evidence in rats to show that when the ratio of n-3 to n-6 acids is modified experimentally, some changes do occur in the electro-retinograms and in behavioural responses. We still do not have sufficient evidence, however, to pinpoint the specific functions for these fatty acids.

The metabolism of the essential fatty acids during the growth and development of animals will be discussed in section 5.4.

Essential fatty acids as precursors for eicosanoids

In section 3.4 we described the biosynthesis of eicosanoids from poly-unsaturated fatty acids of the n-3 and n-6 families and indicated in section 3.4.8 that the balance of eicosanoids produced was important in, for example, maintaining normal vascular function. Several studies have demonstrated that altering the amounts and types of n-6 and n-3 fatty acids in the diet can influence the spectrum of eicosanoids produced. For example, substitution of fish oils in which n-3 polyunsaturated fatty acids predominate for diets in which linoleic acid (n-6) is the main polyunsaturated fatty acid (as typified by the UK diet) results in changes in plasma and platelet fatty acid profiles from arachidonic to eicoasapentaenoic acid as the predominant polyunsaturated fatty acid and a reduction in the formation by platelets of thromboxane A_2, an eicosanoid that stimulates their aggregation (Table 5.7).

Table 5.7 Eicosanoid formation by human cells isolated from subjects given diets of differing fatty acid composition

		Predominant dietary polyunsaturated fatty acids	
Cell type	*Eicosanoid*	*n–6*	*n–3*
Platelets	Thromboxane B_2 (units)	100	30–50*
	Thromboxane B_3 (units)	ND	10
Endothelial cells	Prostaglandin PGI_2 (ng)	146	236*
	Prostaglandin PGI_3 (ng)	ND	134

ND = not detectable. * $p < 0.05$. Source: Weber *et al.* (1986) *Progress in Lipid Research,* **25**, 275.

The physiological effects of eicosanoids are so potent in such minute quantities that they need to be generated locally and destroyed immediately after they have produced their effect by enzymes that convert them into inactive metabolites. The excretion of these metabolites in the urine has been used as a quantitative method for estimating the daily production of eicosanoids and assessing the quantities required by the body. The mechanism that ensures efficient local production is the release of the precursor polyunsaturated fatty acids from membrane phospholipids by specific phospholipases (sections 3.4.9 and 7.2) and transfer to the enzymes of eicosanoid biosynthesis, which are also located within the membrane. The membrane phospholipids can therefore be regarded as a vast body store of essential fatty acids that are immediately available for eicosanoid biosynthesis when required. As they are depleted, they must be replaced by new polyunsaturated fatty acids biosynthesized by the pathways illustrated in Figure 5.2. This in part explains the dynamic turnover of fatty acids in biological membranes, although eicosanoid formation represents a very small proportion of fatty acid turnover.

Several aspects of eicosanoid metabolism and function, however, need urgent research. These include: (a) the mechanism by which the relative proportions of the different eicosanoids are regulated; (b) the significance of changes of dietary fatty acid composition for whole body eicosanoid production; (c) the quantitative significance of the different pathways and sites of synthesis; and (d) the quantitative relationships between the requirements for essential fatty acids, which are measured in grams and the daily production of eicosanoids, which is measured in micrograms.

What are the quantitative requirements for essential fatty acids in the diet?
In assessing nutritional requirements, a distinction needs to be made between physiological requirements and dietary requirements. The primary task is to try to estimate physiological requirements; that is, the need for the substance or substances within the body for essential metabolic functions. Dietary requirements will normally be greater than this because of the need to take into account the fact that absorption is normally less than 100% efficient and there are subsequent losses in metabolic pathways.

At the present time it is impossible to give more than a rough estimate of requirements for EFA. Firstly, as will be clear from the foregoing discussion, we are not sure of all the functions of EFA and the extent to which different EFA have distinct functions. There may be different levels of requirements for different functions. For example, there may be a relatively low level of requirement (as for most vitamins) which is necessary to prevent the overt signs of EFA deficiency, such as the skin lesions and growth retardation. For most species, including man, this seems to be around 1% of dietary energy as linoleic acid. In man, the requirement for linolenic acid is less certain and certainly well below 1% of energy. At a higher level, there may be a less easily definable quantity required for maintaining an optimal balance of membrane fatty acids, preserving a reservoir of precursors for eicosanoid production and for maintaining optimal plasma lipoprotein concentrations (section 5.5). The absolute amounts required at this level will be influenced by the amounts and types of other fats in the diet because of the competition for metabolic pathways between EFA and non-EFA (Figure 5.2).

Secondly, experimental data are in general available only for laboratory animals. Thus, man's requirements can often only be implied. Thirdly, it seems that requirements change with age and with physiological conditions. Thus the requirements for the fetus differ from those of the neonate, which are in turn different from those of adolescents, pregnant and lactating women and the elderly. In man, there is no sound way of approaching the problem of requirements experimentally. We must gradually build up a picture from signs and symptoms occurring under conditions of known intake of EFA. The observations of hospital patients on total parenteral nutrition are helpful in building up this picture. Another approach is the so-called factorial method in which we try to calculate the needs at each growth phase. For example an expert Food and Agriculture Organization/ World Health Organization committee has tried to make estimates for requirements in pregnancy based on the amounts of EFA known to accumulate in the fetus and maternal adipose tissue. Similarly, lactational requirements have been estimated from the average content of human milk and the volumes of milk produced. On this basis, the additional requirements of pregnant and lactating mothers would be increased by between 1 and 2% of dietary energy. The problem with this sort of approach is that it ignores the capacity of organisms to adapt their metabolism to become more efficient during periods of physiological demand. Thus estimates of linoleic acid requirements have been put at between 1 and 3% of dietary energy, depending on physiological state, but in our opinion are more likely to be closer to 1%. Average intakes of linoleic acid in the UK are 12 g per day (5% of energy) and in the USA 15 g per day (9% of energy). Dietary intakes are thus 2–3-fold in excess of requirements. Nevertheless, for reasons discussed in section 5.5.4, it is generally considered that there are distinct advantages in having a greater quantity of polyunsaturated fatty acids in the diet than are needed simply to prevent EFA deficiency.

(b) The fat soluble vitamins

Human requirements for the essential fatty acids are measured in grams per day; those for the fat soluble vitamins, in micrograms per day. It was realized early this century that minute amounts of substances other than proteins, carbohydrates, fats and minerals are needed in the diet to sustain growth and reproduction and maintain health. These accessory food factors appeared to be either fat soluble or water soluble and were given the name vitamins. They were divided into fat soluble A and water soluble B, but eventually it was realized that both these classes were mixtures of chemically unrelated compounds. The fat soluble vitamins, found in mainly fatty foods, are now classified as vitamins A, D, E and K.

Vitamin A
The chemical name for vitamin A is all-*trans*-retinol (Figure 5.6). Retinol itself is found only in animal fats. Plant materials, such as dark green leaves and vegetable oils, like palm oil, contain a precursor, β-carotene (provitamin A), which can serve as a source of retinol in the body.

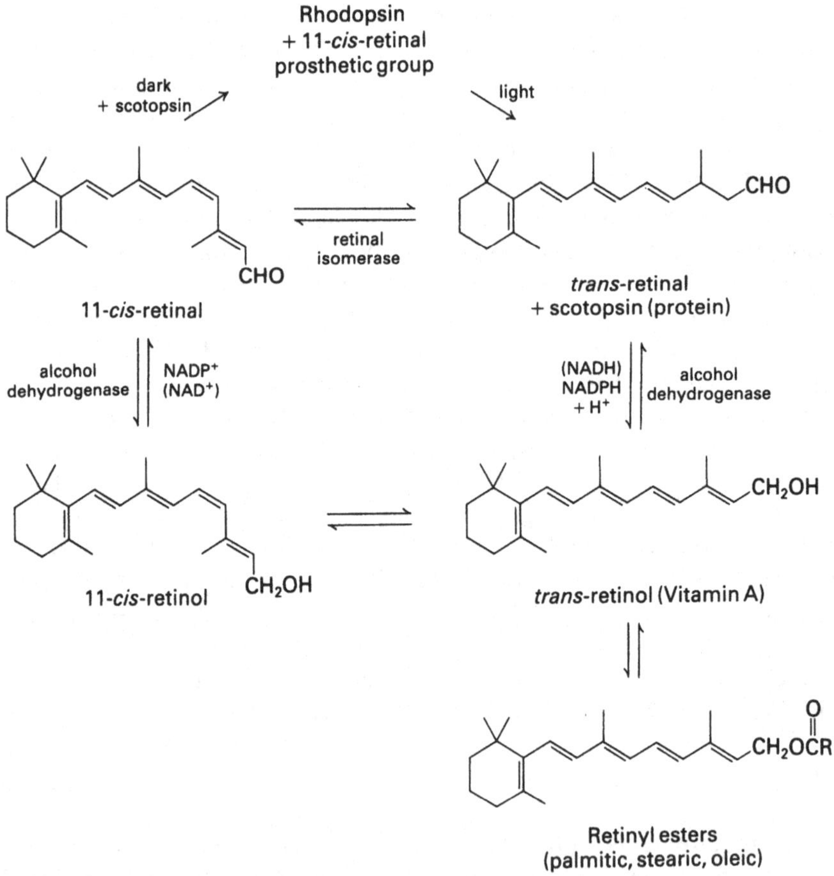

Figure 5.6 Metabolic transformations of vitamin A derivatives related to vision.

The vitamin is important in animal growth, reproduction, the maintenance of function of mucous membranes and vision, but only in the latter case can its role be adequately defined at the molecular level. The enzymic oxidation product of retinol, 11-*cis*-retinal (Figure 5.6) complexes with a protein, opsin, in the retinal rods to form the conjugated protein, rhodopsin. The visual process consists of a series of reactions triggered by the photochemical isomerization of 11-*cis*-retinal to all-*trans*-retinal.

The major forms of vitamin A (apart from carotene) are the esters of retinol with long chain fatty acids. Compared with the normal pattern of fatty acids in tissue lipids, they are a relatively select group and much more saturated. Palmitic acid predominates, together with smaller quantities of stearic and oleic acids. When the esters reach the lumen of the small intestine, they are almost completely hydrolysed, absorbed into the intestinal cells and reesterified. They are transported as components of the chylomicrons to the liver where they are stored almost entirely as retinyl palmitate, regardless of the composition of the dietary esters. Modification of dietary retinyl esters to such a specific fatty acid composition requires, as in the case of the cholesterol esters, specific hydrolytic and synthetic enzymes. These occur mainly in the liver, intestine and the retina of the eye.

There are two forms of the hydrolytic enzyme, one of which is specific for short chain esters like retinyl acetate, even though this ester does not occur naturally! The other has maximum activity with retinyl palmitate as substrate but also hydrolyses other long chain esters. As in the hydrolysis of cholesteryl esters, the enzyme is not just a non-specific esterase, but has quite definite specificity for retinyl esters. In vitamin A deficiency, the activity of the enzyme increases one hundred fold. The esterification enzyme resembles the low energy cholesteryl esterase in that neither ATP nor coenzyme A appear to take part in the reaction; nor are free fatty acids or acyl-CoA thiolesters incorporated into retinyl esters. One of the major problems in this area of research is to identify the acyl donor, which may be, as in plasma cholesteryl ester biosynthesis, a phospholipid.

The physiologically active form of vitamin A is the alcohol retinol, and its transport to target tissue from the liver requires its binding to a specific transport protein, retinol binding protein (RBP). At the target tissue, the RBP-retinol complex binds to specific cell surface receptors before discharging its retinol.

Vitamin A and pro-vitamin A are not widely distributed in foods. The main sources are green vegetables, carrots, liver, milk, butter and margarine. In the UK, all margarine for retail sale is required by law to contain about the same amount of vitamin A (added as synthetic retinol or the pro-vitamin A, β-carotene) as butter. Fish liver oils are by far the most concentrated sources. In contrast, mammalian storage fats, such as lard and dripping, contain none. The British diet contains, on average, about twice the recommended daily amount (RDA) of vitamin A, with two thirds coming from retinol itself and about a third from carotene. (The RDA is not a minimum requirement, but the amount estimated to cover the needs of most of the population.) This happy situation is not shared by many parts of the world, since vitamin A deficiency is widespread, the most afflicted countries being Bangladesh, India, Indonesia and the Philippines. The problem also exists, though to a lesser degree, in many African and Central or South American countries.

Vitamin A is essential for vision and the most tragic manifestation of vitamin A deficiency is blindness in young children. The first effects are seen as severe eye lesions, a condition known as xerophthalmia which is eventually followed by keratomalacia with dense scarring of the cornea and complete blindness. Xerophthalmia is considered to be one of the four commonest preventable diseases in the world. Although there are large scale programmes for the supplementation of children's diets with vitamin A, these are difficult to implement successfully and the World Health Organization considers that if the consumption of green leafy vegetables and suitable fruits by young children could be substantially increased, there is every reason to believe that the problem would be solved. There might be a case for trying to increase the overall fat content of the diet, too. Here, we are more concerned with solving problems of economics, distribution and with changing local eating habits than with biochemistry and nutrition, where the knowledge is already to hand.

Epidemiological evidence shows that people with above average blood retinol concentrations or above average β-carotene intakes have a lower than average risk of developing some types of cancer (section 8.10).

The idea is prevalent amongst many laymen that because vitamins are essential for health, the more that are consumed, the more effective will they be. No more than 0.75 mg vitamin A is required by the average person per day. Whereas an excess of most water-soluble vitamins is excreted from the body, this is not so with fat-soluble vitamins. Vitamin A, if taken in excess, will accumulate in the liver and eventually destroy it. Xavier Mertz, the Antarctic explorer, was forced, during an expedition in 1912, to eat the raw livers of his dogs for sustenance. He died from the excessive effects of vitamin A which resulted in the complete peeling of the skin from his body. Recent cases of vitamin A toxicity have been reported as a result of people following unusual diets requiring the ingestion of abundant quantities of carrot juice.

Vitamin D

Vitamin D is a general name for a family of sterols with anti-rachitic properties. The only one of significance is the one now called vitamin D_3 or cholecalciferol (Figures 5.7 and 7.20). It is present in only a few foods (Table 5.8), the richest sources being the liver oils of fish.

Cholecalciferol is produced by the ultraviolet irradiation of 7-dehydrocholesterol, a sterol widely distributed in animal fats including the skin surface lipids. Many people get little or no vitamin D from their diet but obtain all they require from the action of the ultraviolet rays in sunlight on

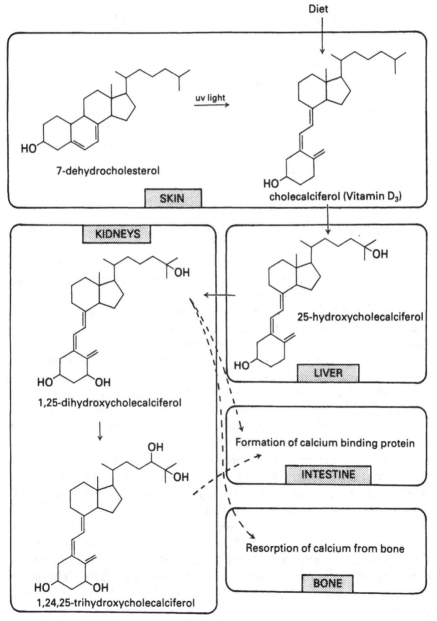

Figure 5.7 Metabolism of vitamin D.

Table 5.8 Vitamin D content of some raw foods (μg/100 g)

Milk (liquid, winter)	0.01
Milk (liquid, summer)	0.03
Milk (UHT)	0.02
Milk (evaporated)*	2.91
Cheese (Cheddar)	0.26
Eggs	1.75
Liver	0.75
Herring, kipper	22.40
Salmon (canned)	12.50
Sardines (canned)	7.50
Butter	0.76
Margarine*	7.94
Cod liver oil	212.50

* Includes added vitamin D.

the 7-dehydrocholesterol in the skin. For this reason, many nutritionists have argued that vitamin D should hardly be called a vitamin at all. Two groups of people, however, may have a special need to obtain vitamin D from the diet. In the first group are children and pregnant and lactating women whose requirements are particularly high. In the second group are people who are little exposed to sunlight, such as the housebound elderly and people in far northern lattitudes or those who wear enveloping clothes. Dark-skinned immigrants to Northern Europe are especially vulnerable. Infants and children who obtain too little vitamin D develop rickets, with deformed bones that are too weak to support their weight. The reason why, in the UK and some other countries, vitamin D preparations are provided for children and pregnant women, and margarine is fortified with it, is because these degenerative changes soon become permanent if supplementation is not begun early enough.

A major role of vitamin D is to maintain the concentrations of calcium and phosphorus in the blood, primarily by enhancing the absorption of dietary calcium from the alimentary tract and regulating the interchange of calcium between blood and bone. It is likely that there are other cellular roles as yet little understood. The active form of vitamin D responsible for the hormone-like functions described above is 1,25-dihydroxycholecalciferol. Cholecalciferol, whether absorbed from the diet or synthesized in the skin, is first hydroxylated in the liver to 25-hydroxycholecalciferol. This is the main form of the hormone circulating in the blood, bound to a sterol binding protein. A further hydroxylation occurs in the kidneys to give 1,25-

dihydroxycholecalciferol. These hydroxylations are catalysed on the endoplasmic reticulum by mixed function monooxygenases that require NADPH and molecular oxygen. The biochemistry of vitamin D is summarized in Figure 5.7. A major function of the 1,25-dihydroxy-vitamin D_3 is to bind to the nucleus of enterocytes where it stimulates the biosynthesis of a calcium binding protein that is involved in calcium absorption and cellular transport. It is now becoming appreciated, however, that the hormone plays a wider role, probably in virtually every cell in the body, although these functions are not well understood.

Like vitamin A, vitamin D is toxic in high doses. Too high an intake causes more calcium to be absorbed than can be excreted, resulting in excessive deposition in, and damage to, the kidneys.

Vitamin E

Substances with vitamin E activity occur in a number of different molecular forms, the major one being α-tocopherol (Figure 5.8). In 1922, the Americans, Evans and Bishop, discovered that vitamin E prevented sterility in rats reared on fat-deficient diets fortified with vitamins A and D. Because vitamin E is so widespread in foods and like all other fat soluble vitamins is stored in the body, deficiency states are rarely if ever seen, a possible exception being in premature infants with low fat stores. The richest sources are vegetable oils, cereal products and eggs.

α-Tocopherol is present in the lipid bilayers of biological membranes (section 6.5) and may play a structural role there. It is known to be a powerful antioxidant and prevents the oxidation of unsaturated lipids (sections 3.3.4, 8.11 and 8.15). The products of lipid peroxidation can cause damage to cells if the oxidative process is not kept in check (section 8.11) and such damage appears to be exacerbated in animals fed diets deficient in vitamin E.

It is a generally held view that dietary requirements should be considered in relation to the polyunsaturated fatty acid content of the diet rather than in absolute amounts. A ratio of vitamin E to linoleic acid of about 0.6 mg g^{-1} is generally recommended. In general, those vegetable oils containing high concentrations of polyunsaturated fatty acids are sufficiently rich in vitamin E to give adequate protection.

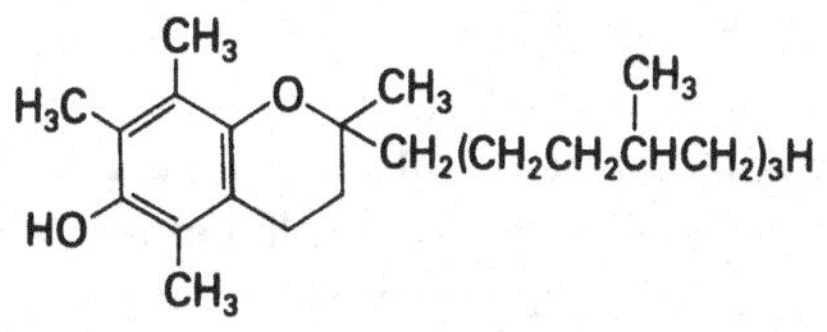

Figure 5.8 α-Tocopherol (vitamin E).

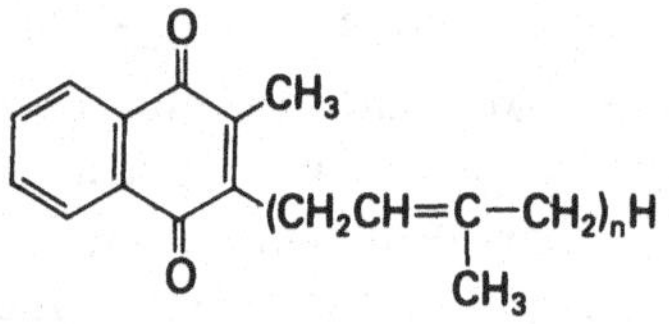

Figure 5.9 Menaquinones (vitamin K).

Vitamin K

Vitamin K activity is shared by a group of chemical compounds called the menaquinones (Figure 5.9). They are produced in green leafy vegetables or by bacteria. Indeed, the reason why no well-defined human deficiency has been described may be because our intestinal bacteria are capable of providing for our entire needs.

Vitamin K is necessary for the normal clotting of blood and a deficiency would prolong the time taken for blood to clot.

Deficiencies of vitamin K and indeed all the other fat soluble vitamins are more likely to occur as a result of impairment in fat absorption than from dietary insufficiency. This could occur when the secretion of bile salts is restricted (as in biliary obstruction), when sections of the gut have been removed or damaged by surgery or in diseases, such as tropical sprue and cystic fibrosis, that are associated with poor intestinal absorption. Even when normal absorptive mechanisms are functioning well, some fat is necessary in the diet to improve the absorption and utilization of fat soluble vitamins. There is little evidence, however, that, within the normal range of fat intakes, the amount of dietary fat significantly affects the utilization of fat soluble vitamins.

5.2.3 Palatability: lipids play an important role in enhancing the flavour and texture of foods

Whatever its nutrient composition might be, food has no nutritive value if it is not eaten. In countries where choice of food is abundant, palatability is a major factor in determining food choice and, therefore, nutrient intakes. Indeed there is evidence that price and the appearance and taste of food are more influential in determining food choice than interest in its nutritive value or healthiness although the latter is now beginning to assume more importance in some countries than hitherto.

Fats contribute to palatability mainly in two ways. Firstly by responses to their texture in the mouth (sometimes called mouthfeel) and secondly by olfactory responses, namely taste in the mouth and aroma or odour in the nose; together, these are called flavour.

Fat is, on the one hand, a source of taste and aroma compounds and, on the other, a medium that regulates the distribution of these compounds between water, fat and vapour phases, which influence their perception by the sense organs. Some flavour compounds result from the decomposition of lipids by lipolysis, oxidation (section 3.3.4), and microbial or thermal degradation. These processes may produce free fatty acids, aldehydes, ketones, lactones and other volatile compounds (Figure 5.10).

Odour

When low molecular weight volatile compounds interact with receptors in the nose, they give rise to the sensation of odour or aroma. This sensation depends on features of the molecular structure of the odour compounds and on their partitioning between the fat and the vapour phases. Thus, short chain length fatty acids (e.g. in butter) have a more intense odour than longer chain acids; those with chain lengths greater than about 10 carbon atoms have little odour because they are not sufficiently volatile. The fat content of the food has an important influence on aroma. A high fat content favours partition into the fat phase and may slow down the loss of volatiles, which are generally hydrophobic. Sometimes an advantage of a high fat

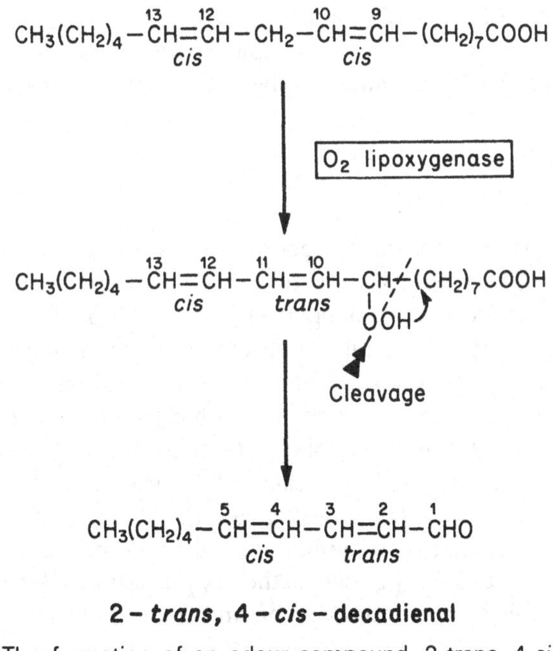

2 – *trans*, 4 – *cis* – decadienal

Figure 5.10 The formation of an odour compound. 2-*trans*, 4-*cis*-decadienol is responsible for some of the flavour of oxidized soyabeans.

content may be in lowering the intensity of the sensation of an unpleasant odour. For all odour and flavour compounds there is a threshold concentration below which there is little sensation and above which the characteristic flavour is apparent. Frequently the perception of a particular odour is quite different (and often more unpleasant) at high compared to low concentrations and the effective concentrations will be much influenced by the fat content of the foods. If food products do not contain much lipid, then there will be an early release of aroma compounds as the food is eaten. The nature and timescale of the perceived sensation will be quite different, and possibly less pleasant, than when a fatty food is eaten.

Taste

Taste compounds, which may elicit acid, sweet, salty or bitter tastes, must pass into the aqueous phase before they are sensed by the taste receptors in the mouth. The aqueous phase is composed of the water from the food mixed with saliva in the mouth. Thus, factors such as water solubility, pH and salt formation, which influence the partitioning of taste compounds between aqueous and fat phases, influence flavour perception. For example, the presence of fat may retard their passage into saliva and thus limit the perception of taste. The way in which fat and water are distributed in a food, and the proportions of each, influence flavour perception. Many foods have a poor flavour if they contain too little fat (cheese is a good example). The effect, while very obvious to the consumer, is not fully understood scientifically but may be related to the state of dispersion of the fat.

Texture

Fat may also have an important effect on the texture of food as perceived during eating. The physical state of the fat is important. Pure oil is unpleasant to swallow for most people, while an emulsion may be perceived as pleasantly creamy. An oil-in-water emulsion gives quite a different impression from a water-in-oil emulsion of the same chemical composition. Thus, full fat milk, cream and butter each have their individual sensory characteristics. A high content of solid fat (e.g. in chocolate) gives a cooling effect on the tongue during eating because of the heat needed for melting. If the fat does not melt completely during eating, however, an impression of stickiness results. Firmness and graininess are other attributes affected by solid fat content. People vary greatly in their appreciation of fat in relation to palatability. In this respect, composition and state of dispersion of the fat may be crucial.

5.3 ASSIMILATION OF LIPIDS BY THE BODY

5.3.1 Digestion. Before dietary fats can be taken up and used by the body, they must first be broken down into their component parts by a variety of digestive enzymes

The bulk of dietary fat is provided by the triacylglycerols, which must be extensively hydrolysed to their constituent fatty acids before they can be assimilated by the body.

In most adults the process of fat digestion is very efficient and the hydrolysis of triacylglycerols is accomplished almost entirely in the small intestine by a lipase secreted from the pancreas. At birth, the newborn animal has to adapt to the relatively high fat content of breast milk after relying mainly on glucose as an energy substrate in fetal life. It is presented with two major problems in fat digestion: the pancreatic secretion of lipase is rather low and the immature liver is unable to provide sufficient bile salts to solubilize the digested lipids. These problems are even more acute in the premature infant. Yet the newborn baby can digest fat, albeit less efficiently than the older child or adult. This is now attributed primarily to the activity of a lipase secreted from the serous glands of the tongue (lingual lipase) which is carried into the stomach where hydrolysis occurs, without the need for bile salts, at a pH of around 4.5–5.5. The secretion is probably stimulated both by the action of sucking and the presence of fat in the mouth, although the evidence for this was obtained from experiments with rats rather than human babies. The products are mainly 2-monoacyglycerols, diacyl-glycerols and non-esterified fatty acids, the latter being relatively richer in medium chain length fatty acids than the original acylglycerols. There is also evidence that a lipase present in human breast milk contributes to fat digestion in the newborn. It is interesting that the milk fat of most mammals is relatively rich in medium chain length fatty acids rather than the usual 16–20C compounds. The relative ease with which lipids containing medium chain fatty acids can be absorbed certainly helps lipid uptake in babies.

As the baby is weaned on to solid food, the major site of fat digestion is shifting from the stomach to the duodenum. The stomach still has a role to play since its churning action creates a coarse oil-in-water emulsion, stabilized by phospholipids. Also, proteolytic digestion in the stomach serves to release lipids from the food particles where they are generally associated with proteins as lipoprotein complexes. The fat emulsion that enters the intestine from the stomach is modified by mixing with bile and pancreatic juice. Bile supplies bile salts, which in man are mainly the glycine and taurine conjugates of tri- and di-hydroxycholanic acids, formed from

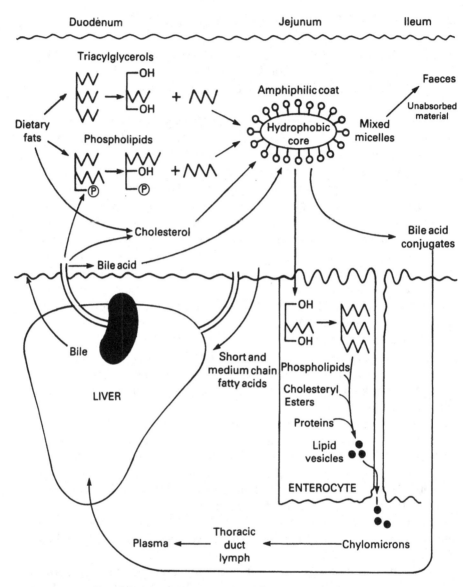

Figure 5.11 The digestion and absorption of fat.

cholesterol in the liver (section 7.5.6) and phospholipids. Much of the intestinal phospholipid in man is of biliary origin and is estimated at between 7–22 g day^{-1} compared with a dietary contribution of 4–8 g day^{-1}. Biliary secretion is enhanced as the amount of fat in the diet increases. Pancreatic juice supplies enzymes that catalyse the hydrolysis of fatty acids from triacylglycerols, phospholipids and cholesterol esters (Figure 5.11).

Pancreatic lipase itself catalyses the hydrolysis of fatty acids from positions 1 and 3 of triacylglycerols to yield 2-monoacylglycerols; there is very little hydrolysis of the fatty acid in position 2 and very little isomerization to the 1-monoacylglycerols. In ruminant animals, the complex population of microorganisms contains lipases that split triacylglycerols completely to glycerol and non-esterified fatty acids. Some of these are fermented to acetic and propionic acids which are absorbed directly from the rumen and carried to the liver, where they are substrates for gluconeogenesis. A proportion of the remaining long chain unsaturated fatty acids undergo several metabolic transformations catalysed by enzymes in the rumen microorganisms (Figure 3.15) before passing into the small intestine where they are absorbed. Principal among these is hydrogenation in which double bonds are reduced by a process that is strictly anaerobic (section 3.2.5). During hydrogenation, the double bonds are isomerized from the *cis* to the *trans* geometrical configuration. Positional isomerization also occurs, in which the double bonds, both *cis* and *trans*, migrate along the carbon chain. The result is a complex mixture of fatty acids, generally less unsaturated than the fatty acids in the ruminant's diet and containing a wide spectrum of positional and geometrical isomers.

Pancreatic lipase attacks triacylglycerol molecules at the surface of the large emulsion particles (the oil-water interface) but before lipolysis can occur, the surface and the enzyme must be modified to allow interaction to take place. Firstly, bile salt molecules accumulate on the surface of the lipid droplet, displacing other surface active constituents. As amphiphilic molecules they are uniquely designed for this task since one side of the rigid planar structure of the steroid nucleus is hydrophobic and can essentially dissolve in the oil surface. The other face contains hydrophilic groups that interact with the aqueous phase (Figure 5.12). The presence of the bile salts donates a negative charge to the oil droplets, which attracts a protein to the surface. This protein has a molecular mass of 10 kDa and is called colipase. The function of the colipase is to attract and anchor the pancreatic lipase to the surface of the droplets. Thus, bile salts, colipase and pancreatic lipase interact in a ternary complex which also contains calcium ions that are necessary for the full lipolytic activity.

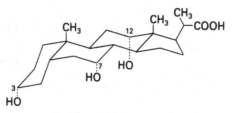

Figure 5.12 The structure of a bile acid (cholic acid).

Phospholipase A_2 (section 7.2) hydrolyses the fatty acid in position 2 of phospholipids, the most abundant being phosphatidylcholine. The enzyme is present as an inactive proenzyme in pancreatic juice and is activated by the tryptic hydrolysis of a heptapeptide from the N-terminus. The major digestion products that accumulate in intestinal contents are lysophospholipids. Any cholesteryl ester entering the small intestine is hydrolysed by a pancreatic cholesteryl ester hydrolase.

As digestion progresses, the large emulsion particles, which may be around 1 μm in diameter, decrease in size as the digestion products pass into large molecular aggregates called mixed micelles (Figure 5.13). The components of these mixed micelles have been classified either as insoluble swelling amphiphiles (Type II) or soluble amphiphiles (Type III) according to their physical properties and the way in which they interact with water, by the Swedish biochemist, Borgstrom who has done much pioneering work on the physical chemistry of fat absorption (Table 5.9). The main components are monoacylglycerols, lysophospholipids, fatty acids which leave the surface of the lipid particles to be incorporated into the micelles. In ruminants, there is very little monoacylglycerol present in the digestion mixture, since the microbial lipases completely hydrolyse dietary triacylglycerols to fatty acids and glycerol, the latter being further fermented. The mixed micelles in ruminants are therefore composed largely of non-esterified fatty acids, lysophospholipids and bile salts.

Fatty acids are in the form of soluble amphiphiles since the pH in the proximal part of the small intestine, where digestion takes place, has risen from the acid values in the stomach to around 5.8–6.5. The partition of long chain fatty acids into the micellar phase is favoured by the gradual increase in pH that occurs in the luminal contents as they pass into the more distal parts of the small intestine. Bile salts are also incorporated into micelles. The presence of these soluble amphiphiles helps to incorporate very

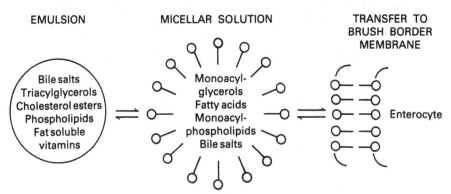

Figure 5.13 Role of mixed micelles in fat absorption.

Table 5.9 Characteristics of the components of mixed micelles

	Mode of interaction with:	
Types of components	*water*	*bile salt solution*
1. *Insoluble non-swelling amphiphiles*		
Triacylglycerols	Insoluble oil or	Low micellar
Diacylglycerols	crystals orientate	solubility
Long chain protonated	at interfaces.	Excess forms oil
fatty acids	Form stable	or crystals
Sterols	monolayers	
Vitamins A, D, E, K		
2. *Soluble swelling amphiphiles*		
Phospholipids	Solubility very low.	Mixed micelles
Glycolipids	Swell to form	with a salt to
Monoacylglycerols	liquid crystals.	bile salt ratio
'Acid soaps'	Form stable mono-	> 0.5. Excess
	layers at interfaces	forms liquid
		crystals
3. *Soluble amphiphiles*		
Soaps of long chain fatty acids	*Form* molecular	Mixed micelles
Monoacylphosphatidyl choline	solutions, micellar	
Most synthetic detergents	solutions or liquid	
Bile salts	crystalline phase	

insoluble non-polar molecules like cholesterol and the fat soluble vitamins into the micelles and aid their absorption. The picture of the luminal lipid digestion contents being distributed between coarse triacylglycerol emulsion particles and small mixed micelles is probably a grossly oversimplified one. The American, Carey and his team have described a hierarchy of lipid particles, in which non-micellar liposomes (section 8.5) with diameters larger than the mixed micelles may also play an important role.

Although it is possible to demonstrate differences in the digestibility of different types of fatty acids, long chain unsaturated fatty acids being somewhat better digested than saturated ones, there is so much excess capacity in the healthy human gut to digest fat and solubilize the digestion products that such compositional differences are probably rarely of physiological significance.

5.3.2 Absorption: intraluminal phase. The first phase of fat absorption is passage of digestion products into the absorbing cells

Lipid absorption in man occurs largely in the jejunum. The principal molecular species passing across the brush border membrane of the enterocyte are the monoacylglycerols and non-esterified long chain fatty acids. The bile salts themselves are not absorbed in the proximal small intestine but pass on to the ileum where they are absorbed and recirculated in the portal blood to the liver and then to the bile for re-entry at the duodenum.

The digestion products encounter two main barriers to their absorption. The first is the layer of unstirred water at the surface of the microvillus membrane and this is thought to be the rate-limiting step in the uptake process. The second barrier is the microvillus membrane (brush-border membrane) itself. As mentioned in section 6.5.2, this membrane has a particularly robust structure. After many years of research, it is still not entirely clear how the lipid digestion products cross this barrier.

5.3.3 Absorption: intracellular phase. In the second phase of fat absorption, absorbed products are resynthesized in the enterocytes

For efficient absorption into the enterocytes to occur, it is essential that an inward diffusion gradient of lipolysis products is maintained. Two cellular events ensure that this occurs. First, the fatty acids entering the cells bind to a fatty acid binding protein (or Z protein) of molecular mass 12 kDa. The protein binds long chain unsaturated fatty acids in preference to saturated fatty acids and this may explain the more rapid absorption of oleic than stearic acid that is observed. Up to this stage, the absorption process has not been dependent on a source of energy. The next phase which removes fatty acids, thereby maintaining a gradient, is the energy dependent re-esterification of the absorbed fatty acids into triacylglycerols and phospholipids.

The first step in re-esterification is the activation of fatty acids to their acyl-CoA thiolesters (section 3.2.1). Again, the preferred substrates are the long chain fatty acids. In man and other simple stomached animals, the major acceptors for esterification of acyl-CoAs are the 2-mono-acylglycerols, which together with the non-esterified fatty acids are the major forms of absorbed lipids. Resynthesis of triacylglycerols, therefore, occurs mainly via the monoacylglycerol pathway (section 4.4.2(a)). In ruminant animals, the major absorbed products of lipid digestion are glycerol and non-esterified fatty acids and resynthesis occurs via the glycerol phosphate pathway (section 4.4.1(a)) after phosphorylation of glycerol catalysed by glycerol kinase. It should be noted, however, that lipids usually form a minor part of ruminant diets, so that glucose is probably the major precursor of glycerol phosphate in the ruminant enterocyte.

The main absorbed product of phospholipid digestion is monoacyl-phosphatidylcholine (lysophosphatidylcholine). A fatty acid is re-esterified to position 1 to form phosphatidylcholine by an acyl transferase located in the villus tips of the intestinal brush border. The function of this phospho-lipid will be to stabilize the triacylglycerol-rich particles, or chylomicrons, exported from the cell as described later. It is thought that the phosphatidyl-choline used for the synthesis and repair of membranes in the enterocytes (cells with a rapid turnover) is synthesized by the classical CDP-choline pathway (section 7.1) in cells at the villus crypts.

The absorption of cholesterol is slower and less complete than that of the other lipids, about half of the absorbed sterol being lost during desquamation of cells. Most of the absorbed portion is esterified either by reversal of cholesteryl esterase or via acyl-CoA: cholesterol acyl transferase (section 7.5). The latter enzyme is induced by high concentrations of dietary cholesterol. During fat absorption, the biosynthetic activity of the enterocyte is geared to packaging the resynthesized absorbed lipids in a form that is stabilized for transport in the aqueous environment of the blood. Within minutes of absorption products entering the enterocyte, fat droplets can be seen within the cysternae of the smooth endoplasmic reticulum, where the enzymes of the monoacylglycerol pathway are located. The rough endoplasmic reticulum is the site of the synthesis of phospholipids (section 7.1) and apolipoproteins (section 5.3.5(b)) which provide the coat that stabilizes the lipid droplets. These gradually increase in size, pinch off from the endoplasmic reticulum to form lipid vesicles and fuse with the Golgi apparatus. The latter also provides carbohydrate moieties for the apolipoproteins (section 5.3.5(b)) and acts as a vehicle for the transport of the particles, which are now fully formed chylomicrons, to the lateral surface of the enterocyte. The final phase of transport from the cells involves fusion with the membrane and secretion into the intercellular space by a process known as exocytosis (Figure 5.11).

Chylomicrons are the main route for the transport of long chain fatty acids. Those with chain lengths of less than twelve carbon atoms are absorbed in the non-esterified form, passing directly into the portal blood and are metabolized directly by β-oxidation in the liver. There are several reasons for this partition. Firstly, short and medium chain fatty acids are more readily hydrolysed from triacylglycerols and since they occupy mainly position 3, are not retained in the 2-monoacyglycerols. Secondly, they are more likely to diffuse into the aqueous phase rather than the mixed micelles and for this reason are more rapidly absorbed. We can calculate that about 4 g day^{-1} short and medium chain fatty acids enter the diet from dairy products or foods that incorporate coconut or palm kernel oils. Some very short chain fatty acids derived by microbial fermentation of non-starch polysaccharides (dietary fibre) in the colon may also be absorbed and contribute to lipid metabolism.

5.3.4 Malassimilation of lipids: failure to digest or absorb lipids properly can arise from defects in the gut and several other tissues

Failure to assimilate lipids of dietary origin into the body may arise from defects in digestion (maldigestion) or absorption (malabsorption).

Maldigestion can occur because of incomplete lipolysis. Thus, pancreatic insufficiency, which may result from pancreatitis, pancreatic tumour or in diseases of malnutrition such as kwashiorkor, can lead to a failure to secrete enough lipase or the production of an enzyme with reduced activity. Alternatively, the lipase may be fully functional but a failure to produce bile (generally arising from hepatic insufficiency) may result in an inability to effect the micellar solubilization of lipolysis products. This, in turn, feeds back to cause an inhibition of lipolysis. Gastric disturbances that result in abnormal acid secretion also inhibit pancreatic lipase and, furthermore, gastric problems may cause poor initial emulsification of the lipid in the stomach, further reducing the efficiency of digestion. Thus, maldigestion seems to arise from defects in a variety of organs contributing to different aspects of the digestive process.

Malabsorption may occur, even when digestion is functioning normally, due to defects in the small intestine affecting the absorptive surfaces. There may be a variety of causes, some common ones being bacterial invasion of the gut or sensitization of the gut to dietary components such as gluten, as in coeliac disease. Malabsorption syndromes (often called sprue) are characterized by dramatic changes in the morphology of the intestinal mucosa. The epithelium is flattened and irregular and atrophy of the villi reduces the absorbing surface. Tropical sprue is a prevalent disease in many countries of Africa and Asia.

A common feature of all fat malabsorption syndromes is a massively increased excretion of fat in the faeces (steatorrhoea) which arises not only from unabsorbed dietary material but also from the bacterial population that usually proliferates in the gut and the breakdown of cells. The bacteria undoubtedly affect the composition of the excreted fat. For example, a major component (absent from the diet) of faecal fat, 10-hydroxystearic acid, was shown by tracer studies to be formed by bacteria from stearic acid. This is a normal component but found in particularly high concentration in patients with steatorrheoa.

Malabsorption of fat can also occur in a number of inherited disorders in which the biosynthesis of different apoproteins in the enterocytes is impaired. Without the apoproteins, the stabilization of the lipid droplets cannot occur and the fat cannot be transported out of the cell. Triacylglycerols then begin to accumulate in the enterocyte.

Patients with poor fat absorption are very much at risk from deficiencies of energy, fat soluble vitamins and of essential fatty acids (section 5.2.2). The clinical management of fat malassimilation is facilitated by replacing normal dietary fats by medium chain triacylglycerols (MCT). This product is composed largely of triacylglycerols with C8 and C10 saturated fatty acids, refined from coconut oil and can be purchased as a cooking oil or as a fat spread. The medium chain fatty acids are rapidly hydrolysed and efficiently absorbed into the portal blood, thereby bypassing the normal absorptive route of long chain fatty acids and chylomicron formation (see also comments on infant digestion in section 5.3.1).

5.3.5 Transport of lipids in the blood: plasma lipoproteins

(a) Introduction: lipoprotein classes. Plasma lipoproteins represent a continuum of lipid-protein complexes in which the ratio of lipid to protein, and hence their density, varies. These can be conveniently divided into groups according to density

The biological problem of how to transport water-immiscible lipids in the predominantly aqueous environment of the blood has been solved by stabilizing the lipid particles with a coat of amphiphilic compounds: phospholipids and proteins. The resulting particles are the lipoproteins and you have already been introduced to them when we discussed the export of newly absorbed fat from the enterocytes. The protein moieties are known as apolipoproteins and, as will be discussed later, have much more than a stabilizing role. They also confer specificity on the particles, allowing them to be recognized by specific receptors on cell surfaces, thereby regulating their metabolism.

There are several types of lipoproteins with differing chemical compositions, physical properties and metabolic functions (Table 5.10) but their common role is to transport lipids from one tissue to another to supply the lipid needs of different cells. The different types may be classified in a number of ways depending on their origins, their major functions, their composition, physical properties or method of isolation. Lipoproteins differ according to the ratio of lipid to protein within the particle as well as having different proportions of lipids: triacylglycerols, esterified and non-esterified cholesterol and phospholipids. These compositional differences influence the density of the particles and there is a strong relationship between biological function and the broad density classes into which they fall. It is convenient, therefore, to make use of density to separate and isolate lipoproteins by ultracentrifugation and it is now usual to classify plasma lipoproteins into different density classes. From lowest to highest density, these are: chylomicrons, very low density lipoproteins (VLDL), low density lipoproteins (LDL) and high density lipoproteins (HDL). As density increases, particle size decreases and so does the ratio of lipid to protein and the ratio of triacylglycerols to phospholipids and cholesterol (Table 5.10). It is important to emphasize that the classes are not homogeneous: there is a

Table 5.10 Composition and characteristics of the human plasma lipoproteins

	Chylo-microns	VLDL	LDL	HDL
Protein (% particle mass)	2	7	20	50
Triacylglycerols (% particle mass)	83	50	10	8
Cholesterol (% particle mass) (free + esterified)	8	22	48	20
Phospholipids (% particle mass)	7	20	22	22
Particle mass (daltons)	$0.4\text{--}30 \times 10^6$	$10\text{--}100 \times 10^6$	$2\text{--}3.5 \times 10^6$	$1.75\text{--}3.6 \times 10^5$
Density range (g/ml)	>0.95	0.95–1.006	1.019–1.063	1.063–1.210
Diameter (nm)	>70	30–90	18–22	5–12
Apolipoproteins	A_1, B–48 C_1, C_2, C_3	B–100, E	B–100	A_1, A_2

wide variety of particle sizes and chemical compositions within each class and, therefore, a certain amount of overlap between them. It is also important to realize that they are not molecules in the normal sense. They are aggregates of individual lipid and protein molecules with a degree of structural organization (section 5.3.5(k)). We should not, therefore, talk about molecular mass to describe their mass but use the term particle mass.

In the following sections, we will first outline the types of peptides present in lipoproteins and then briefly describe the broad classes of lipoproteins, summarizing their nomenclature, composition, functions and biosynthesis. This will be followed by a more detailed description of the metabolic interrelationships between lipoproteins and how their metabolism is integrated within the body. Methods of isolation and analysis will be described and the section will end with a summary of their clinical relevance, leading to a discussion of the importance of lipoprotein metabolism in some metabolic diseases.

(b) The apolipoproteins: the protein moieties serve to help solubilize the lipid. They also provide specificity and direct the metabolism of the lipoproteins

The protein moieties of lipoproteins fulfil two main functions: firstly, they provide a means of solubilizing the lipid particles and maintaining their structural integrity (section 5.3.5(k)). Secondly, they are important in identifying the lipoprotein and directing its metabolism in specific ways. The main metabolic functions of the apolipoproteins are shown in Table 5.11. The term apoprotein was first used in 1963 to describe the protein moiety of a number of delipidated lipid–protein complexes and the more specific term apolipoprotein will be used here. It has now become accepted to use a series of letters A–E to identify apolipoproteins but it soon became apparent that most of these could be divided further into several sub-classes (Table 5.11). These are usually referred to in abbreviated form, thus: $apoA_1$, $apoC_3$, etc.

The complete amino acid sequences of $apoA_1$, A_2, A_4, C_1, C_2, C_3 and E are now known. Specific regions of these proteins which contain some of the structural and functional determinants of the lipoprotein particles have been synthesized and tested for biological activity. Knowledge of the detailed structure of the polypeptides has recently been enormously extended by the increasing use of the monoclonal antibody technique.

Much interest has been centred on the properties of apoB since it is common to the triacylglycerol-rich VLDL and cholesterol-rich LDL, is important in the recognition of lipoproteins by cell surface receptors (section 5.3.5(h)) and is now being used as a marker in vascular disease (section 5.5.3). Its characterization has been hampered by its insolubility, its

Table 5.11 Characteristics of the human apolipoproteins

Shorthand name	Molecular mass (daltons)	Amino acid residues	Function	Major sites of synthesis
A$_1$	28,000	243	Activates LCAT*	Liver Intestine
A$_2$	17,000	154	Inhibits LCAT? Activates hepatic lipase	Liver
B–48			Cholesterol clearance	Intestine
B–100	350–550,000		Cholesterol clearance	Liver
C$_1$	6,605	57	Activates LCAT?	Liver
C$_2$	8,824	79	Activates LPL**	Liver
C$_3$	8,750	79	Inhibits LPL? Activates LCAT?	Liver
E	34,000	279	Cholesterol clearance	Liver

* LCAT: Lecithin-cholesterol acyltransferase.
** LPL: Lipoprotein lipase.

susceptibility to degradation and its propensity to aggregate. The peptide can be solubilized by trypsin and detergent treatment. The amino-terminal residue is glutamic acid and the carboxyl terminal residue serine. ApoB has considerable β-structure in addition to regions of random coil and α-helix. The content of β-structure depends on the lipid content and the temperature at which the observations are made. ApoB is a glycoprotein linked to glucosamine through asparagine residues. Between one third to one half of the carbohydrate is released by trypsin treatment and, therefore, probably not buried in the lipid core.

ApoB is now known to exist as two variants of different molecular masses (approximately 100 and 48 kDa) designated as apoB$_H$ (heavy) and apoB$_L$ (light), now more usually referred to as apoB-100 and apoB-48 respectively. Human VLDL are believed to contain almost exclusively the heavy variant while rat VLDL contains both variants. The apoB-48 may be synthesized in a phosphorylated form which is dephosphorylated after excretion.

Naturally occurring variants of apolipoproteins have been identified that result in specific metabolic disorders (apolipoproteinopathies).

(c) Chylomicrons are rich in triacylglycerols and transport lipids of dietary origin

Chylomicrons are the largest and least dense of the lipoproteins and their function is to transport lipids of exogenous (or dietary) origin. Their size depends on factors such as the rate of lipid absorption and the type of dietary fatty acids that predominate. Thus larger chylomicrons are produced after the consumption of large amounts of fat, at the peak of absorption or when apolipoprotein synthesis is limiting. When the fatty acids are largely unsaturated, the size tends to be larger than those in which saturated fatty acids predominate.

Because of their role in transporting absorbed dietary fat, the principal components are triacylglycerols, with small amounts of phospholipids and proteins, sufficient to cover the surface (Table 5.10). The core lipid also contains some cholesteryl esters and minor fat soluble substances absorbed with the dietary fats: fat soluble vitamins, carotenoids and possibly environmental contaminants such as traces of pesticides.

Shortly after the consumption of a fatty meal, the presence of chylomicrons is very apparent in a sample of plasma (lipaemia). The very large particles scatter light giving the plasma an opalescent appearance. The particles can be isolated as a floating layer after a short low speed centrifugation and can be withdrawn from the top of the tube with a Pasteur pipette for analysis.

Several apoproteins are present in the surface layer: $apoA_1$ (15–35%); $apoA_4$ (10%); apoB (10%); the apoC group (45–50%) and apoE (5%). The A group apolipoproteins are synthesized on the endoplasmic reticulum of the intestinal epithelial cells, whereas apoC and apoB are acquired from other lipoproteins once the chylomicrons have entered the blood.

(d) Very low density lipoproteins (VLDL) are also rich in triacylglycerols but transport lipids of mainly endogenous origin

These lipoproteins, like chylomicrons, contain predominantly triacyl-glycerols (Table 5.10) and their function is to transport triacylglycerols of endogenous origin, synthesized mainly in the liver or intestine. In the older literature, they are sometimes referred to as pre-β-lipoproteins, since they were found to migrate just ahead of β-globulins on electrophoresis of plasma. VLDL are spherical particles with a core consisting mainly of triacylglycerols and cholesteryl esters, with cholesterol, phospholipids and protein mainly on the surface. Do not be alarmed if you find that different authors give surprisingly different data on the dimensions and compositions of VLDL and other lipoproteins. This is because, as stated earlier, there are no sharp dividing lines between the classes; they form a continuum of particle sizes and characteristics. The composition of the classes analysed will depend on the method of isolation, the species of animal and its nutritional and physiological state.

VLDL may be conveniently isolated for compositional, structural and metabolic studies by flotation in the ultracentrifuge, although other methods are available (section 5.3.5(j)). The amount of apoB per VLDL particle is independent of the particle mass and is the same as that in LDL particles for reasons that will be apparent when we discuss metabolic relationships. In contrast, the amounts of $apoC_1$, C_2, C_3 and apoE are variable and decrease relative to apoB as particle density increases.

The major site of synthesis of VLDL is in the liver, although some VLDL are produced in the enterocytes. The source of carbon is mainly glucose derived from dietary carbohydrate and converted into the lipid precursors, glycerol-3(sn)-phosphate via the glycolytic pathway and long chain fatty acids via the malonyl-CoA pathway (section 3.2.2). Nevertheless, some VLDL fatty acids are derived from circulating albumin-bound NEFA, which in turn, may originate from lipolysis of adipose tissue triacylglycerols, intravascular lipolysis of triacylglycerol-rich lipoproteins or directly absorbed medium chain fatty acids. Factors affecting hepatic triacylglycerol biosynthesis and its control are discussed in section 4.6. As in chylomicron biosynthesis in the intestine, nascent VLDL particles originate in the smooth endoplasmic reticulum and acquire phospholipids and apoproteins from the rough endoplasmic reticulum. The resulting vesicles move to the Golgi apparatus where some of the apoproteins are glycosylated. The Golgi vesicles then migrate to the cell surface where VLDL are exported by exocytosis into the space of Disse.

In contrast to nascent chylomicrons, which acquire apoC and apoE only after they reach the plasma, VLDL receive their full complement of apoproteins in the hepatocyte or enterocyte.

(e) Low density lipoproteins (LDL) are major carriers of plasma cholesterol in man

In older literature these lipoproteins were referred to as β-lipoproteins because they migrated with β-globulins on electrophoresis of plasma proteins. Their role is to transport cholesterol to tissues where it may be needed for membrane structure or conversion into various metabolites such as steroid hormones. LDL provide the major carriers of plasma cholesterol in man although this is not the case for all mammalian species (section 5.3.5(i)).

LDL are normally isolated from plasma by ultracentrifugation at salt densities between 1.019 and 1.063 g ml^{-1}. Each lipoprotein particle contains the same mass of apoprotein B but each differs with respect to the amount of bound lipid. An average composition is shown in Table 5.10.

LDL are largely derived from VLDL by a series of degradative steps that remove triacylglycerols, resulting in a series of particles that contain a progressively lower proportion of triacyglycerols and are correspondingly richer in cholesterol and phospholipids. These intermediate particles are termed intermediate density lipoproteins (IDL) and the above reactions take place, first in the blood capillaries associated with adipose tissue and latterly in the liver. During the transformations, the apoB component remains with the LDL particles and the apoC and E components are progressively lost. ApoB has an important role in the recognition of LDL by cells since it must interact with specific cell surface receptors before the LDL particle can be taken up and metabolized by the cell. Other receptors (for example on macrophages) recognize modified LDL and are responsible for the degradation of LDL particles that cannot be recognized by normal cell surface LDL receptors. ApoE plays a role in receptor binding while the C group apolipoproteins are involved in reactions by which the particles are sequentially degraded by lipases.

(f) High density lipoproteins (HDL) are involved in the transport of excess cholesterol to the liver for reprocessing

In older literature, these lipoproteins were referred to as α-lipoproteins because they migrated with α-globulins on electrophoresis of plasma proteins. A major role is to carry cholesterol from peripheral cells to the liver, a process generally termed reverse cholesterol transport (section 5.3.5(h)). Table 5.10 summarizes their composition and properties. HDL are usually divided into two subclasses: HDL_2 and HDL_3 because rate zonal centrifugation gives rise to a bimodal distribution in the HDL region, whereas other lipoprotein classes form a continuum. The distinction between HDL_2 and HDL_3 may have metabolic significance in that HDL_2 appears to have a stronger inverse relationship with cardiovascular disease than HDL_3 (section 5.5.3). The major apolipoproteins of HDL are $apoA_1$ and $apoA_2$ but the surface coat also contains some apoC, apoE and a protein unique to HDL – apoD.

Unlike triacylglycerol-rich lipoproteins, nascent HDL have not been identified within subcellular compartments. Much of our knowledge is obtained from observations of nascent HDL in liver perfusates. At this early stage, they are disc-shaped rather than spherical and consist of a bilayer of mainly phosphatidylcholine with $apoA_1$ and apoE at the margins of the disc. Similar particles are also found in intestinal lymph. After excretion into the plasma, HDL acquire additional surface components: phospholipids, cholesterol and apoproteins by transfer from chylomicrons and VLDL

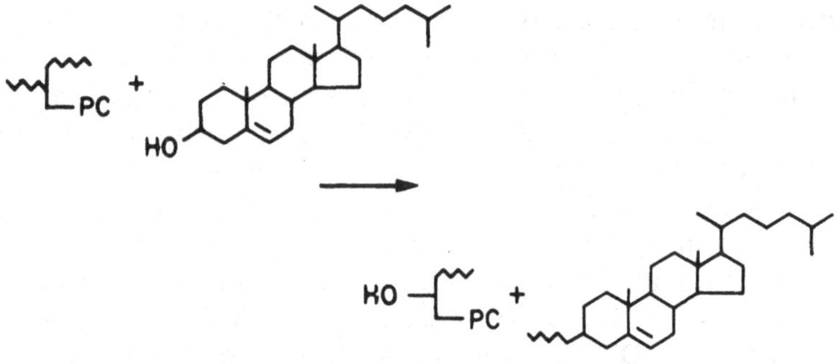

Figure 5.14 The lecithin-cholesterol acyltransferase (LCAT) reaction.

during their catabolism by lipoprotein lipase. ApoA, much of which is synthesized in the intestine, and some apoC, are transferred during the breakdown of chylomicrons. Further apoC is transferred from VLDL breakdown products. In plasma, a subfraction of HDL, containing $apoA_1$ and apoD becomes associated specifically with an enzyme: lecithin-cholesterol acyltransferase (LCAT; Figure 5.14) which is synthesized in the liver and excreted into plasma. The function of this enzyme is to transfer a fatty acid from phosphatidylcholine to the cholesterol in the surface of the disc-shaped HDL to form cholesteryl ester which transfers into the core of the HDL particle. The other product, lysophosphatidylcholine is transferred to plasma albumin from which it is rapidly removed from blood and reacylated. This redistribution of lipid in the HDL particle converts the particles from discoidal to spherical in shape. This sequence is part of the process called reverse cholesterol transport (section 5.3.5(h)).

(g) Metabolic relationships between the lipoproteins: apolipoproteins interact with enzymes or receptors on cell surfaces resulting in degradation of the lipoproteins and the uptake of the products into cells. These processes may not result in complete degradation but rather in the transformation of one type of lipoprotein into another

After the consumption of a meal containing an appreciable amount of fat, the absorbed, resynthesized triacylglycerols are transported in the blood as chylomicrons. (It is important to realize that small amounts of lipid-carrying particles also circulate when no fat is being absorbed from the diet. Under these conditions the chylomicron fatty acids are derived from biliary phospholipids or the lipids of cells shed from the gut mucosa and the VLDL fatty acids from NEFA transported to the liver from adipose tissue.)

The chylomicrons initially bind to the enzyme lipoprotein lipase (LPL) on the endothelial surfaces of blood capillaries in muscle and other organs, but primarily adipose tissue. This enzyme catalyses the hydrolysis (lipolysis) of triacylglycerols, releasing fatty acids (Figure 5.15). ApoC$_2$ plays a key role in activating LPL. Indeed, it is absolutely essential for activity as demonstrated by the failure of people with apoC$_2$ deficiency to hydrolyse circulating lipoprotein triacylglycerols. The peptide seems to facilitate interaction of the enzyme with the lipoprotein interface and to modulate the specificity of the enzyme so that it catalyses hydrolysis of long chain rather than short chain fatty acids. The hydrolysis is accomplished in only a few minutes ($T_{1/2}$ of chylomicron clearance is 2–3 minutes). At this stage, the A group apolipoproteins and the remaining apoC are transferred to HDL, as are the phospholipids. The remaining particle, although retaining the same basic structure, contains fewer triacylglycerols, is enriched in cholesterol esters and is known as a chylomicron remnant. These particles are no longer able to compete effectively for lipoprotein lipase and circulate in the plasma to be taken up by liver cells (hepatocytes) by a receptor-mediated process. This means that specific receptors on the surface of the hepatocytes recognize and bind to the apoE component of the remnant. The whole receptor-remnant complex then pinches off from the membrane and is taken up into the cell (internalized) where it is degraded by lysosomal enzymes which catalyse virtually the complete hydrolysis of the lipid and protein components. The uptake of remnants is inhibited by the C group apolipoproteins, which thereby prevent the premature uptake of small unhydrolysed chylomicrons by the remnant receptor.

The regulation of lipoprotein lipase itself is crucial to the control of lipoprotein metabolism in different tissues in the body. The enzyme is synthesized in the parenchymal cells of the tissues and secreted into the capillary endothelium where it is bound to the cell surface by sulphated glycosaminoglycans. The activity of the enzyme is regulated by diet and hormones, of which the most important seems to be insulin. After consuming a meal, when the supply of energy may exceed the body's immediate needs, the secretion of insulin ensures that the adipose tissue enzyme is active and the muscle enzyme suppressed. In a state of fasting, the adipose tissue lipoprotein lipase activity is suppressed and the hormone-sensitive lipase switched on allowing the mobilization of NEFA (sections 4.5 and 4.6.3). At the same time, the muscle lipoprotein lipase activity is elevated so that fatty acids from circulating lipoproteins can be used as fuel. In lactation, the synthesis of lipoprotein lipase seems to be regulated by prolactin which promotes the utilization of chylomicron triacylglycerol fatty acids for milk fat synthesis.

In Man, up to 300 g of chylomicron triacylglycerols can be hydrolysed by lipoprotein lipase each day, although less than 1% of dietary lipids may be found in the blood at any given time.

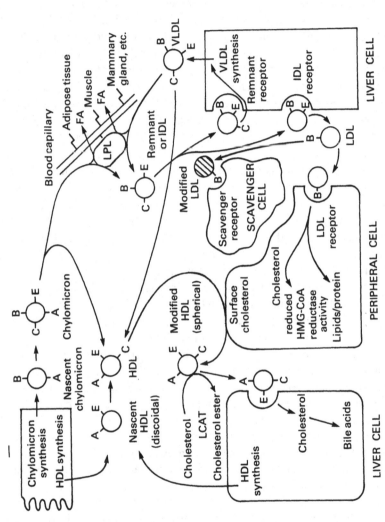

Figure 5.15 Metabolism of lipoproteins.

The catabolism of VLDL is basically the same as for the chylomicrons. They bind to endothelial lipoprotein lipases and the triacylglycerols are hydrolysed. The larger the particles (and, therefore, the more apoC they contain) the greater the rate of hydrolysis. Phospholipids and remaining apoproteins are transferred to HDL while apoB and apoE are retained. The resulting particles, depleted of triacylglycerol and richer in cholesteryl esters, are termed VLDL remnants or, more usually, intermediate density lipoproteins (IDL) since further degradation results in the formation of LDL. In many species, most of the IDL are taken up by the liver by receptor-mediated endocytosis in a manner analogous to chylomicron remnants, but in man, almost a half is normally processed further to yield LDL.

As we pointed out in section 5.3.5(a), lipoproteins are not discrete particles of fixed size but form a distribution of sizes; thus the sizes and densities of IDL or remnant particles overlap those of the precursor VLDL or the product LDL but can be distinguished by their mobility on electrophoresis, presumably because of their different apolipoprotein composition.

The final processing step of the triacylglycerol-rich, apoB-containing lipoproteins is the further loss of triacylglycerols and phospholipids (catalysed by hepatic lipases) and of apoE, to form LDL which retain only apoB-100.

In contrast to IDL and remnants, which are removed almost entirely in the liver, LDL are less efficiently removed by hepatic receptors (although it can occur) and are principally removed by specific extra-hepatic LDL receptors (Figure 5.16). This mechanism was first described by the American

CONTROL OF CHOLESTEROL UPTAKE AND SYNTHESIS

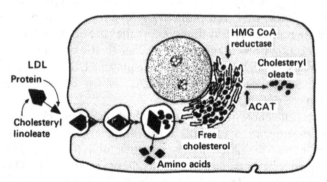

Figure 5.16 The receptor-mediated mechanism for the uptake of low density lipoproteins by cells. Redrawn from M.S. Brown and J.L. Goldstein (1976) *Science*, **191**, 150, with permission.

biochemists Goldstein and Brown and the work, with its enormous implications for human disease (section 5.5), was recognized by the award of the Nobel Prize in 1985. The detailed mechanism and its ramifications are described in highly readable form in many reviews by these authors (see references and section 7.5).

The distribution of LDL to various tissues may depend on the rate of transcapillary transport as well as the activities of the LDL receptors on the cell surface. Adipose tissue and muscle have few LDL receptors and take up LDL only slowly, whereas adrenal gland (important in the synthesis of steroid hormones derived from cholesterol) has a highly fenestrated epithelium with abundant receptors and takes up LDL avidly. Much of our understanding of the pathway of receptor-mediated uptake of LDL has been derived by studying cultured cells, mainly fibroblasts, smooth muscle cells and macrophages. As far as is known the behaviour of the LDL receptor *in vivo* is generally consistent with that observed in cultured cells.

The LDL receptors function in a manner similar to that described earlier for the hepatic remnant receptor. Both apoE and apoB contain a recognition site for the LDL receptor. The human LDL receptor is a transmembrane protein of 839 amino acid residues, that is synthesized in the rough endoplasmic reticulum and inserted at random in the plasma membrane. It then migrates laterally in the plane of the membrane until it reaches a pit that is coated with the protein, clathrin; about 80% of the LDL receptors are concentrated in these coated pits which cover about 2% of the surface of the cell. A region of negatively charged residues binds electrostatically with a positively charged region in apoE. This apolipoprotein binds at several receptor sites, whereas apoB interacts with only a single site. (LDL contain only apo-B and no apo-E, whereas HDL_1 contain only apo-E. HDL_1, also bind to the LDL receptor, which is now often referred to as the apo-B,E receptor.) Once bound, the receptor-LDL complex is internalized by endocytosis and its component parts degraded by lysosomal enzymes. LDL receptors are recycled within minutes of endocytosis and are reutilized many times before they are eventually catabolized. The LDL receptors are regulated according to the amount of incoming LDL. Normally, a high concentration of plasma LDL will suppress the number of specific cell surface receptors, which are increased under conditions in which LDL are depleted.

Inside the cell, cholesteryl esters are hydrolysed by cholesteryl ester hydrolase. Incorporation of cholesterol into the endoplasmic reticulum membranes serves to inhibit hydroxymethylglutaryl-CoA reductase, the rate limiting enzyme in cholesterol biosynthesis (section 7.5). Thereby, an abundant supply of cholesterol in the plasma is able to suppress its own endogenous biosynthesis and ensure that excessive amounts do not accumulate. The importance of this pathway for cholesterol homeostasis is highlighted by the example of familial hypercholesterolaemia. This is a disease in which a defective gene results in the lack of the LDL receptor. The same condition occurs in an animal model, the Watanabe heritable hyperlipidaemic rabbit. In both conditions, the fractional catabolic rate of LDL is significantly decreased; plasma LDL concentrations are characteristically elevated. The implications for vascular disease are discussed in section 5.5.3.

About 65% of LDL uptake is receptor-mediated. Some LDL in liver and other tissues are taken up by endocytosis that does not involve receptors, although this is less efficient.

The laboratory of the American biochemist Mahley has been in the forefront of work aimed at modifying lipoproteins in various ways to try to probe the structural features needed for receptor uptake. This has involved reacting the lipoproteins with a number of chemical groups, acetyl, acetoacetyl, maleyl, succinyl and others. Acetyl and some other groups convert the ε-amino groups of lysine in apoB-100 to a neutral or negatively charged moiety that results in its inability to be recognized by the appropriate receptor. Many modified lipoproteins are, however, recognized by a so-called scavenger receptor on some cells such as macrophages (section 5.5.3).

(h) Reverse cholesterol transport: HDL pick up cholesterol that is excess to requirements and transport it to the liver for further metabolism

A major function of HDL is to remove unesterified cholesterol (where it may have accumulated in cell membranes and plasma lipoproteins) and transport it to the liver where it can be degraded and utilized for, among other things, the synthesis of the bile acids. This process is now referred to as reverse cholesterol transport.

A key step in this process is catalysed by the enzyme lecithin : cholesterol acyltransferase (LCAT, section 7.5). The enzyme catalyses the transfer of a fatty acid from phosphatidylcholine to cholesterol to form a cholesterol ester (Figure 5.14). In human plasma, LCAT is associated with a subfraction of HDL that contains apoA$_1$ (but not apoA$_2$) and apoD. The phospholipid substrate for the reaction is present in the HDL particle, having been transferred from chylomicron remnants or IDL during the degradation of chylomicrons and VLDL respectively (Figure 5.15). The cholesterol substrate is derived from the surfaces of the plasma lipoproteins or plasma membranes of cells. Molecules of the substrate cholesterol and the product cholesteryl ester exchange readily between plasma lipoproteins and between lipoproteins and cell membranes. This exchange is often mediated by transfer proteins (cholesteryl ester transfer protein, CETP) similar to the phospholipid transfer protein discussed in section 5.3.5(l). Rat, dog and pig plasma, however, contain little transfer protein activity, so that the cholesteryl esters remain mainly with the HDLs.

LCAT, by consuming cholesterol, promotes its net transfer from non-hepatic cells into plasma and from other lipoproteins to the site of esterification. Evidence comes from experiments *in vitro* as well as *in vivo*. LCAT can be inhibited *in vitro* by compounds such as dithiobis (2-nitrobenzoic) acid (DTNB). Under these conditions net transport of cholesterol from cells ceases. In patients with a genetic deficiency of LCAT, free cholesterol accumulates in cells. Plasma from such people is ineffective in promoting net transport of cholesterol from cellular membranes. One product, cholesteryl ester, is redistributed among plasma lipoproteins. The other, lysophosphatidylcholine, is transferred to albumin from which it is rapidly removed and recycled.

Molecules of cholesteryl ester transferred to lipoproteins containing apoB-100 or apoE are taken up by the liver, thereby completing the process of reverse cholesterol transport.

(i) Species differences in lipoproteins

In some species, like man, guinea pig and pig, lipoproteins of the LDL type, in which apolipoprotein B predominates, account for more than 50% of the total substances of density $< 1.21 \, \text{g ml}^{-1}$. They are the LDL mammals. In the vast majority of mammals, however, HDL are the predominant class and may account for up to 80% of plasma substances of density $< 1.21 \, \text{g ml}^{-1}$. Herbivorous species, with the exception of guinea pigs, camels and rhinos, and carnivores are HDL mammals. It is worth noting that although rats are most frequently the animal of choice for the study of lipid biochemistry in the research laboratory, their lipoprotein pattern is of the HDL type and very different from that of man. Caution needs to be exercised in extrapolating results on experimental animals to the human situation.

(j) Isolation and analysis of lipoproteins: a combination of ultra-centrifugation, gel filtration, precipitation and electrophoresis is sufficient to isolate and identify the different lipoprotein classes

Isolation

To study the chemical composition and the physical properties of lipoproteins it is necessary to isolate them from plasma and separate them from each other. This will also be useful in many metabolic studies. There are three basic methods for their isolation: separation in the ultracentrifuge; gel filtration or precipitation. Ultracentrifugation is costly and time consuming but enables relatively large quantities of reasonably discrete classes to be isolated. Gel filtration seems to cause less damage to the proteins and may be the method of choice when compositional and structural integrity is of utmost importance. Affinity columns employing antibodies to specific apolipoproteins are useful for isolation on a small scale. The fastest and cheapest method is precipitation by employing a combination of different reagents, although some damage to lipoprotein structure may have to be accepted.

The first important step in lipoprotein isolation is to obtain a good sample of blood under conditions that ensure minimal damage to the lipoprotein. Unless serum is to be used, an anticoagulant is generally included such as heparin or EDTA. The latter has the advantage of chelating heavy metals (e.g. Cu^{2+}) that might catalyse the autoxidation of unsaturated fatty acids and cholesterol and inhibits phospholipases that may alter the structure of the lipoproteins. Many careful workers also include protease inhibitors to maintain the integrity of the apolipoproteins.

Schemes for the separation of lipoproteins by these different methods are illustrated in Figures 5.17–5.19.

Analysis
Lipoproteins can be analysed after separation by any of the methods described for their preparation. When only small quantities are available for analysis, however, electrophoresis is the general method of choice. Particularly useful is gradient gel electrophoresis which consists of the migration of charged particles through a polyacrylamide gel of increasing concentration. As the concentration of the gel increases, the pore size

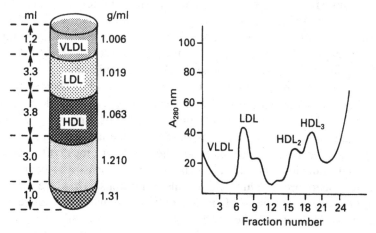

Figure 5.17 Separation of plasma lipoproteins by gradient ultracentrifugation.

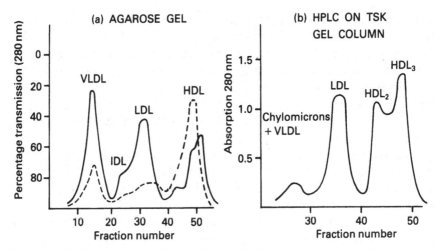

Figure 5.18 Examples of separations of plasma lipoproteins by column chromatography. Reproduced with kind permission of Academic Press and (a) Dr L.L. Rudel and (b) Dr I. Hara, from *Methods in Enzymology* (1986) **129**, (a) p. 53, Figure 2; (b) p. 66, Figure 3.

decreases and there is a differential retardation of particles, eventually reaching an exclusion limit that is a function of the size and shape of the particles.

Lipoproteins can be localized on gels by staining either for protein or lipid moieties. Proteins are stained with Coomassie blue, although the lipoproteins should be free from plasma proteins before separation. Quantitation is achieved by densitometry. Lipids can be stained with oil red O and in this case plasma proteins do not interfere. Other stains can also be used, such as periodic acid-Schiff reagent for carbohydrate moieties. If the lipoproteins are radioactively labelled they can also be visualized on film by autoradiography. Immunoblotting can be used for the localization of specific apolipoproteins if suitable antibodies are available.

Lipoproteins can be analysed by their lipid content. For cholesterol, the classical Liebermann-Burchard colour reaction gives a reliable means of quantitation which is amenable to automated methods. Triacylglycerols can be quantitated by fluorescence, a method that depends on the presence of the glycerol moiety. Enzymic methods, however, are most often used in modern laboratories since they are usually more sensitive, have better specificity, need small volumes and mild conditions. They are also well suited to modern methods of automation, often without the need for extraction or hydrolysis. The 1980s have seen dramatic advances in lipid laboratory automation, spurred on by the needs of clinical screening, and

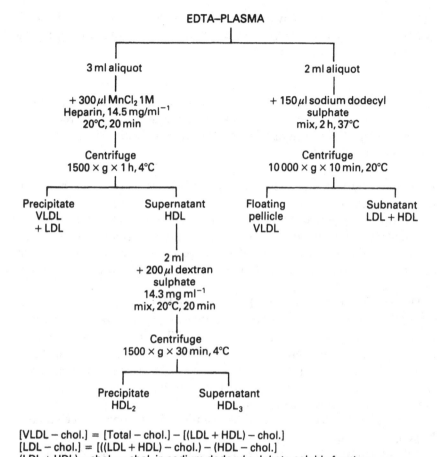

EDTA–PLASMA

3 ml aliquot

+ 300 µl MnCl$_2$ 1M
Heparin, 14.5 mg/ml^{-1}
20°C, 20 min

Centrifuge
1500 × g × 1 h, 4°C

Precipitate
VLDL
+ LDL

Supernatant
HDL

2 ml
+ 200 µl dextran
sulphate
14.3 mg ml^{-1}
mix, 20°C, 20 min

Centrifuge
1500 × g × 30 min, 4°C

Precipitate
HDL$_2$

Supernatant
HDL$_3$

2 ml aliquot

+ 150 µl sodium dodecyl
sulphate
mix, 2 h, 37°C

Centrifuge
10 000 × g × 10 min, 20°C

Floating
pellicle
VLDL

Subnatant
LDL + HDL

[VLDL – chol.] = [Total – chol.] – [(LDL + HDL) – chol.]
[LDL – chol.] = [((LDL + HDL) – chol.) – (HDL – chol.]
(LDL + HDL) – chol. = chol. in sodium dodecyl sulphate-soluble fraction

Figure 5.19 Separation of plasma lipoproteins by precipitation techniques.

much of the drudgery has been removed by highly accurate repetitive pipetting devices and computer-aided programming of the analyses and presentation of results.

Delipidation

Detailed studies of the apolipoproteins require the removal of the lipid from the lipoproteins. This can be achieved by extraction of the lipoproteins either direct from plasma or in the precipitated form with ethanol–diethyl ether mixtures (3 : 1, v/v). The more powerful lipid solvent, chloroform–methanol (2 : 1, v/v) results in a poorly soluble apolipoprotein. With ethanol–ether, however the advantage of having a soluble apoprotein is partly offset by the loss of 20% of VLDL protein. Lyophilization before extraction has several advantages. The sample size is reduced; small peptides are not soluble in the organic phase and the removal of structural water weakens non-polar interactions.

(k) Structure of lipoproteins: a combination of powerful physical techniques including NMR, ESR, X-ray and neutron scattering and electron microscopy have given insights to the molecular structures of lipoprotein particles

It is only during the 1980s that great strides have been made in understanding lipoprotein structure. These have been due to the development of techniques for pulling apart and reassembling the particles and improved techniques for investigating the interactions between the component lipids and proteins. Knowledge of the detailed structure of apolipoproteins (section 5.3.5(b)) has been the most recent and crucial development.

The most studied lipoproteins have been the HDL and, to a lesser extent, the LDL because they have a more specific and tightly organized structure than the triacylglycerol-rich lipoproteins. HDL has a readily soluble apolipoprotein component and model studies of the interaction of these peptides with phospholipid and neutral lipid mixtures have been invaluable. In contrast, the insolubility of the LDL apoproteins has restricted progress in the study of their structure.

Investigation of LDL structure has employed a number of techniques. Negative staining is a simple and useful way to examine particle size distribution and morphology. Other electron microscopic techniques of freeze-etching and freeze-fracture can provide complementary information on gross structure. To probe the interior of the particles, more discriminating techniques such as nuclear magnetic resonance (NMR) and other spectroscopic methods must be used. Using NMR, chemical groups can be distinguished by their chemical shifts, for example the $(CH_2)_n$ and terminal CH_3 groups of the hydrocarbon chains and the $N^+(CH_3)_3$ headgroup of the polar lipids (Figure 5.20). ^{13}C-NMR provides an additional technique to probe the environment of carbon–carbon groups in both the lipids and proteins. It is useful for measuring the state of molecules within their native environments but is relatively insensitive. A more sensitive technique but one that suffers from the fact that it requires an external probe (a spin-label) which is bulky and may itself perturb the natural structure is electron spin resonance (ESR). In combination with X-ray, neutron scattering and calorimetric studies, such techniques provide detailed information about lipid–protein interactions in lipoproteins. Further information on these techniques can be found in the references.

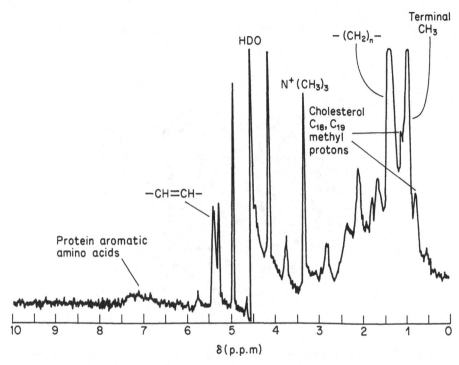

Figure 5.20 Nuclear magnetic resonance (NMR) spectrum (220 MHz) of HDL_2 in deuterium oxide at 50°C. Reproduced with kind permission of Dr R.B. Leslie.

Using these methods, it has been demonstrated that apolar lipids associate at the hydrophobic surfaces of amphipathic α-helical regions of specific amino acid sequences of apolipoproteins while phospholipids and water interact at polar surfaces. Interaction of the major apolipoproteins of HDL with lipids stabilizes the α-helical structure and increases its content. In contrast, as the lipid content of LDL increases, the β-structure of apoB increases.

Out of these studies, models for the structures of HDL and LDL have emerged. LDL is envisaged as having a core composed of neutral lipids, mainly cholesterol esters and triacylglycerols, with apoB at the surface, interacting with both surface components, mainly phospholipids and unesterified cholesterol, and the core lipid (Figure 5.21). It is interesting to note that a more complex model has been proposed by Finer and his colleagues which envisages some protein at the core and a trilayer of lipids: internal phospholipid–cholesterol ester–external phospholipid. They proposed this structure to account for the finding that, according to the NMR signals derived, the proteins and some of the phospholipids were very immobile, which the authors interpreted as indicating an internal location.

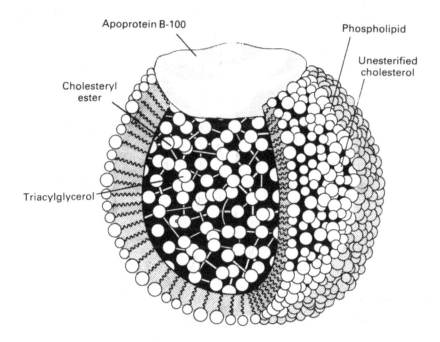

Figure 5.21 A model of the structure of a low density lipoprotein. Redrawn with kind permission of M.S. Brown and J.L. Goldstein (1984) *Scientific American*, **251** (5), 60. © 1984 Scientific American Inc.

In the case of HDL, there are subtle changes of structure associated with changing lipid patterns as the particles evolve. Soluble HDL apolipo-proteins interact with well-defined synthetic phospholipids to form a disc-like structure similar to that observed *in vivo* (section 5.3.5). In the absence of neutral lipids, the structural organization is dominated by the bilayer formation of the phospholipids. As neutral lipids (cholesterol esters and triacylglycerols) are incorporated into the core, the particles assume a spherical morphology that is governed by the hydrophobic interactions of the core lipids.

(l) Other lipoproteins

It would be a mistake to think that plasma lipoproteins were the only examples of lipid–protein complexes in the body. Many other lipid transport proteins exist and, while they are not usually referred to as lipoproteins, they nevertheless have analogous roles.

Serum albumin, for example, binds many types of molecules including non-esterified fatty acids and is, indeed, the main transporter of NEFA, released by lipases from adipose and other tissues into the blood. There are three classes of binding sites that bind respectively, 2, 5 and 20 molecules of fatty acids. These sites have a decreasing affinity for binding NEFA in the above order and there are striking differences between the binding constants for different fatty acids at each site. The complex is completely water-soluble and is often used as a tool for introducing fatty acids as substrates into an experimental system (either *in vivo* or *in vitro*) to overcome problems of water insolubility.

Other fatty acid binding proteins, such as the Z-protein, have been described in sections 3.3.1 and 5.3.3. Their function is thought to be to transport fatty acids and acyl-CoAs within cells rather than in the blood. This is important, not only because of the problem of water insolubility but because of the disruptive detergent effects of these substances in the free state.

Phospholipids may be transported within cells by binding to specific phospholipid exchange proteins (PLEP) whose function may be to introduce lipids into cell membranes and organelles remote from their sites of synthesis. For example, mitochondria do not possess enzymes for the complete biosynthesis of phospholipids *de novo* and may acquire their characteristic lipid composition by exchange from endoplasmic reticulum via the PLEP. Carrier proteins also exist for sterols, including vitamin D, and for retinol. These are referred to in the appropriate sections.

Table 5.12 Lipoproteins of egg yolk

A. Lipid composition (g/100 g lipoprotein)

Lipoprotein	Protein	Phospho-lipid	Cholesterol	Cholesterol ester	Triacyl-glycerol
High density (lipovitellin)	78	12	0.9	0.1	9
Low density	18	22	1.8	0.2	58

B. Fatty acid composition (g/100 g total fatty acids)

Fatty acid	Whole egg	Low density lipoprotein
16:0	29	32
16:1	4	8
18:0	9	8
18:1	43	45
18:2	11	7
Others	4	0

Biological membranes can legitimately be regarded as insoluble lipoproteins and this subject is treated in detail in Chapter 6. The brain contains a large proportion of lipid that is bound both in membranes and as poorly characterized lipoprotein complexes. Finally, eggs contain a high proportion of lipids present in the form of both high density lipovitellin and low density lipoproteins. The lipid composition of the latter is illustrated in Table 5.12.

5.4 LIPIDS IN GROWTH AND DEVELOPMENT. DURING GROWTH, THE SIZE AND NUMBERS OF CELLS IN TISSUES ARE INCREASING RAPIDLY, MAKING LARGE DEMANDS ON THE SUPPLY OF LIPIDS FOR MEMBRANE SYNTHESIS AND OTHER FUNCTIONS

From conception, the cells of the growing fetus need to incorporate lipids into their rapidly proliferating membranes. The fetus is totally dependent on the placental transfer of substrates from the mother's circulation, which could then be elaborated into lipids in at least four different ways:

1. biosynthesis *de novo* from glucose in fetal tissues;
2. incorporation of fatty acids transferred from maternal to fetal circulation;
3. incorporation of fatty acids from circulating maternal lipoproteins after release by a placental lipoprotein lipase; and
4. biosynthesis of lipids in the placenta itself and their transfer to the fetal circulation.

Most authors agree that glucose is a major substrate for the fetus which possesses the enzymic activities for its conversion into fatty acids and for the production of the glycerol moiety of glycerides via glycerol phosphate.

Clearly animals are incapable of synthesizing all their lipid requirements. A crucial question in developmental biology is how essential fatty acids are acquired and metabolized at this early stage. There are several ways in which the fetus could obtain the essential fatty acids it needs for its membranes but their relative importance has not been fully established in man (Figure 5.22):

1. Precursor and product EFA could be obtained directly from maternal plasma, transferred across the placenta and used directly by fetal tissues.
2. Precursor EFA could be taken up by the placenta and converted into product EFA in the placenta before transfer to fetal tissues, or
3. Precursor EFA could be transferred across the placenta and converted into product EFA in fetal tissues.
   (We are using the term precursor EFA to refer to linoleic (n-6) and α-linolenic (n-3) acids in species other than obligate carnivores,

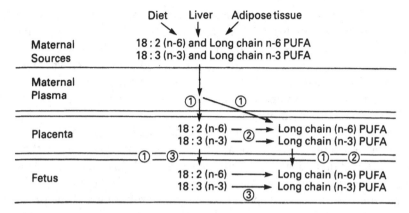

Figure 5.22 Origins and metabolism of polyunsaturated fatty acids in the fetus. The numbers refer to the three alternatives outlined in the text. The sources of maternal plasma fatty acids could be dietary (as chylomicrons) or from the liver (as VLDL) or adipose tissue (as albumin-bound NEFA).

while product EFA are their long chain polyunsaturated fatty acid metabolites.)

The placenta of most mammals, including human beings, is permeable to non-esterified fatty acids although there are large differences between species in rates of transfer. A common finding is a higher concentration of arachidonic acid in fetal than in maternal plasma. Whether this results from metabolism of linoleic acid in the placenta itself or particularly rapid passage of arachidonic acid is still to be resolved but there are again clear differences between species. The rat appears to rely more on maternal transfer than on placental synthesis and although the human placenta is capable of forming arachidonic from linoleic acid, the activity is limited. Arachidonic acid is selectively incorporated and trapped into placental phospholipids for export to the fetal circulation. Crawford's group in London, England, has coined the term biomagnification for a process in which the proportion of product EFA increases in phospholipids progressing from maternal blood, to cord blood, fetal liver and fetal brain.

The study of fetal conservation of EFA in ruminants is extremely rewarding since the availability of EFA to the mother is severely limited by rumen hydrogenation (section 3.2.5). The triene/tetraene ratio in fetal lamb tissues is about 1.6, a value that in simple stomached animals would be associated with extensive signs of EFA deficiency. By 10 days after birth, the ratio has fallen to 0.4 and by 30 days to 0.1. These values are well within the normal range despite the extremely low concentration of linoleic acid in ewe's milk (0.5% of energy). Ruminants are, therefore, able to conserve

EFA with supreme efficiency. Sheep placenta transfers linoleic acid at a relatively slow rate but has a very high Δ6-desaturase activity by comparison with non-ruminants, providing the major source of arachidonic acid for the fetus. This metabolite is concentrated into phosphoglycerides whereas the linoleic acid precursor is in higher concentration in the triacylglycerols. This molecular compartmentation has the effect of conserving arachidonic acid and directing it into membranes.

In some species, reserves of fatty acids are already being built up in the fetus. For example, the development of human fat cells begins in the last third of gestation and at birth a baby weighing 3.5 kg has, on average, 560 g adipose tissue. Guinea pigs are also born with a large amount of adipose tissue whereas, in contrast, pigs, cats and rats are born with little or none. There is a remarkable parallelism between the placental permeability to non-esterified fatty acids and the tendency to accumulate adipose tissue by the fetus, suggesting that an important source of fetal fat reserves is derived from circulating maternal lipids. The adipose tissue composition of the fetal and new-born fat will, therefore, reflect the fatty acid composition of the maternal diet as illustrated in Table 5.13.

Concentrations of all lipoprotein classes increase in the maternal circulation during pregnancy, a process that is mediated by the sex hormones. A lipoprotein lipase has been discovered in human placenta which is consistent with the hypothesis that during hyperlipidaemia of pregnancy, release of fatty acids by placental lipoprotein lipase could generate substrates for synthesis of lipids by the fetus. The transfer of intact lipoproteins by a receptor-mediated pathway has not been demonstrated but cannot be ruled out.

Table 5.13 Influence of the fatty acid composition of the mother's dietary fat on the fatty acid composition of adipose tissue in new-born guinea-pigs

Fatty acid	Dietary fat			
	Maize oil		Beef tallow	
	maternal diet	adipose tissue	maternal diet	adipose tissue
	(g/100 g total fatty acids)			
16:0	13	23	25	26
18:0	3	6	18	9
18:1	26	28	36	42
18:2	54	33	7	16
18:3	2	5	2	6
Others	2	5	12	1

A large proportion of the product EFA synthesized or accumulated during the perinatal period is destined for the growth of the brain, 50% of whose acyl groups may consist of the long chain polyunsaturated fatty acids 20:4 (n-6), 22:4 (n-6), 22:5 (n-6) and 22:6 (n-3). As with adipose tissue, there are large species differences in the time at which birth occurs in relation to the extent of brain development. The peak rate of brain development occurs in guinea pigs in fetal life; in the rat, postnatally; while in man and pig it reaches a peak in late gestation and continues after birth. It has been suggested that transfer from the placenta is the major source of long chain polyunsaturated fatty acids for the human fetal brain but the experimental difficulties of demonstrating this in man are enormous. Because of their remarkable similarity to man in the timing of brain development and their lipid composition, pigs have been used as models. Long chain derivatives of linoleic acid (the n-6 family) increase in brain from mid-gestation to term. Little linoleic acid, however, accumulates until birth, when the concentration increases three-fold while product EFA remain constant. By labelling experiments with $1-^{14}C$ linoleic acid *in vivo*, it has been shown that linoleic acid is metabolized to long chain polyunsaturated fatty acids by piglet brain and liver throughout the perinatal period. The contribution of the liver was many fold greater than the brain at all stages. Whether fetal tissues can supply all the needs of the nervous system for long chain polyunsaturated fatty acids without the need for the maternal transfer of intact product EFA is still to be resolved.

An outstanding feature of the composition of brain phospholipids is its remarkable consistency, irrespective of species or diet. The concentrations of the precursor EFA are extremely low (18:2, n-6, 0.1–1.5% and 18:3, n-3, 0.1–1.0%) while arachidonic (20:4, n-6) and docosahexaenoic (22:6, n-3) acids predominate at 8–17% and 13–29% respectively in all species. This contrasts with the liver lipids where there is much greater variation between species. The precursor EFA are present in much greater concentrations than they are in brain and there are major differences in the product EFA. For example, 22:5 is the major n-3 fatty acid in the liver lipids of ruminants and other herbivores while 22:6 predominates in the carnivores and omnivores. Fatty acids of the n-6 family usually predominate in liver phosphoglycerides, even when the overwhelming dietary intake is in favour of the n-3 fatty acids. Thus, zebra and dolphin, both species that have an overwhelming excess of n-3 fatty acids in the diet, attain a preponderance of n-6 acids in the liver phosphoglycerides (Table 5.14).

It is fortunate that the development of adipose tissue lipids in man can be studied with relatively simple biopsy techniques. Detailed temporal studies of human brain lipids are more problematical and highlight the need for a good animal model. The pig, which is inadequate as a model for the temporal aspects of adipose tissue development, is useful as a model for

Table 5.14 Principal polyunsaturated fatty acids of the liver phospholipids of zebra and dolphin

	(g/100 g total fatty acids)	
	Zebra	Dolphin
Liver fatty acids		
18 : 2(n-6)	40	1
20 : 4(n-6)	5	10
18 : 3(n-3)	4	6
22 : 6(n-3)	1	1
n-6/n-3	9 : 1	1.6 : 1
Food fatty acids		
n-6/n-3	1 : 3	1 : 20
	(mainly 18C)	(mainly 20, 22C)

brain development as described above. Such studies are important since the increasing expertise in sustaining life in premature infants has bequeathed us the problem of how to feed them to achieve normal development. These children may be born well before brain development has reached its peak. It is crucial, therefore, that the lipids in the infant's food are able to contribute to the brain growth that would have occurred in the fetus. Any long chain essential fatty acids that the fetus could not elaborate for itself and had to be supplied by the mother's circulation or the placenta will now have to be contributed by the milk. Thus, appropriate feeding strategies will require greater knowledge of the relative roles of the mother, placenta and the fetus or new-born's own tissues in supplying long chain polyunsaturated fatty acids.

At birth, quite large changes in lipid metabolism occur. Whereas the fetus relied extensively on glucose and had elaborated the pathways of lipid synthesis from this substrate, the sole source of nutrition for the new-born is milk from which about 50% of energy comes from fat. The enzymes of fatty acid synthesis are suppressed and the baby's metabolism becomes geared to using fat directly from the diet. Human milk contains quite a high proportion of long chain polyunsaturated fatty acids and this may be a response to the needs of the still developing brain and nervous tissue. There are, however, wide differences in milk fat composition depending on the mother's diet (Table 5.15) so that large variations in composition must be tolerated without ill effects. Despite the low content of polyunsaturated fatty acids in cow's milk fat, there is no evidence that babies given mainly formulas based on cow's milk develop less well than those primarily breast-fed. Nevertheless, manufacturers have made many modifications to the fat in baby formulas to mimic more closely human milk fat and some have had influences on body composition (Table 5.16).

Table 5.15 Influence of diet on the fatty acids of human milk

| | (g/100 g total fatty acids) | |
	Subject A	Subject B
Energy intake (MJ)	11.7	10.5
Energy as fat (%)	7	51
Fatty acid		
12:0	17	6
14:0	13	7
16:0	18	24
16:1	1	2
18:0	6	7
18:1	40	40
18:2	1	11
Others	4	3

Table 5.16 Influence of the type of fat in infant formulas on adipose tissue composition of babies

		(g/100 g total fatty acids)				
		British *			Dutch **	
Fatty acid	Fat in formula	Adipose tissue		Fat in formula	Adipose tissue	
		Birth	6–12 months		Birth	6–12 months
14:0	11.5	3.8	8.5	0	3.3	4.8
16:0	30.0	48.9	32.0	10.7	48.5	29.7
18:0	14.3	4.1	5.6	2.0	3.8	2.9
16:1	2.0	12.6	7.0	0	15.2	13.8
18:1	31.1	29.6	43.4	27.2	29.0	40.8
18:2	1.8	1.0	3.5	58.2	2.9	8.0

* British babies were given a formula in which the fat component came from cow's milk.
** Dutch babies were given a formula in which the fat component was provided by a vegetable oil.

5.5 DISEASES INVOLVING CHANGES OR DEFECTS IN LIPID METABOLSIM

Earlier in this chapter we described deficiency diseases that arise because the diet contains insufficient of several types of lipids that the body is unable to synthesize for itself. Here, the discussion is devoted to diseases that involve a defect in a metabolic pathway, usually because an enzyme is missing or not functioning correctly. We have classified them into: transport diseases, which involve defects in lipoprotein metabolism; and diseases of complex multifactorial aetiology in which changes in lipid metabolism may play a direct, but more usually an indirect role. The disorders in question may be primary or secondary. Primary disorders are normally inherited and result from a defect in a gene coding for a specific enzyme or receptor protein or the failure of a regulatory gene, not necessarily in lipid metabolic sequences. They are frequently referred to as inborn errors of metabolism. Secondary lipid disorders occur when lipid metabolism is altered as a result of another recognizable disease (e.g. diabetes or obesity), which, if treated, will lead to a normalization of the lipid patterns. Although these diseases are not primarily the result of a nutritional deficiency, nutrition may often play an important role in controlling, modifying or even reversing the worst effects of the disease and may sometimes be the only way of allowing the patient to lead a normal life.

Lipid storage diseases, which usually involve the inability to break down a particular lipid so that it accumulates in tissues, are described in section 8.14.

5.5.1 Lipid transport diseases: as a result of genetic defects resulting either in faulty apoproteins or their cellular receptors, the circulating concentrations of lipoproteins can be abnormally high, or, more rarely, low. High concentrations of plasma lipoproteins are, in turn, associated with a number of diseases including diabetes and ischaemic heart disease

(a) Primary hyperlipoproteinaemias

The primary hyperlipoproteinaemias (sometimes referred to as familial or essential hyperlipoproteinaemias or hyperlipidaemias) are normally classified for convenience in five categories according to a system introduced by the American clinician and biochemist, Frederickson. It must be emphasized, however, that because of the gross heterogeneity of the lipoproteins themselves, not all types of abnormality fall into these neat categories. There are some blurred edges that need to be clarified by specific clinical investigations. In general, each category is characterized by the elevation of the plasma concentration of specific classes of lipoproteins, resulting in particularly high concentrations of triacylglycerols or cholesterol or both. A summary of these different disorders, the primary defects, clinical features and treatments is presented in Table 5.17. It is interesting to note that an important clinical feature of abnormally high blood lipid concentrations, skin xanthomata, was first recognized as long ago as 1827 in an illustration in Rayer's *Traite theoretique et pratique des maladies de la peau.*

The underlying causes are either the absence of a specific enzyme (e.g. lipoprotein lipase in Type I); the absence of a specific apolipoprotein (e.g. apoC$_2$ in another form of Type I); the absence, reduction or impairment of function of specific cell surface receptors (e.g. the LDL receptor in Type IIa); the overproduction of specific apolipoproteins (e.g. apoB in Type IIb); or the overproduction of lipids in the liver (e.g. triacylglycerols in Types IV and V). Different gene defects may produce the same lipoprotein pattern although by different mechanisms. Alternatively, factors such as body weight and dietary composition may modify the phenotypic expression of a particular genotype.

The diagnosis of primary hyperlipoproteinaemia can usually be confirmed, after exclusion of secondary causes, by an investigation of medical history, analysis (by electrophoresis and determination of blood lipids) of lipoprotein patterns and screening of near relatives. Further ambiguities may be removed by such procedures as the measurement of post-heparin plasma lipolytic activity or assay of LDL receptor function in cultured fibroblasts or blood lymphocytes.

Analysis of genetic polymorphisms by recombinant DNA techniques is now being used to unravel the basis for many of these disorders. Some patients with familial hypercholesterolaemia (FH; normally Type IIa) have a mutation in the gene encoding the LDL receptor, although the mutations are not always the same. For example, in some there is a five kilobase pair deletion of DNA of the gene coding sequence, leading to the production of a truncated LDL receptor protein that is unable to attach to the cell membrane and, therefore, internalize the receptor-LDL complex (section 5.3.5). A restriction fragment-length polymorphism revealed by the restriction enzyme Pvu has been shown to occur commonly enough to be of value in the presymptomatic diagnosis of FH, as alleles of the LDL receptor gene coinherit with FH.

(b) Secondary hyperlipoproteinaemias

Several diseases result in disturbances of plasma lipoprotein metabolism which are reversed upon treatment of the primary disease. Diabetes and obesity are discussed in more detail later. Because of the prominent role of

Table 5.17 Diseases involving abnormal plasma lipoprotein concentrations – the lipoproteinaemias

Classification	Characteristic lipoprotein pattern	Major lipids involved	Treatment	Clinical features
Hyperlipoproteinaemias				
Type I	Raised chylomicrons in fasting plasma	Triacylglycerols Free cholesterol	Restricted fat diet; replace long chain fats by medium chain triacylglycerols (MCT)	Rather rare; usually diagnosed before age 10
Type II	Raised LDL but the LDL are normal in composition	Cholesterol esters	Restrict dietary cholesterol by limiting the dietary intake of egg yolks, liver, dairy products. Substitute poly-unsaturated oils, PUFA margarines, skimmed milk	A common disorder; very strong associations with premature ischaemic heart disease (IHD); children who are homozygotes for the disorder exhibit features of IHD in the first decade of life; xanthomas (massive accumulation of cholesterol in the skin) and corneal arcus (white ring in the eyes) are common features of the disease; occurs both as a heritable disorder and secondary to hypo-thyroidism
Type III	Raised abnormal LDL concentrations	Cholesterol esters Triacylglycerols	(i) Restrict cholesterol intake (ii) Reduce weight (iii) Dietary composition should be protein: carbohydrate: fat = 20 : 40 : 40	The third most common disorder after types II and IV; accompanied by extensive vascular disease

			(iv) Drug treatment	
Type IV	Elevated VLDL concentrations	Triacylglycerols	(i) Weight control (ii) Avoidance of excessive dietary carbohydrate (iii) Hypolipidaemic drugs	Associated with abnormal glucose tolerance and a family history of diabetes; obesity is extremely common; not associated with IHD to the same extent as type II; occurs both as a heritable disorder and secondary to diabetes pancreatitis, etc.
Type V	Elevated chylomicrons and VLDL	Triacylglycerols Cholesterol esters	(i) Weight reduction (ii) Low energy diet not rich in either carbohydrate or fat; difficult to achieve	Very rare
Hypolipoproteinaemias (very rare)				
Familial LDL deficiency	Deficiency or complete absence of LDL; poor ability to form chylomicrons after a fatty meal	Cholesterol esters Triacylglycerols	Limit long chain saturated fat intake and replace by MCT with some PUFA vegetable oils	Neuromuscular disturbances; retinal changes; red blood cell abnormalities; steatorrhea (bulky and excessively fatty stools)
Familial HDL deficiency (Tangier disease)	Abnormally low HDL concentrations	Cholesterol Phospholipids	—	Enlarged tonsils, spleen, liver and lymph nodes; accumulation of lipids in reticulo-endothelial tissues

the liver in lipoprotein biosynthesis and catabolism, diseases affecting liver function are likely to influence plasma lipoproteins. Thus, chronic liver failure may result in deficiencies in lipases involved in the catabolism of remnant lipids and LCAT deficiency which can result in hypertriacyl-glycerolaemia and hypercholesterolaemia. Alcohol abuse can be one such cause of liver failure with wide ranging results. Its effect on stimulating the synthesis of triacylglycerols has already been referred to in section 4.6.2, More recent evidence suggests a consistent positive relationship between alcohol intake and HDL concentrations although the mechanism of this effect is unknown. Chronic renal failure is also associated with the increased synthesis and decreased catabolism of VLDL.

Hyperlipoproteinaemias, both primary and secondary, are mainly important insofar as they are powerful risk factors for cardiovascular disease (section 5.5.3).

(c) Hypolipoproteinaemias

These conditions in which lipoprotein concentrations are below normal or there is a complete absence of one lipoprotein class are much rarer than those characterized by raised lipoprotein concentrations (Table 5.17).

5.5.2 Obesity and diabetes: obesity or the excessive storage of triacylglycerols in adipose tissue occurs when energy intake exceeds energy expenditure over a significant period of time. Diabetes is a disorder of metabolic control arising either through lack of insulin (Type I) or insensitivity of tissues to insulin (Type II)

These two conditions will be considered together since recent research has confirmed important relationships between the two which will have a profound influence on our thinking concerning the so-called diseases of affluence.

Obesity is one of the predominant health problems in affluent societies. It has been the subject of much research but little progress has been made towards its control. The most obvious implication of lipids in obesity is their accumulation in adipose tissue, but this is only the final stage in a series of metabolic changes leading to obesity. The cause is not necessarily to be found in adipose tissue metabolism. Because dietary lipids contain over twice as much energy per gram as carbohydrates or proteins, high fat diets, common in many Western countries, have often been assumed to play a role in the development of obesity. However, some populations may tend to consume high fat diets, yet not have a high prevalence of obesity and some individuals may consume diets containing more than enough fat to satisfy their energy requirements yet not gain weight. Adding fat to the predominantly carbohydrate diets of laboratory rodents results initially in an adjustment by the animal of its voluntary food intake to maintain constant energy intake, but when the level of fat inclusion exceeds a certain amount, this regulatory mechanism starts to break down. Many things affect food intake and palatability (enhanced by the presence of fat (section 5.2.3)) is one of the most important.

In searching for the causes of obesity, energy intake is not the only factor to consider: energy expenditure is at least as important. One of the biggest changes in modern society is a general reduction in the need to expend energy: both work and popular leisure have become increasingly sedentary. Although there seems to be slow decline in average energy intakes in countries such as the UK, for many people this does not offset the reduction in their energy expenditure. Exercise is important not only in stimulating the mobilization of fatty acids from adipose tissue to supply substrates for oxidation. It also influences metabolism in ways which together are conducive to improved health. Muscle tissue that is exercised can extract oxygen more efficiently from the blood, resting heart rate and blood pressure are lower and the capacity of cardiac output and blood flow at the maximum rate of oxygen consumption are greater. Moreover, concentrations of plasma HDL correlate consistently with habitual physical activity (section 5.5.3).

Another component of energy expenditure is termed diet-induced thermogenesis. Thus, energy expenditure increases after a meal because the biochemical reactions involved in the metabolism of food components following the digestion and absorption of fats, proteins and carbohydrates are thermodynamically inefficient. Differences between individuals in metabolic efficiency may partly explain why some individuals gain weight easily while others appear to remain lean even though they apparently consume large quantities of food. It is not yet certain which metabolic pathways are primarily responsible for these individual differences in thermogenic response. Protein turnover is clearly an energy demanding process, while in carbohydrate metabolism, the fructose-6-phosphate/fructose-1,6-diphosphate cycle has been investigated as an energy-wasting substrate cycle.

In lipid metabolism the turnover of triacylglycerols in adipose tissue has been examined as a possible substrate cycle but it is difficult to see how this could be quantitatively important. Much attention has been focused on brown adipose tissue, since it is known to be important in the generation of heat to maintain body temperature in new-born animals (Figure 5.23). Brown adipose tissue mitochondria are able to generate large quantities of heat because electron transport is largely uncoupled from oxidative

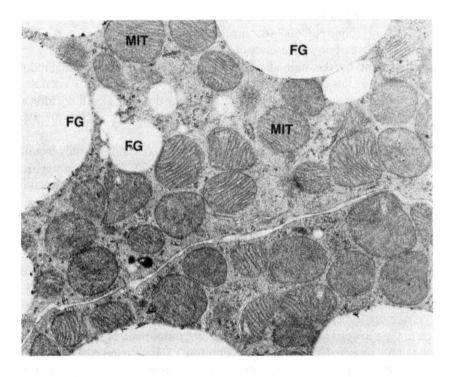

Figure 5.23 Structure of brown adipose tissue as seen by transmission electron microscopy. FG = Fat globule; MIT = mitochondria. Reproduced with kind permission of Dr M. Ashwell.

phosphorylation. Instead of being used to drive ATP synthesis, hydrogen ions leak back into the mitochondria through a channel in a protein known as the GDP-binding protein or uncoupling protein that is unique to this tissue. The amount of this protein has been used as a biochemical index of the potential for thermogenesis. Experiments in which rats were induced to over-feed demonstrated striking enhancement of thermogenesis accompanied by elevated concentrations of GDP-binding protein. It was suggested that this could be an adaptive mechanism whereby some individuals would be able to dispose of excess dietary energy without gaining weight. There is, to date, little or no evidence that this mechanism is important in Man and the search for the biochemical basis for adaptation to excess energy consumption must continue.

Maturity-onset diabetes (now more frequently termed Type II or non-insulin-dependent diabetes mellitus, NIDDM) is associated with obesity, hyperlipoproteinaemia and an increased risk of cardiovascular disease. Lipid metabolism tends to be enhanced and an important characteristic is hyperinsulinaemia. Although the aetiology of this condition is still far from clear, the following sequence is gaining wide acceptance. The initial lesion is a resistance of various peripheral tissues (e.g. muscle, adipose tissue) to the action of insulin; in other words, more hormone is required to produce a given metabolic effect. To overcome this insulin resistance, the pancreatic β-cells are compelled to hypersecrete insulin and hyperinsulinaemia results. After many years of hypersecretion, the pancreatic β-cells eventually fail and insulin deficiency and clinical diabetes (Type I) result. (The term diabetes is a source of some confusion, since it is applied to two separate conditions that are quite different in their associated metabolism. In juvenile-onset diabetes, now more often referred to as Type I or insulin-dependent diabetes mellitus (IDDM), the pancreatic β-cells are unable to produce sufficient, or any, insulin and the person is unable to maintain normal glucose homeostasis. The condition is controlled by injection of appropriate amounts of insulin combined with careful control of diet. Usually when we have referred to diabetes in earlier chapters, the Type I form was implied. It is associated with leanness and a reduced level of fatty acid synthesis, desaturation and esterification. Administration of insulin reverses these metabolic changes. Such changes in lipid metabolism can be studied in animal models for diabetes in which the pancreas has been poisoned by administration of substances such as alloxan or streptozotocin.

It has been a consistent finding that Type II diabetes is almost invariably associated with obesity but of more fundamental interest is its association with a particular distribution of body fat. The excessive adipose tissue characteristic of obesity tends to be distributed in the upper body, including the waist (android or upper body obesity) or the lower body, including the hips and buttocks (gynoid or lower body obesity). The former is more characteristic of men and is more closely associated with Type II diabetes, hyperlipidaemia and other metabolic disorders while the latter is more usually seen in women and has fewer adverse metabolic implications. These relationships are so strong that risk can be assessed by a relatively simple index: the ratio of waist to hip circumference.

The mechanisms whereby body fat distribution may be associated with diabetes independently of overall adiposity are still unresolved but the subject of intense research attention. Subjects with predominantly upper-body fat distribution are more insulin-resistant than subjects equally obese but with a fat distribution predominantly in the lower body. Upper body obesity is associated primarily with fat cell hypertrophy (increased adipocyte size) while lower body fat distribution is associated more with fat cell hyperplasia (increased number of adipocytes). As fat cells enlarge, they become more resistant to insulin. The adipose tissue, however, accounts for only about 5% of total body glucose disposal, so that insulin resistance in this tissue alone should not account for significant increases in total body insulin resistance. It is more likely, therefore, that an interaction between upper body fat distribution and the skeletal muscle response to insulin is the basis for the increased insulin resistance. The nature of this interaction is obscure but it is known that enlarged fat cells have an increased rate of basal lipolysis. Thus, it may be that the resulting increased circulating non-esterified fatty acid concentrations are responsible for the diminution in insulin sensitivity of muscle cells.

5.5.3 Blood vessel diseases: atherosclerosis involves the build-up of deposits in arteries and these are characterized by high concentrations of lipid that derive from the plasma lipoproteins. Lipids are also involved in arterial disease insofar as they are concerned in the formation of thrombi which may lead to the blockage of blood vessels

Cardiovascular disease is a broad term for a number of conditions involving pathological changes in blood vessels associated mainly with the heart and brain. The major cause of death is ischaemic heart disease (IHD) or, as it is often called, coronary heart disease (CHD). In IHD, the coronary arteries become narrowed by an accumulation of deposits to such a degree as to prevent the coronary circulation meeting the metabolic demands of the heart. Local conditions may also increase the likelihood of platelet aggregation and thrombus formation. The immediate cause of death is failure to supply a sufficient proportion of the heart muscle with oxygen. Peripheral vascular disease involves blood vessels in, for example, the limbs and although it is not a major cause of mortality, it is associated with considerable morbidity and distress for the patient (Figure 5.24).

In considering the role of fats, it is useful to consider the disease as consisting of two distinct phases (Figure 5.25). The first phase is atherosclerosis, an irregular thickening of the inner wall of the artery that reduces the size of the arterial lumen. The thickening is caused by the accumulation of plaque, consisting of smooth muscle cells, connective tissue, mucopolysaccharides, fat-filled foam cells, in which the predominant lipid is cholesteryl ester, and deposits of calcium. The artery wall is locally

---

Figure 5.24 An illustration of arterial disease due to atheromatous deposits: (a)–(c) illustrate how the aorto-iliac arteries can be examined for atheroma in the living person by X-rays ('arteriograms'). (a) Slight atheroma with no occlusion; (b) moderate atheroma exhibiting the characteristic 'string of sausages' effect but with no occlusion; (c) one iliac artery has been totally occluded; (d) illustrates an atheromatous aorta examined *post mortem*. Reproduced with kind permission of Dr K.J. Kingsbury.

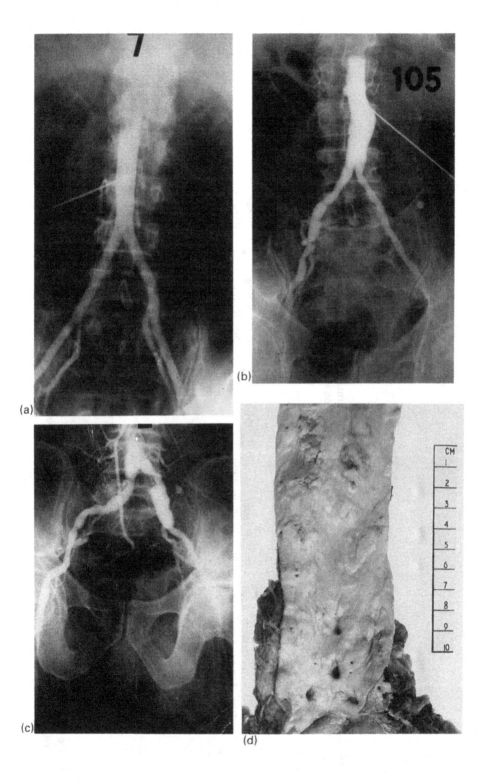

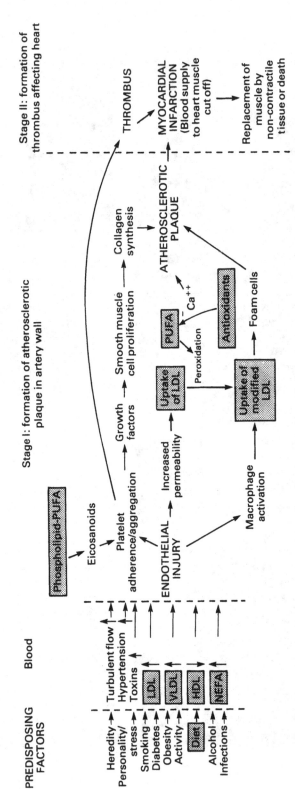

Figure 5.25 Possible processes involved in the development of ischaemic heart disease. Shaded boxes indicate points at which lipids may be involved.

thickened and loses elasticity. Lesions resembling human atherosclerosis in some, but not all, respects have been produced in a variety of animal species by feeding high levels of lipids containing predominantly saturated fatty acids and cholesterol. Using this technique, a prelude to the development of atherosclerosis is hyperlipoproteinaemia and by labelling the fats carried by the lipoproteins with radioactivity, it can be shown that at least some of the lipid deposited in arterial plaque can originate from the plasma lipoproteins. The second phase, and the fatal episode, in IHD is the formation of a clot or thrombus in a coronary artery, usually at the site of an atherosclerotic lesion so that the blood supply to the heart is cut off (myocardial infarction). This leads either to the degeneration of part of the contractile heart tissue and then replacement by non-contractable scar tissue, or else complete cessation of the heart beat. No experimental animal has yet been found in which this phase can be mimicked reproducibly and for this reason experimental study of this disease has always been difficult.

Several hypotheses have been erected to explain the natural history of cardiovascular disease and the one given the most attention has been the lipid hypothesis. Its main postulates are:

1. Hyperlipidaemia is a major cause of the train of events leading to atherosclerotic plaques.
2. A prime cause of hyperlipidaemia is the overconsumption of fat and particularly saturated fatty acids in the diet.

If these postulates are accepted, it follows that:

3. Dietary modification should lead to reduced blood lipid concentrations; and
4. Lowering blood lipid concentrations should reduce morbidity and mortality from IHD.

Evidence for the lipid hypothesis is both epidemiological and experimental. In epidemiology, the investigator observes how disease patterns in a population relate statistically to endogenous and environmental factors in those populations. Thus, several decades ago a close correlation was discovered between the average serum cholesterol of a population and the risk of developing heart disease. These communities also seemed to be ones that had the highest consumption of fat and saturated fatty acids, although there are many exceptions to this rule. In the light of current knowledge, we must specify that the strongest relationship is between plasma LDL cholesterol and IHD, rather than total cholesterol. An important recent epidemiological finding is that the concentration of plasma HDL (especially HDL_2) is inversely related to IHD risk. In other words, HDL and factors that increase its plasma concentration may be protective. HDL concentrations tend to be higher in women than in men and in people who lead active as opposed to sedentary lives.

The association between LDL and the risk of IHD is much weaker for individuals within a population than between populations indicating that other factors of environment and lifestyle may exert a greater influence. Other important risk factors are smoking, hypertension, upper body obesity, Type II diabetes, lack of exercise and personality type and these factors may themselves be independently related to plasma lipoprotein concentrations. The known risk factors, however, can account for only a third of the current incidence of IHD.

It is important to realize that epidemiology can provide only statistical correlations; it can give no information about mechanisms. That requires experimental evidence. Many studies with animals and man confirm that certain saturated fatty acids in the diet elevate plasma LDL. This is mainly confined to lauric, myristic and palmitic acids. Those with lower or higher chain lengths have very little effect on blood cholesterol. Careful evaluation of the scientific evidence indicates that the main role of polyunsaturated fatty acids is to prevent the rise otherwise associated with saturated fatty acids. The effects of monounsaturated fatty acids on blood cholesterol were thought to be neutral until recent more carefully controlled studies indicated that they depress LDL and slightly elevate HDL concentrations.

The next stage, the influence of diet and/or blood lipids on the development of atherosclerosis, has been mainly studied in experimental animals. Feeding high fat diets containing a high proportion of saturated fatty acids, sometimes with and sometimes without cholesterol, can give rise to arterial lesions that some researchers regard as sufficiently similar in their pathology to human lesions to act as a good model. Others point out that the inclusion of cholesterol has often been extreme and not related to normal human diets and point to what they regard as serious differences between the pathology of the lesions induced by cholesterol feeding and those typical of the human disease. The definitive experiment cannot of course be done in man, but some regard the condition of homozygous familial hypercholesterolaemia as a natural experiment that provides the best possible evidence for the lipid hypothesis. Patients may exhibit a greater than three-fold elevation of plasma cholesterol, develop rapidly progressing vascular disease in childhood and show an increased mortality from IHD. Death frequently occurs before the age of 30. That the hyperlipidaemia can be controlled at least in part by diet (drugs are almost invariably needed), and thereby life expectancy can be prolonged, is taken by many researchers to indicate that similar measures would be beneficial for those people in the general population whose blood LDL concentration are on the high side of average. Whereas several dietary trials have managed to reduce blood cholesterol concentrations significantly, few have shown significant effects on IHD mortality and none has significantly influenced total mortality. The lipid hypothesis has to be regarded at the present time as unproven.

The current trend is for more general hypotheses to be erected and it is likely that aspects of the lipid hypothesis will be incorporated into these more general proposals. Of great importance in this regard is the endothelial injury hypothesis. It is proposed that an initial injury to the endothelial lining of the artery occurs in response to a range of chemicals (one of which could be peroxidized LDL), toxins, viral infections or immunological reactions. This gives rise to platelet aggregation. Lipids play a role here through the conversion of polyunsaturated fatty acids into eicosanoids, and the modulation of end-product formation by changes in the balance of fatty acid types, which in turn may be influenced by diet. Platelet aggregation is followed by the release of platelet-derived growth factors that initiate a phase of cell proliferation. This accounts for the smooth muscle cell proliferation characteristic of the arterial lesion.

Another characteristic of the lesion is the presence of lipid-laden foam cells. Whereas it had been formerly assumed that foam cells were derived from smooth muscle cells (mainly because of the sheer numbers of the latter and their known capacity to take up lipid from LDL by the receptor-dependent mechanism), new evidence suggests that many, if not most, of the foam cells derive from circulating macrophages. These adhere to the endothelium and penetrate into the sub-endothelial space. There, they imbibe lipoproteins which have become loaded with cholesterol esters. An important observation is that macrophages take up native LDL at a surprisingly low rate; they do, however, take up modified LDL avidly via the scavenger receptor. One possibility is that lipid peroxidation occurring *in vivo* (perhaps as a consequence of a local deficiency of antioxidant) provides the modified LDL substrates for the scavenger receptor. We know that the transformation of normal into peroxidized LDL is reversed in the presence of sufficient vitamin E. It is also well established that the arterial lesions are characterized by high concentrations of peroxides, but whether this is a cause of the lesions or a result of disrupted metabolism in a dying tissue is uncertain. The only other class of naturally occurring lipoproteins that bind to the macrophage scavenger receptors is the β-VLDL. Although not detectable in normal plasma, these lipoproteins accumulate in the plasma of patients with Type III hyperlipoproteinaemia and of animals and human beings given diets containing abnormally large amounts of cholesterol. This is an exciting area of lipid research which should help further to unravel the very complex problem of the aetiology and eventually the prevention of cardiovascular disease. Another newly developing area that may turn out to be exciting is the discovery that high plasma concentrations of an unique lipoprotein, designated Lp(a), are associated with a 2- to 3-fold increase in risk for cardiovascular disease. Lp(a) has a similar structure to LDL but with a higher density and different electrophoretic mobility. Its functions and origins are as yet unknown but we can expect to see some interesting developments in this field.

5.6 SUMMARY

Lipids are important constituents of the diet, although the proportion of energy provided by lipids varies widely between different human societies and between individuals. Quantitatively, the most important are the triacylglycerols that derive mainly from the adipose tissue of farm animals, the liver and flesh of marine animals and the seed oils of plants. Their fatty acids may be predominantly saturated and monounsaturated, as in ruminant fats and coconut oil, di- and triunsaturated as in many seed oils or highly polyunsaturated as in many marine oils. Their physical form is usually modified during food manufacture and processing, and not infrequently their chemical composition too, for example in the hydrogenation employed in the manufacture of margarine to reduce the degree of unsaturation.

Lipids confer palatability to the diet through the textural properties they give to foods and through low molecular weight flavour and odour compounds produced by enzymic oxidation, heating and fermentation. Triacylglycerols provide a concentrated form of dietary energy and act as vehicles for the fat soluble vitamins. Whereas the human body has the capacity to synthesize for itself saturated and monounsaturated fatty acids and cholesterol, tissues are unable to synthesize linoleic and α-linolenic acids (essential fatty acids) and vitamins A, E and K. The amounts required daily in the diet can be assessed, albeit rather imprecisely. Requirements vary between individuals and depend on the rate of growth, age and physiological states such as puberty, pregnancy and lactation. Linoleic and α-linolenic acids are converted in many tissues into two series of longer chain more highly polyunsaturated fatty acids. These are important in contributing to the structural integrity of membranes and in acting as precursors for a range of oxygenated fatty acids, the eicosanoids. The latter are produced locally, have short half-lives and exert powerful physiological activities such as muscle contraction and platelet aggregation at extremely low concentrations. The metabolism of essential fatty acids is influenced by the amounts and types of other fatty acids in the diet.

Before absorption can take place, dietary fats are digested by lipases in the small intestine. Bile salts aid the solubilization of digestion products, which then diffuse into the enterocytes where triacylglycerols, phosphoglycerides and cholesteryl esters are resynthesized and packaged into large particles termed chylomicrons. These lipid transport particles belong to a group known collectively as plasma lipoproteins. Chylomicrons carry triacylglycerols of dietary origin while very low density lipoproteins carry triacylglycerols synthesized in the liver from dietary carbohydrates. Lipoproteins have a core of neutral lipids stabilized by a coat of amphiphilic lipids and apolipoproteins. The latter also give specificity to the lipoproteins. By interacting with specific cell surface receptors, they direct lipoproteins to specific sites of metabolism. Lipoproteins rich in triacylglycerols are degraded by tissues such as adipose tissue and muscle, depending on the individual's energy status. One product of this breakdown is a smaller lipoprotein, the low density lipoprotein, which is the main carrier of plasma cholesterol to the tissues for membrane synthesis in Man. Other particles, called high density lipoproteins, have the function of removing excess cholesterol from membranes and carrying it to the liver for degradation and further metabolism, for example to bile acids.

An inherited lack of, or deficiency in, cell surface receptors for low density lipoproteins results in a condition, familial hypercholesterolaemia, in which blood cholesterol concentrations are rather high. This condition, if untreated, leads to severe vascular disease and death from ischaemic heart disease. Lipids are involved in several ways. First, one of the characteristics of developing atherosclerotic plaques is an accumulation of lipids, particularly cholesteryl esters, which are derived from plasma lipoproteins; secondly, lipids are involved (because of their role as precursors of eicosanoids) in the formation of thrombi which may block arteries and cause ischaemia. Another risk factor for ischaemic heart disease that involves lipid metabolism is obesity, characterized by an excessive accumulation of adipose tissue. In particular, upper body obesity is also associated with Type II diabetes and hyperinsulinaemia. Hyperlipoproteinaemia is secondary to obesity and diabetes mellitus and if these conditions are treated, blood lipid concentrations return to normal.

Lipids thus play a role in the aetiology of various diseases but also have therapeutic importance, so that control of obesity, diabetes and hyper-lipoproteinaemia should include, among other measures, careful attention to the amount and types of fat in the diet.

REFERENCES

Lipids in foods

Forss, D.A. (1972) Odor and flavor compounds from lipids. *Progress in the Chemistry of Fats and Other Lipids*, **13**, 181–258.

Gunstone, F.D. and Norris, F.A. (1983) *Lipids in Foods*, Pergamon, Oxford.

Gurr, M.I. (1984) *Role of Fats in Food and Nutrition*, Elsevier Applied Science Publishers, London.

Paul, A.A. and Southgate (1980) *McCance and Widdowson's The Composition of Foods*, HMSO, London.

Wiseman, J. (ed.) (1984) *Fats in Animal Nutrition*, Butterworth, London.

Acylglycerol structure and composition

Gunstone, F.D., Harwood, J.L. and Padley, F.B. (1986) *The Lipid Handbook*, Chapman and Hall, London.

Hilditch, T.P. and Williams, P.N. (1964) *The Chemical Composition of Natural Fats*, 4th edn, Chapman and Hall, London.

Stumpf, P.K. and Conn, E.E. (eds) (1980, 1987) *Biochemistry of Plants*, Vol. 4, 9, Academic Press, New York.

Digestion and absorption

Carey, M.C., Small, D.M. and Bliss, C.M. (1983) Lipid digestion and absorption. *Annual Review of Physiology*, **45**, 651–677.

Hamosh, M. (1979) Role of lingual lipase in neonatal fat digestion, in *Development of Mammalian Absorption*, Proceedings of CIBA Symposium No. 70, Excerpta Medica, Amsterdam, pp. 92–98.

Sickinger, K. (1975) Clinical aspects and therapy of fat malassimilation with particular reference to use of medium chain triglycerides, in *Role of Fat in Human Nutrition* (ed. A.J. Vergroesen), Academic Press, London, pp. 116–209.

Tso, P. (1985) Gastrointestinal digestion and absorption of lipid. *Advances in Lipid Research*, **21**, 143–186.

Essential fatty acids and eicosanoids

Holman, R.T. (1970) Biochemical activities of and requirement for polyunsaturated fatty acids. *Progress in the Chemistry of Fats and Other Lipids*, **9**, 607–687.

Perkins, E.G. and Visek, W.J. (eds) (1983) *Dietary Fats and Health*, American Oil Chemist's Society, Champaign.

Needleman, P., Turk, J., Jakschik, B.A., Morrison, A.R. and Lefkowith, J.B. (1986) Arachidonic acid metabolism. *Annual Review of Biochemistry*, **55**, 69–1102.

Sanders, T.A.B. (1988) Essential and trans-fatty acids in nutrition. *Nutrition Research Reviews*, **1**, 57–78.

Zollner, N. (1986) Dietary linolenic acid in Man: an overview. *Progress in Lipid Research*, **25**, 177–80.

Lipoproteins

Brown, M.S. and Goldstein, J.L. (1983) Lipoprotein metabolism in the macrophage: implications for cholesterol deposition in atherosclerosis. *Annual Review of Biochemistry*, **52**, 223–261.

Goldstein, J.L. and Brown, M.S. (1977) The low density lipoprotein pathway and its relation to atherosclerosis. *Annual Review of Biochemistry*, **46**, 897–930.

Segrest, J.P. and Albers, J.J. (eds) (1986). The Lipoproteins. Vol. 128: Separation, structure and molecular biology; vol. 129: Characterization, cell biology and metabolism, in *Methods in Enzymology*, Academic Press, New York.

Lipids in disease

Steinberg, D. (1987) Current theories of the pathogenesis of atherosclerosis, in *Hypercholesterolaemia and Atherosclerosis: Pathogenesis and Prevention* (ed. D. Steinberg and J.M. Olefsky), Churchill Livingston, New York, pp. 5–23.

Stern, M.P. and Haffner, S.M. (1986) Body fat distribution and hyperinsulinaemia as risk factors for diabetes and cardiovascular disease. *Arteriosclerosis*, **6**, 123–130.

Trembath, R.C. and Galton, D.J. (1987) Towards genetic markers for coronary artery disease. *Cardiovascular Medicine*, 16 January.

6 *Lipids in cellular structures*

One of the most important general functions for lipids is their role as constituents of cellular membranes. These membranes not only separate cells from the external environment but also compartmentalize cells and provide a special milieu for many important biochemical processes. The functions of lipids in membranes will be dealt with in Chapter 8; here we shall describe the different types of lipids found and their distribution in various cellular membranes.

6.1 CELL ORGANELLES

Eukaryotic cells characteristically contain a range of different cell organelles. All of these are separated from the cytosol (and usually from each other) by membranes. Some organelles such as mitochondria or chloroplasts are surrounded by two distinct membranes whereas others, such as microbodies, have a single membrane only. The relative numbers and detailed structure of most of these organelles vary from cell to cell. Moreover, in some cases, for example chloroplasts, the morphology will change depending on the developmental stage of the cell and will even be affected by the environment and metabolic state of the organelle.

Even in prokaryotes, many species contain a complex internal membrane structure. This is especially true in photosynthetic organisms such as purple non-sulphur bacteria and cyanobacteria (Figure 6.1). Moreover, in some organisms membrane (or organelle) formation can be induced (Table 6.1) – thus providing a convenient experimental system from which to isolate membranes and in which to study membrane biogenesis.

Since organelles and their membranes carry out such a wide range of different functions, it is scarcely surprising that their lipid compositions vary. Moreover, the percentage of lipid and protein in different membranes can range from 80% lipid in myelin to only about 25% in mitochondria

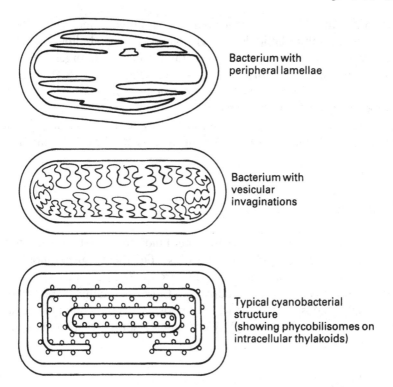

Bacterium with
peripheral lamellae

Bacterium with
vesicular
invaginations

Typical cyanobacterial
structure
(showing phycobilisomes on
intracellular thylakoids)

Figure 6.1 Diagrammatic representation of internal membrane systems of bacteria and cyanobacteria.

Table 6.1 Experimental systems for membrane induction

Organism	Stimulus	Morphological change
Photosynthetic bacteria	Starvation, light	Photosynthetic internal membranes formed
Euglena	Starvation, light	Chloroplasts synthesized
Yeast	Glucose in medium reduced	Mitochondria synthesized
Etiolated plants	Light	Etioplasts converted to chloroplasts
Animals	Fed barbiturates	Liver endoplasmic reticulum proliferates
Animals	Fed clofibrate	Liver peroxisomes proliferate

(Table 6.2). Almost all membrane lipids are amphipathic molecules (section 6.5) such as phospholipids or glycolipids; neutral lipids such as triacylglycerols have little part in membrane structure. In organisms which make or utilize sterols, these tend to be concentrated in the external surrounding membrane (compare chloroplast or mitochondria with erythrocyte or myelin membranes in Table 6.2).

Before describing the composition, structure and function of membranes, it is necessary to detail the molecules which make up the lipid components.

6.2 GLYCEROLIPIDS

Three important classes of membrane lipid are widely distributed – glycerolipids, sphingolipids and steroids. Of these, glycerolipids are quantitatively by far the most important group. They can be conveniently divided into two main groups – those containing phosphorus (phosphoglycerides) and those without phosphorus but containing a sugar constituent (glycosylglycerides). Confusingly, some compounds can be classified as both.

6.2.1 Phosphoglycerides are the major lipid components of most biological membranes

The stereochemistry of phosphoglycerides was discussed in Chapter 1. The phosphoglycerides are a very widespread and diverse group of structures. In most membranes they are the main lipid components and, indeed, the only general exceptions to this statement are the photosynthetic membranes of plants, algae and cyanobacteria and the archaebacterial membranes.

Usually, phosphoglycerides contain fatty acids esterified at the *sn*-1 and *sn*-2 positions of glycerol. They are, thus, diacylphosphoglycerides. These lipids are named after the moiety which is attached to the phosphate esterified at the *sn*-3 position of glycerol. Thus, the compounds can be thought of as derivatives of diacylglycerols in which the hydroxyl on carbon atom 3 is esterified with phosphoric acid which in turn is esterified with a range of molecules – organic bases, amino acids, alcohols.

The simplest phosphoglyceride contains only phosphoric acid attached to diacylglycerol and is called phosphatidic acid (Table 6.3). Where additional 'X' groups are esterified to the phosphate moiety the lipids are called phosphatidyl-X. Major types of such diacylphosphoglycerides are shown in Table 6.3 where relevant comments about their distribution and properties are made also.

Table 6.2 Lipid composition of different membranes

	Membrane (% total lipid)				
	Chloroplast (spinach)	Protoplast (B.megaterium)	Mitochondrion (rat)	Erythrocyte (rat)	Myelin (rat)
Lipid: protein	1:1	1:3	1:3	1:3	3:1
Phospholipid	12	48	90	61	41
PC	tr	0	40	34	12
PE + PI + PS	tr	19	41	11	26
PG	12	26	—	—	—
DPG	—	3	7	—	—
SPH	—	—	2	16	3
Glycolipid	80	52	—	11	42
MGDG	41				
DGDG	23				
SQDG	16				
Sterol. sterol ester	tr	0	tr	28	17
Acylglycerols	—	—	10	—	—
Pigments	8	—	—	—	—

PC = phosphatidyl choline; PE = phosphatidyl ethanolamine; PS = phosphatidyl serine; PG = phosphatidyl glycerol; SPH = sphingomyelin; MGDG = monogalactosyldiacylglycerol; DGDG = digalactosyldiacylglycerol; SQDG = sulpholipid; DPG = diphosphatidylglycerol.

Table 6.3 Structural variety of different diacyl-glycerophospholipids

X	Name of phospholipid	Source	Remarks
H	Phosphatidic acid	Animals, higher plants, microorganisms	Only minute amounts found. Main importance as a biosynthetic intermediate
$OH \cdot CH_2CH_2N^+(CH_3)_3$ Choline	Phosphatidyl choline (lecithin)	Animals. First isolated from egg yolks. Higher plants. Rare in microorganisms	Most abundant animal phospholipid
$OH \cdot CH_2CH_2NH_3^+$ Ethanolamine	Phosphatidyl-ethanolamine	Animals, higher plants, micro-organisms	Widely distributed and abundant. Major component of old 'cephalin' fraction. N-acetyl derivatives in brain; fatty amides in wheat flour, peas
$OH \cdot CH_2CH \cdot NH_3^+$ $\mid$ COO^- Serine	Phosphatidyl serine	Animals, higher plants, microorganisms	Widely distributed but in small amounts. Minor component of old 'cephalin' fraction. Serine as L isomer. Lipid usually in salt form with K^+, Na^+, Ca^{2+}
myo-Inositol	Phosphatidyl inositol	Animals, higher plants, microorganisms	The natural lipid is found as a derivative of myo-inositol-1-phosphate only
Inositol-4-phosphate	Phosphatidyl inositol phosphate (Diphosphoinositide)	Animals, trace in yeast, plants	Mainly nervous tissue but also plasma membranes of other cells

Inositol-4,5-bisphosphate	Phosphatidyl inositol bisphosphate (Triphosphoinositide)	Animals, trace in yeast, plants	Distribution as above. Both compounds have very high rates of turnover
	Phosphatidyl inositol mannoside x = 0, monomannoside x = 1, dimannoside etc.	Microorganisms (*M. phlei, M. tuberculosis*)	

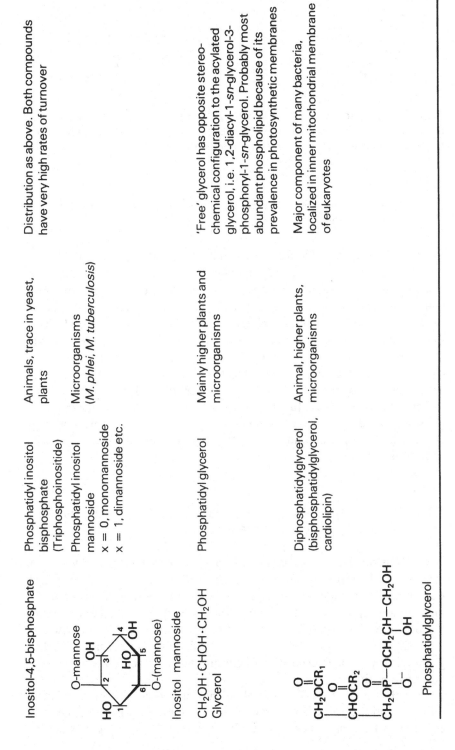

Inositol mannoside

$CH_2OH \cdot CHOH \cdot CH_2OH$
Glycerol

Phosphatidyl glycerol	Mainly higher plants and microorganisms	'Free' glycerol has opposite stereo-chemical configuration to the acylated glycerol, i.e. 1,2-diacyl-1-*sn*-glycerol-3-phosphoryl-1-*sn*-glycerol. Probably most abundant phospholipid because of its prevalence in photosynthetic membranes
Diphosphatidylglycerol (bisphosphatidylglycerol, cardiolipin)	Animal, higher plants, microorganisms	Major component of many bacteria, localized in inner mitochondrial membrane of eukaryotes

Phosphatidylglycerol

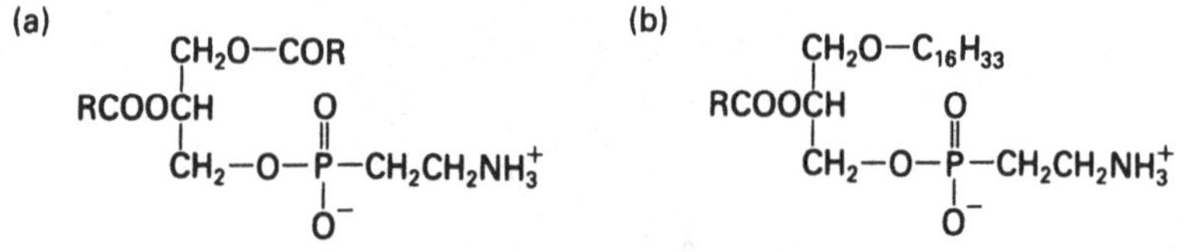

(a)

$$CH_2O-COR$$
$$RCOO\overset{|}{C}H \qquad O$$
$$\overset{|}{C}H_2-O-\overset{\overset{\displaystyle O}{\|}}{\underset{\underset{\displaystyle O^-}{|}}{P}}-CH_2CH_2NH_3^+$$

(b)

$$CH_2O-C_{16}H_{33}$$
$$RCOO\overset{|}{C}H \qquad O$$
$$\overset{|}{C}H_2-O-\overset{\overset{\displaystyle O}{\|}}{\underset{\underset{\displaystyle O^-}{|}}{P}}-CH_2CH_2NH_3^+$$

Figure 6.2 Glycerophosphonolipids, (a) is phosphatidylethylamine while (b) is an analogue which is a derivative of chimyl alcohol rather than a diacylphosphonolipid. It is named 1-hexadecyl-2-acyl-*sn*-glycero(3)-2-phosphonoethylamine.

6.2.2 Phosphonolipids constitute a rare class of lipids found in a few organisms

In 1953, Rouser and his associates first identified a phosphonolipid in biological extracts – from a sea anemone. Work since has shown that such lipids, which contain a C–P bond, are significant constituents of lower animals such as molluscs, coelenterates and protozoa. Phosphonolipids can be detected in bacteria and mammals but only in very low quantities. Two types of glycerophosphonolipids have been found (Figure 6.2) and there are also other structures which contain a sphingosine (and, sometimes, a galactosyl) residue instead of glycerol. Usually, organisms tend to accumulate phosphonolipids based on either glycerol or on sphingosine rather than both types. In protozoa such as *Tetrahymena*, the glycerophosphonolipids are concentrated in the ciliary membranes. This is perhaps because the phosphonolipids are particularly resistant to chemical as well as enzymic attack.

6.2.3 Glycosylglycerides are particularly important components of photosynthetic membranes

Glycolipids which are based on glycerol have been found in a wide variety of organisms. However, whereas in animals they are only found in very small quantities they are major constituents of some microorganisms and are the main lipid components of the photosynthetic membranes of algae (including the blue-greens or cyanobacteria) and plants (Table 6.4). Their structure is analogous to that of glycerophospholipids with the sugar(s) attached glycosidically to the *sn*-3 position of glycerol and fatty acids esterified at the other two positions.

Since the membranes of higher plant chloroplasts are the most prevalent on land, and the photosynthetic membranes of marine algae are the most common in the seas and oceans, it follows that these glycosylglycerides are quantitatively the most important membrane lipids in Nature, in spite of the sparse attention paid to them in most standard biochemistry textbooks. The two galactose-containing lipids – monogalactosyldiacylglycerol and digalactosyldiacylglycerol – represent about 40% of the dry weight of photosynthetic membranes of higher plants. The 1-position of the galactose ring has a β-link to glycerol whereas, in digalactosyldiacylglycerol there is an $\alpha,1 \rightarrow 6$ bond between the sugars. Whereas galactose is almost the only sugar found in the glycosylglycerides of higher plants, other sugars such as glucose may be found in algae, particularly marine species. In bacteria several combinations of residues may be found in diglycosyldiacylglycerols (Table 6.5). The most common combinations are two glucose, two galactose or two mannose residues linked $\alpha,1 \rightarrow 2$ or $\beta,1 \rightarrow 6$. Such glycosylglycerides do not form a large proportion of the total lipids in bacteria but are found more frequently in the Gram-positives or photosynthetic Gram-negatives. In addition, bacteria may contain higher homologues with up to seven sugar residues.

Apart from the galactose-containing lipids, a third glycosylglyceride is found in chloroplasts (Table 6.4). This is the plant sulpholipid. It is more correctly called sulphoquinovosyldiacylglycerol and contains a sulphonate constituent on carbon 6 of a deoxyglucose residue. This sulphonic acid group is very stable and also highly acidic so that the plant sulpholipid is a negatively charged molecule in Nature. Although this sulpholipid occurs in small amounts in photosynthetic bacteria and some fungi, it is really characteristic of the photosynthetic membranes of chloroplasts and cyanobacteria.

All of the three chloroplast glycolipids usually contain large amounts of α-linolenic acid (Table 6.6). In fact, galactosyldiacylglycerol may have up to 97% of its total acyl groups as this one component in some plants. The reason for this exceptional enrichment is not known although speculations have been made (section 8.4). Moreover, a unifying theory connected with photosynthesis is not possible since many cyanobacteria do not make α-linolenate and marine algae contain little of the acid. In the plant sulpholipid the most usual combination of acyl groups is palmitate/α-linolenate. Interestingly, unlike animal lipids (section 6.2.1) most palmitate is esterified at the *sn*-2 position and most of the α-linolenate at the *sn*-1. Since the position at which saturated and unsaturated fatty acids are attached to the glycerol can have a marked effect on their melting points (section 8.2), this may have functional significance. Alternatively, the different distribution of acyl moieties in plants and animals may be related to specific interactions with membrane proteins (sections 6.5.5 and 8.4), or merely because of the specificity of the enzymes which make the lipids (sections 4.4.7 and 7.1.2).

Bacteria contain a number of phosphatidylglycolipids. These compounds are confined to certain types of Gram-positive organisms such as the

Table 6.4 Major glycosylacylglycerols of plant and algal photosynthetic membranes

Common name	Structure and chemical name	Source and fatty acid composition
Monogalactosyl diacylglycerol (MGDG)	1,2-diacyl-[β-D-galactopyranosyl-(1' → 3)]-*sn*-glycerol	Especially abundant in plant leaves and algae; mainly in chloroplast. Contains a high proportion of polyunsaturated fatty acids. *Chlorella vulgaris* MGDG has mainly 18 : 1, 18 : 2 when dark grown but 20% 18 : 3 when grown in the light. *Euglena gracilis* MGDG has 16 : 4. Spinach chloroplast MGDG has 25% 16 : 3, 72% 18 : 3. Also found in the central nervous systems of several animals in small quantity
Digalactosyl diacylglycerol (DGDG)	1,2-diacyl-[α-D-galactopyranosyl-(1' → 6')- β-D-galactopyranosyl-(1' → 3)]-*sn*-glycerol	Usually found together with MGDG in chloroplasts of higher plants and algae. Not quite so abundant as MGDG. Also has high proportion of polyunsaturated fatty acids, especially 18 : 3. In both lipids the glycerol has the same configuration as in the phospholipids. The naming is confusing; many authors still use the D-glycerol system and you will see the galactolipids written: β-D-galactopyranosyl-(1 → 1')-2',3'-diacyl-D-glycerol (MGDG) and α-D-galactopyranosyl-(1 → 6)-β-D-galactopyranosyl-(1 → 1')-2',3'-diacyl-D-glycerol (DGDG)

Plant sulpholipid
(sulphoquinovosyl-
diacylglycerol;
SQDG)

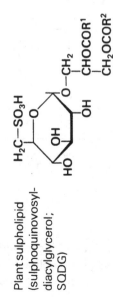

1,2-diacyl-[6-sulpho-α-D-quinovopyranosyl-(1′ → 3)]-sn-glycerol

D-quinovose is
6-deoxy-D-glucose.
Note the carbon-
sulphur bond

Usually referred to as a 'sulpholipid' as
distinct from a 'sulphatide' which is reserved
for cerebroside sulphates. Found in leaves
and algae. Contains more saturated fatty
acids (mainly palmitic) than the galactolipids,
e.g. spinach leaf sulpholipid has: 27% 16 : 0,
39% 18 : 2, 28% 18 : 3. Glycerol has the same
configuration as in the galactolipids and
phospholipids. The alternative name would
be 6-sulpho-α-D-quinovopyranosyl-(1 → 1′)-
2′,3′-diacyl-D-glycerol

Table 6.5 Some glycosylglycerides found in bacteria

Glyceride	Structure of glycoside moiety	Occurrence
Monoglucosyldiacylglycerol	α-D-Glucopyranoside	*Pneumococcus, Mycoplasma*
Diglucosyldiacylglycerol	β-D-Glucopyranosyl-(1 → 6)-*O*-β-D-glucopyranoside	*Staphylococcus*
Diglucosyldiacylglycerol	α-D-Glucopyranosyl-(1 → 2)-*O*-α-D-glucopyranoside	*Mycoplasma, Streptococcus*
Dimannosyldiacylglycerol	α-D-Mannopyranosyl-(1 → 3)-*O*-D-mannopyranoside	*Microccus lysodeikticus*
Galactofuranosyldiacyl-glycerol	β-D-Galactofuranoside	*Mycoplasma, Bacteroides*
Galactosylglucosyldiacyl-glycerol	α-D-Galactopyranosyl-(1 → 2)-*O*-α-D-glucopyranoside	*Lactobacillus*
Glucosylgalactosylglucosyl-diacylglycerol	α-D-Glucopyranosyl-(1 → 6)-*O*-α-D-galactopyranosyl-(1 → 2)-*O*-α-D-glucopyranoside	*Lactobacillus*

Table 6.6 Fatty acid compositions of glycosylglycerides in two plants

Plant leaf		(% total fatty acids)				
		16 : 0	16 : 3	18 : 1	18 : 2	18 : 3
Spinach	MGDG	trace	25	1	2	72
('16 : 3-plant')	DGDG	3	5	2	2	87
	SQDG	39	–	1	7	53
Pea	MGDG	4	–	1	3	90
('18 : 3-plant')	DGDG	9	–	3	7	78
	SQDG	32	–	2	5	58

'16 : 3-plants' contain hexadecatrienoate in their monogalactosyldiacylglycerol while '18 : 3-plants' contain α-linolenate in its stead. The reason for this is provided by the differences in fatty acid metabolism between these two types of plants.
(See Harwood (1989) in the reference list to Chapter 7.)

streptococci or mycoplasmas. In addition, different species of algae and bacteria (including archaebacteria) contain various sulphoglycolipids – usually with the sulphur present in a sulphate ester attached to the carbohydrate moiety.

6.3 SPHINGOLIPIDS

The lipids described in sections 6.2.2 and 6.2.3 were based on a glycerol backbone. However, another important group of acyl lipids have sphingosine-based structures. Both glycolipids and phospholipids are found, with some compounds capable of dual classification, i.e. phospholipids which contain sugar residues.

Sphingosine (D-erythro-2-amino-*trans*-4-octadecane-1,3-diol or 4D-sphingenine) is a long chain amino alcohol (Figure 6.4). Several other sphingosyl-alcohols are also found in naturally-occurring sphingolipids (e.g. phytosphingosine, dihydrosphingosine, Figure 6.4). Both the amino and the alcohol moieties of sphingosine can be substituted to produce the various sphingolipids (Figure 6.3).

Attachment of an acyl group to the amino group yields a ceramide. This acyl link is resistant to alkaline hydrolysis and, therefore, can be easily distinguished from the *O*-esters found in glycerol-based acyl lipids. The simplest glycosphingolipids are the monoglycosyl ceramides or cerebrosides. In animals galactosylcerebroside is the most common. Further attachment of hexosides to glucosylcerebroside yields the neutral glycosylsphingolipids or neutral ceramides (Table 6.7). These are usually written with a shorthand nomenclature, e.g. Glc-Gal-Gal-Glc-Cer would be glucosyl(1 → 4)-galactosyl(1 → 4)galactosyl(1 → 4)glucosyl-ceramide. The deacylated product of galactosylcerebroside, *O*-sphingosyl-galactoside is called psychosine.

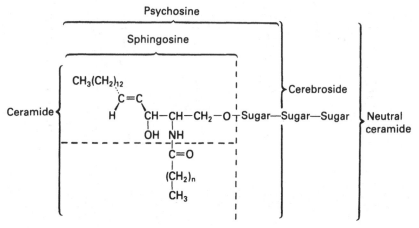

Figure 6.3 Basic structure of sphingolipids.

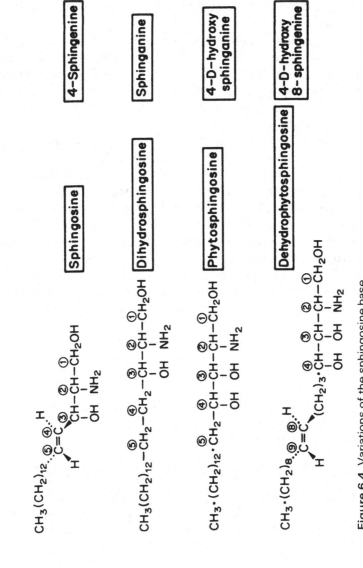

Figure 6.4 Variations of the sphingosine base.

Table 6.7 The structure of some glycosyl ceramides
[Cer = ceramide; glc = glucose; gal = galactose; GalNAc = N-acetyl galactosamine]

Accepted nomenclature	Old trivial name	Structure	Description	General remarks
MONOGLYCOSYL CERAMIDES	Cerebroside	Cer-gal Cer-glc In all glycosyl ceramides and gangliosides there is an O-glycosidic linkage between the primary hydroxyl of sphingosine and C-1 of the sugar	'Cerebroside' originally used for galactosyl ceramide of brain but now widely used for monoglycosyl ceramides. Sugar composition depends largely on tissue. Brain cerebroside mainly galactoside while serum has mainly glucose. In animals the highest concentration is in brain. Monogalactosyl ceramide is largest single component of myelin sheath of nerve. Intermediate concentrations in lung and kidney. Also found in liver, spleen, serum with trace amounts in almost all tissues examined	*Fatty acids:* cerebrosides containing galactose are characterized by having large concentrations of (a) hydroxy acids, (b) long chain odd and even fatty acids, in comparison with other lipids. The hydroxy acids include α-hydroxy acids formed by α-oxidation. This oxidation mechanism is probably also responsible for odd chain acids. Typical acids are: 22 : 0 (behenic); 24 : 0 (lignoceric); 24 : 1 (nervonic); α-OH-24 : 0 (cerebronic)
Sulphatide	Cerebroside sulphate	Cer-gal-3-sulphate	Very generally distributed like cerebrosides. Fatty acid and base composition similar	

Table 6.7 Continued

Accepted nomenclature	Old trivial name	Structure	Description	General remarks
DIGLYCOSYL CERAMIDES	Cytosides			*Base*: C_{18}-sphingosine is dominant. Smaller amounts of C_{18}-dihydrosphingosine (especially leaves, wheat flour) C_{18}-phyto- and dehydrophyto-sphingosine also occur (see also Fig. 6.4)
(a) Lactosyl Ceramide	Cytolipin H	Cer-glc-(4 ← 1)-gal	Major diglycosyl ceramide. Widely distributed. Key substance in glycosyl ceramide metabolism. It may accumulate, but as it is a precursor for both ceramide oligosaccharides or gangliosides it may be present in trace quantities only	
(b) Digalactosyl ceramide		Cer-gal-(4 ← 1)-gal	Minor diglycosyl ceramide. Found especially in kidney (human and mouse)	

TRIGLYCOSYL CERAMIDES

Digalactosyl glucosyl ceramide	Cer-glc-(4←1)-gal (4←1)-gal	Source: kidney, lung, spleen, liver. Most analyses have been on human tissue	In general, each *organ* has a dominant type of glycolipid but its nature may depend also on the *species*, e.g. monoglycosyl ceramides are dominant in brain; trihexosides and aminoglyco-lipids in red cells. Modern analytical techniques are revealing that most types are widespread among tissues. Cerebrosides are widespread but minor components of higher plants, the best characterized being those of bean leaves and wheat flour. Glucose is probably the only sugar. Sphingolipids are rare in microorganisms

TETRAGLYCOSYL CERAMIDES

(a) Aminoglycolipid globoside	Cer-glc-(4←1)-gal-(4←1)-gal-(3←1)-β-N-acetyl galactosamine	Most abundant lipid in human erythrocyte stroma
(b)	Cer-glc-(4←1)-gal-(4←1)-gal-(3←1)-α-N-acetyl galactosamine	So-called 'Forssman Antigen'
(c) Asialoganglioside	Cer-glc-(4←1)-gal-(4←1)-galNAc-(3←1)-gal	Basic ganglioside structure without N-acetyl neuraminic acid (sialic acid) residues. Intermediate in ganglioside biosynthesis (rat, frog, brain)

Table 6.8 Structure of some gangliosides

Class	Structure	Wiegandt shorthand	Svennerholm shorthand	Composition and occurrence
MONOSIALO-GANGLIOSIDE	(a) Cer-glc-(4 ← 1)-gal-(4 ← 1)-galNAc-(3 ← 1)-gal β β β $\begin{pmatrix} 3 \\ \uparrow \\ 2 \end{pmatrix}$ NANA	G$_{GNT}$1	G$_{M1}$	*Major bases:* C_{18} and C_{20}-Sphingosines. Minor amounts of dihydroanalogues *Fatty acids:* Large amounts of 18 : 0 (86–95% in brain) *Occurrence:* Mainly in grey matter of brain but also in spleen, erythrocytes, liver, kidney. Modern analytical techniques have shown them to be present in a much wider range of tissues than
Tay–Sachs ganglioside	(b) Cer-glc-(4 ← 1)-gal-(4 ← 1)-galNAc β β $\begin{pmatrix} 3 \\ \uparrow \\ 2 \end{pmatrix}$ NANA	G$_{GNTrII}$1	G$_{M2}$	
Haematoside	(c) Cer-glc-(4 ← 1)-gal-(3 ← 2)-NANA β β	G$_{Lact}$1	G$_{M3}$	previously realized. Main gangliosides of human brain are G$_{GNT}$1, 2a, 2b, 3a.

DISIALO-GANGLIOSIDE	(a) Cer-glc-$\overset{\beta}{(4\leftarrow1)}$-gal-$\overset{\beta}{(4\leftarrow1)}$-galNAc-$(3\leftarrow1)$-gal $\begin{pmatrix}3\\\uparrow\\2\end{pmatrix}$ $\begin{pmatrix}3\\\uparrow\\2\end{pmatrix}$ NANA NANA	$G_{GNT}2a$ G_{D1a}	Gangliosides appear to be confined to the animal kingdom. In man, cattle, horse, main ganglioside outside brain is $G_{Lact}1$. N-glycolyl-neuraminic acid is chief sialic acid in erythrocyte and spleen ganglioside of horse and cattle. *Physical properties*: Insoluble in non-polar solvents; form micelles in aqueous solution
	(b) Cer-glc-$\overset{\beta}{(4\leftarrow1)}$-gal-$\overset{\beta}{(4\leftarrow1)}$-galNAc-$(3\leftarrow1)$-gal $\begin{pmatrix}3\\\uparrow\\2\end{pmatrix}$ NANA-$(8\leftarrow2)$-NANA	$G_{GNT}2b$ G_{D1b}	
TRISIALO-GANGLIOSIDE	Cer-glc-$\overset{\beta}{(4\leftarrow1)}$-gal-$\overset{\beta}{(4\leftarrow1)}$-galNAc-$(3\leftarrow1)$-gal $\begin{pmatrix}3\\\uparrow\\2\end{pmatrix}$ $\begin{pmatrix}3\\\uparrow\\2\end{pmatrix}$ NANA-$(8\leftarrow2)$-NANA NANA	$G_{GNT}3a$ G_{T1}	

Abbreviations in column 2 are the same as in Table 6.7. NANA = N-Acetylneuraminic acid (sialic acid). Wiegandt abbreviations: G = ganglioside. Subscript denotes sialic-free oligosaccharide. Arabic numeral gives number of sialic residues. T = basic tetraose; Tr = triose; Lact = lactose. Trisaccharides and disaccharides that can be derived from the basic tetraose are distinguished by I, II, etc., I describing the sugar that originated from the non-reducing end of the tetrasaccharide. a, b refer to positional isomers with respect to sialic residues.

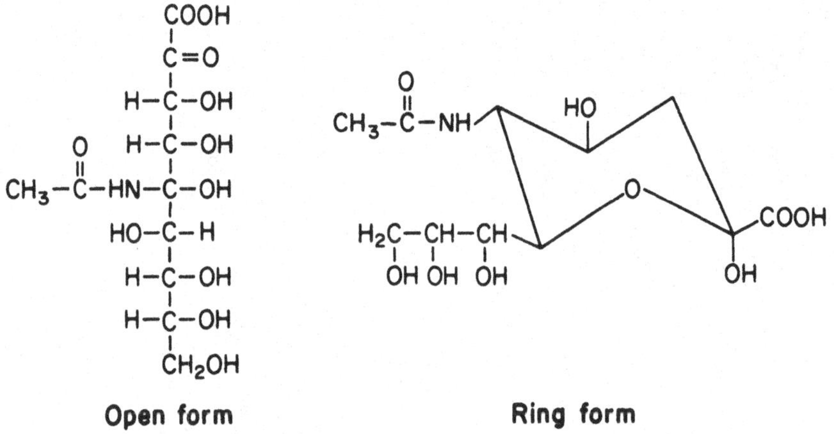

Open form   **Ring form**

Figure 6.5 *N*-acetylneuraminic acid (NeuAc) or sialic acid.

Some glycosphingolipids contain one or more molecules of sialic acid linked to one or more of the sugar residues of a ceramide oligosaccharide. These lipids are called gangliosides (Table 6.8). Sialic acid is *N*-acetyl neuraminic acid (NANA; NeuAc; Figure 6.5). Glycolipids with one molecule of sialic acid are called monosialogangliosides; those with two sialic residues are disialogangliosides and so on. Disialogangliosides may have each sialic moiety linked to a separate sugar residue or both sialic acids may be linked to each other and one of them linked to a central sugar residue. Because of their complex structure and, hence, cumbersome chemical names, many shorthand notations have been employed. One of the most commonly used is that introduced by the Swedish scientist, Svennerholm. The parent molecule is denoted as G_{MI} and other derived structures are shown in Table 6.8. The subscripts M, D, T and Q refer to monosialo-, disialo-, trisialo- and quatra(tetra)sialo-gangliosides respectively. A more recent nomenclature system, due to the German, Wiegandt, is also included in Table 6.8. The latter method has the advantage that, once the symbols have been learnt, the structure can be worked out from the shorthand notation and it can be applied to non-sialic acid containing glycolipids as well. However, it has yet to supersede the Svennerholm notation which will be used in this book.

Instead of glycosylation of the alcohol moiety of the sphingosine base, esterification with phosphocholine can take place. The phospholipid produced thus is called sphingomyelin. Three structures constitute the main bases in animals with dihydrosphingosine and sphingosine being present in the sphingomyelin of most tissues while phytosphingosine is found in kidney. This base is also prominent in phytoglycolipid which is a sphingolipid exclusive to plants which contains inositol as well as several other monosaccharide residues.

Physical techniques have revolutionized the analysis and identification of sphingolipids. In many cases a combination of gas-liquid chromatography, mass spectrometry and nuclear magnetic resonance spectroscopy provide sufficient information to elucidate a complete structure. The position of attachment of the sugar residues can be worked out from the use of glycosidase enzymes or by permethylation analysis. On the other hand the molecular weight and sequence of some quite complex glycosphingolipids can often be established in one step by the use of the gentler techniques of mass spectrometry such as fast-atom bombardment.

As a generalization, sphingolipids have a rather specific distribution in cells where they are concentrated in the outer (external) leaflet of the plasma membrane.

6.4 STEROLS

6.4.1 Major sterols

Cholesterol and the functionally related sterols of fungi and plants are significant components of many organisms especially of their external cellular membranes. However, they are not needed by all types of organisms nor, indeed, are they components (or, therefore, necessary) for all classes of membranes even in organisms with an appreciable sterol content. Sterols are common in eukaryotes but rare in prokaryotes. Whereas vertebrates

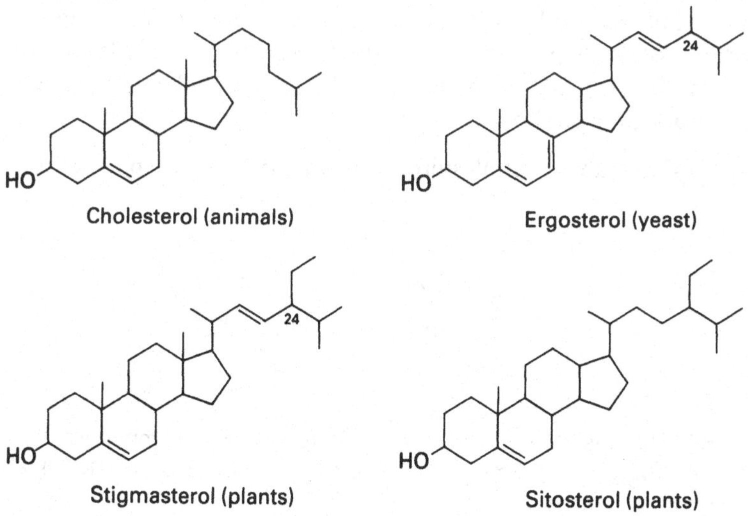

Figure 6.6 Major types of membrane sterols.

synthesize cholesterol, invertebrates rely on an external sterol supply. Yeast and fungi, in the main, have side-chain alkylated compounds while plants and algae contain sitosterol and stigmasterol as their most abundant sterols (Figure 6.6). However, it should be emphasized that, as with most situations in biology, too many generalizations should not be made. For example, cholesterol which in many animal tissues comprises over 95% of the sterol fraction cannot be regarded exclusively as the animal sterol. Thus, marine invertebrates, whether or not they are sterol auxotrophs, discriminate much less than the mammalian intestine in the sterols that they absorb and, therefore, have compositions dependent on the general availability of sterols. Furthermore, whereas plants typically have C_{24}-ethyl sterols such as stigmasterol, some species of red algae contain only cholesterol.

It is believed generally that the main function for membrane sterol is in the modulation of fluidity. The latter is mediated by the interaction of sterol with the glycerolipid components. For that purpose, cholesterol seems to have been optimally designed. Any changes in the structure appear to reduce its effect on membranes (Figure 6.7). Points to note are:

1. All *trans*-fusion of the rings gives a planar molecule capable of interacting on both faces.
2. The 3-hydroxy function permits orientation of cholesterol in the bilayer.
3. In naturally occurring sterols, the side chain is in the $(17\beta, 20R)$ configuration which is thermodynamically preferred and permits maximal interaction with phosphoglycerolipids.
4. Methyl groups at C10 and C18 are retained and provide bridgeheads.
5. An unmodified isooctyl side chain renders the core of the bilayer relatively fluid (c.f. substitutions at C24 in ergosterol or stigmasterol).
6. The tetracyclic ring is uniquely compact and rigid. Other molecules of comparable hydrophobicity are much less restrained conformationally. Because of these properties such sterol molecules are able to separate or laterally displace both the acyl chains and polar head groups of membrane phospholipids.

As mentioned above, cholesterol is, by far, the most important sterol in

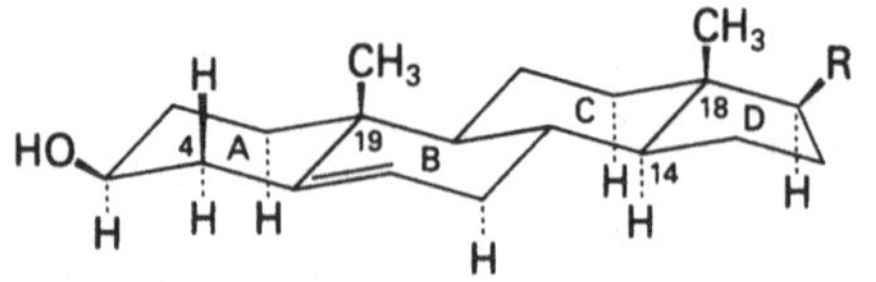

Figure 6.7 The fused ring structure of sterols showing important features. Reproduced with kind permission of Professor K. Block from (1983) *Critical Reviews in Biochemistry*, **17**, 47–92.

mammalian tissues. So far as plants and algae are concerned the major structures are sitosterol (~70%), stigmasterol (~20%), campesterol (~5%) and cholesterol (~5%). Yeasts can accumulate large amounts of sterols (up to 10% of the dry weight) and phycomycetes contain almost exclusively ergosterol. This compound is the major sterol of other yeasts and mushrooms except the rust fungi in which it is absent and replaced by various C29 sterols. With very few exceptions, sterols are absent from bacteria.

The 3-hydroxyl group on ring A can be esterified with a fatty acid. Sterol esters are found in plants as well as animals. Although sterols are present in most mammalian body tissues, the proportion of sterol ester to free sterol varies markedly. For example, blood plasma, especially that of humans, is rich in sterols and like most plasma lipids they are almost entirely found as components of the lipoproteins (section 5.3.5). About 60–80% of this sterol is esterified. In the adrenals, too, where cholesterol is an important precursor of the steroid hormones, over 80% of the sterol is esterified. By contrast, in brain and other nervous tissues, where cholesterol is an important component of myelin, virtually no cholesterol esters are present.

The 3-hydroxyl group can also form a glycosidic link with the 1-position of a hexose sugar (usually glucose). Sterol glycosides are widespread in plants and algae and the 6-position of the hexose can be esterified with a fatty acid to produce an acylated sterol glycoside.

6.4.1 Other steroids

A full description of the various types of steroids, their metabolism and function, is beyond the scope of this book. Cholesterol is the essential precursor for bile acids, corticoids, sex hormones and vitamin D-derived hormones, as has been established well for all vertebrates. Their synthesis is outlined briefly in section 7.5.6.

6.5 MEMBRANE STRUCTURE

6.5.1 Early models already envisaged a bilayer of lipids but were uncertain about the location of the proteins

It has been known for a long time that membranes were composed basically of lipid and protein. The question was how were these constituents arranged in the membrane and how could their special properties be reconciled with their possible physiological functions.

In 1925, the Dutch workers, Gorter and Grendel extracted the lipids from erythrocytes and calculated the area occupied by the lipid from a known

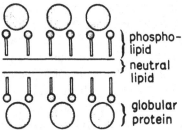

(a) Gorter and Grendel's bimolecular lipid membrane. Circles represent the polar ends of the molecule while the bars represent the long chain hydrocarbon moieties.

(b) The Danielli-Davson Model. Two monolayers of phospholipid are on either side of a layer of neutral lipid of unspecified thickness. The polar ends of the phospholipid molecules are associated with a monolayer of globular protein.

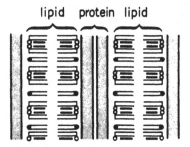

(c) Myelin membrane according to Finean. Lipid bimolecular layers contain phospholipid (⇌) glycolipid (⟹) and cholesterol (—)

An example of a Robertson "Unit Membrane"

Figure 6.8 Early models of membrane structures.

number of cells when it was spread as a monolayer on a *Langmuir Trough*. There was sufficient, they claimed, to surround a red cell in a layer two molecules thick (Figure 6.8a). (Actually these workers made two mistakes which happened to exactly cancel each other out. First, they didn't know how hard to compress the monolayer and hence estimate the area accurately. Secondly, they made no allowance for the contribution of proteins). Nevertheless, the estimates made by Gorter and Grendel turned out to be quite accurate and were accepted at the time to indicate that a bimolecular lipid leaflet surrounded cells. This organization which provides a permeability barrier between various cellular compartments was further refined by Danielli and Davson (Figure 6.8b). They were unaware of Gorter and Grendel's paper and proposed a similar model in 1935 but realized that the measured surface tension of a membrane was too low to be accounted for by lipid and proposed that a layer of protein was present at each surface.

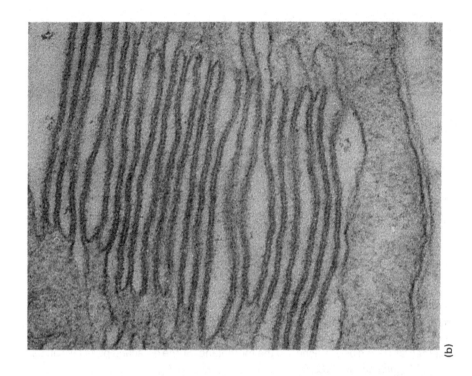

Figure 6.9 *Legend overleaf.*

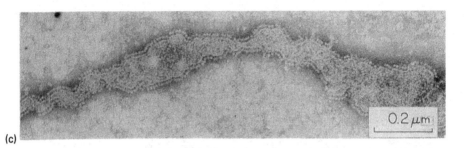

(c)

Figure 6.9 Examples of some different electron microscopic techniques used for examining membrane preparations. (a) An electron micrograph of a 2% dispersion of dioleoyl phosphatidylcholine in water. This is an example of the 'freeze-etching' technique and illustrates the way in which phosphatidylcholine molecules take up a lamellar configuration when dispersed in water. Magnification ×98 000. (b) An electron micrograph of the chloroplast lamellae of a green narcissus petal. The specimen was fixed in glutaraldehyde osmium, embedded in epon and post-stained with lead citrate. This is an example of the 'positive staining' technique and clearly illustrates the typical 'unit membrane' feature of two dark lines separated by a light band. Magnification ×142 500. (c) An electron micrograph of beef heart mitochondria membranes. This is an example of the 'negative staining' technique and illustrates the regular array of globular lipoprotein particles (known as 'elementary particles') attached to the membrane. Some detached particles can be seen. Magnification ×93 000.

With the advent of electron microscopy came pictures of membranes which seemed to support the Danielli model. For the purposes of electron microscopy, the specimen has to be dehydrated, stained (usually with osmium tetroxide or potassium permanganate – so-called positive staining) and embedded in a plastic material such as an epoxy resin. Thin sections are then cut and examined under the microscope. Much of the interpretation of the structures observed depends on how much shrinkage occurs during dehydration of the sample, and what regions of the membrane take up the stain; it is now generally accepted that osmium tetroxide accumulates at the polar regions of lipid and protein. In just about every membrane examined in this way a triple layered structure (two dark lines on either side of a light band, see Figure 6.9(b)), was seen on the micrographs and on this basis Robertson put forward the unit membrane hypothesis which held that every membrane had a basic structure consisting of a bimolecular lipid leaflet sandwiched between a layer of protein one one side and glyco-protein on the other (Figure 6.8c). Further support for the hypothesis came from X-ray studies on myelin by Finean. Myelin is an ideal material for this purpose; it can be easily isolated in a pure state, has a simple composition and regular repeating features which provide good diffraction patterns. Most of the evidence, in fact, supporting the unit membrane hypothesis comes from studies with myelin. To sum up the evidence in favour of this model: (1) the physical-chemical studies which we discussed earlier show that lipids can take up a lamellar configuration; (2) observations of a ubiquitous three-layered pattern in fixed, stained tissue sections under the electron microscope are in keeping with this idea; (3) X-ray diffraction patterns of myelin are also in accord with the model. There are several limitations, however. Examination of Figure 6.8c will show that electrostatic bonding of lipid to protein is implicit in the unit membrane model. Both membrane lipids and membrane proteins are on the whole rather acidic and the weight of evidence at present suggests that hydrophobic bonding is the more important. Furthermore, extraction of more than 90% of the lipid from the mitochondria results in a membrane preparation which has a normal appearance under the electron microscope; this fact has presented something of a puzzle for those looking for a major role for lipids in membranes.

In Table 6.2 we have tried to indicate that from the point of view of composition at least, there is no such thing as a typical membrane. With such widely differing compositions it is unreasonable to expect a universal structure. Myelin, whose structure lent most support to the unit membrane theory, seems to be the most atypical.

6.5.2 The lipid-globular protein mosaic model now represents the best overall picture of membrane structure

Implicit in the bimolecular leaflet model is the idea that the protein is spread as an extended sheet (β-conformation) over the ionic heads of the phospholipids and that the binding between phospholipid and protein is essentially electrostatic. A number of observations about membrane lipids and proteins are not consistent with this model as it stands.

First, the protein component of many membranes represents well over 50% of the bulk of the membrane material. The average amino acid composition of membrane proteins shows no marked preponderance of ionic or hydrophobic residues. Thus, if the protein were extended as a sheet, a significant proportion of the hydrophobic groups would have to be in contact with the water.

Secondly, increasing the ionic strength of the medium surrounding the membrane does not dissociate a large fraction of membrane proteins from lipids as would be expected if electrostatic interactions were predominant.

Thirdly, when the lipid moiety is removed, the proteins are not very soluble in water; in fact they tend to interact hydrophobically with one another.

Lastly, most membrane lipids are zwitterionic rather than having a net charge and would have no strong electrostatic interactions. Those that are charged, tend to be acidic and would react with predominantly basic proteins which are not common in membranes.

Subsequent observations such as the known rapid lateral diffusion of lipid and protein in the plane of the membrane and knowledge that proteins are often inserted into and through the lipid matrix have been added to the points made above. These considerations have been allowed for in the proposal by Singer and Nicholson in 1972 of their fluid mosaic model for membrane structure (Figure 6.10). In the diagram an asymmetric lipid bilayer forms the basis of the membrane structure with proteins spanning the membrane or embedded into the hydrophobic core region.

Two techniques in the electron microscopist's armoury have helped to encourage a re-appraisal of membrane structure. One is negative staining, in which the sample is not fixed and embedded but dispersed in an aqueous solution of the negative stain (phosphotungstate) and dried down on a support film (Figure 6.9c). Strain accumulates in the hydrophilic regions. The other method, drastically different from the staining methods and therefore useful to give independent corroboration, is the freeze-etching technique (Figure 6.9a). These approaches have indicated that certain membranes are composed of globular sub-units. These have been interpreted as lipoprotein particles with lipid hydrocarbon chains and protein apolar side chains in the centre. We should be aware of a profound change of approach of this model, in that we are now dealing with individual lipoprotein particles in contrast to separate and continuous lipid and protein phases. Mitochondria can be fragmented into sub-units – particles having the enzymic activities of part of the electron transport chain – by ultrasonic disruption or by treatment with detergents which loosen the bonds between sub-units. When the detergent's concentration is reduced by dilution or dialysis, the individual sub-units appear to re-aggregate into structures which have all the appearance of membranes under the electron microscope. The sub-units theory gained support not only from new techniques in electron microscopy but from the physical-chemical findings discussed earlier that lipids can adopt phases other than lamellar.

6.5.3 No one model can yet be considered as a true picture of a natural membrane. Transitions between different conformations may occur within a single membrane

The electron microscopist Sjöstrand argues for at least two broad types of

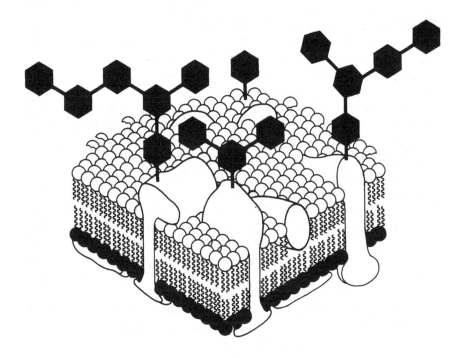

Figure 6.10 The Singer and Nicholson fluid-mosaic proposal for membrane structure. The topography of membrane protein, lipid and carbohydrate in the fluid mosaic model of a typical eukaryotic plasma membrane. Phospholipid asymmetry results in the preferential location of phosphatidylethanolamine and phosphatidyl-serine in the cytosolic monolayer. Carbohydrate moieties on lipids and proteins face the extracellular space. Reproduced with kind permission of Dr P.R. Cullis and the Benjamin/Cummings Company, from *Biochemistry of Lipids and Membranes*, Benjamin/Cummings, Menlo Park, USA.

membranes on grounds of dimensions, ultrastructure and function. The first type would be exemplified by plasma membranes. They are 90–100 Å thick and almost certainly have the bimolecular lipid leaflet structure. These membranes shield the interior of the cell from the outside and impose restrictions on the penetration of ions and molecules into, and out of, the cell. As far as structure is concerned, at least, myelin would belong to this group. To the second type belong the cytoplasmic membranes such as the mitochondria or the endoplasmic reticulum with a thickness of 50–70 Å. They can be considered primarily as metabolic and as the location of the cell's multienzyme systems. Of course, the distinction cannot be pushed to extremes, for the plasma membrane has considerable metabolic activity, while the mitochondrion has to regulate the flow of metabolites between the cytoplasm and the intramitochondrial space.

We should emphasize that some of the models described here represent rather extreme views of membrane structure. In between, there are many different possibilities and the student is encouraged to use the references at the end of the chapter for further details.

The ability of many membrane lipids to form the basic bilayer structure is caused by a number of properties, the most important of which is their **amphipathic** character. Amphipathicity is caused by the lipids having a polar or hydrophilic head group region and a nonpolar or hydrophobic part. Thus, such molecules will naturally orientate themselves to ensure that the polar groups associate with water molecules while the hydrophobic tails interact with each other. Provided the molecule in question is roughly cylindrical in dimension and has no net charge then bimolecular planar leaflets will be the most stable configuration in an aqueous system. However, it is also true that many major lipid components of membranes do not form bilayers when isolated and placed into aqueous systems *in vitro*. Some of the packing characteristics of different lipids and the structures they form are shown in Figure 6.11. Israelachvili, in a masterly exposition in 1980, has discussed the theoretical shapes of these structures and the consequences for cells (see references). In some naturally-occurring membranes, a very high proportion of the lipid is capable of adopting the hexagonal II phase (inverted micelles) rather than bilayers. Thus, many bacteria contain unsaturated phosphatidylethanolamine and diphosphatidylglycerol in large amounts while chloroplast thylakoids contain up to 50% of their lipid as monogalactosyldiacylglycerol. Clearly, such membranes provide good permeability barriers with the vast bulk of the lipid arranged in a bilayer *in vivo*. Therefore, other factors such as the ionic environment, interactions with other lipids and the contributions of membrane proteins are obviously important.

The fluidity of membranes is dependent on the nature of the hydrophobic moieties of lipids. When lipids are isolated the acyl chains of each type undergo transitions from a viscous gel to a fluid state at a certain temperature. Above the transition temperature, the molecules exist in a liquid-crystalline state where the acyl chains, but not the head group, are fluid. This is the normal state of membrane lipids and ensures proper functioning of the membrane proteins. When the temperature of a cell is lowered, three phenomena can contribute to impairment of membrane function. First, ice crystals formed in the aqueous compartments can cause physical damage. Second, transition of the acyl chains into the gel phase leads to changed enzyme activity and altered transport. Third, temperature lowering can lead to the membrane lipids becoming phase separated.

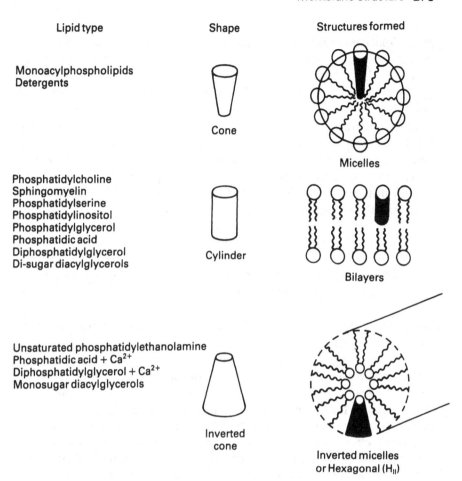

Lipid type	Shape	Structures formed

Monoacylphospholipids
Detergents — Cone — Micelles

Phosphatidylcholine
Sphingomyelin
Phosphatidylserine
Phosphatidylinositol
Phosphatidylglycerol
Phosphatidic acid
Diphosphatidylglycerol
Di-sugar diacylglycerols — Cylinder — Bilayers

Unsaturated phosphatidylethanolamine
Phosphatidic acid + Ca^{2+}
Diphosphatidylglycerol + Ca^{2+}
Monosugar diacylglycerols — Inverted cone — Inverted micelles or Hexagonal (H_{II})

Figure 6.11 Lipid shapes and their packing characteristics.

Concentration of particular lipid types by this phase separation can cause the formation of non-bilayer structures. Lipid adaptations to environmental temperatures are discussed in section 8.2.

6.5.4 Further remarks on the lipid composition of membranes

In section 6.1, we emphasized that different membranes contain quite different lipid compositions. Because such compositions are maintained generally within quite strict limits, one must presume that a membrane's complement of lipids is particularly adapted to its needs. However, the overall lipid composition of a membrane is itself a simplification which needs some qualification. First, individual lipid classes (e.g. phosphatidylcholine) consist of a large number of molecular species where the combinations and positional distribution of acyl chains differ. The nature of these individual molecular species is important for the physical properties, membrane function and further metabolism of the lipids. Second, lipids are not distributed evenly in membranes. For example, all naturally-occurring membranes which have been examined shown sidedness, i.e. the two halves of the bilayer are dissimilar. In addition, lateral segregation has been shown in a few instances or implied from the specific association of certain lipids with particular membrane proteins.

6.5.5 Transbilayer lipid asymmetry is an essential feature of all known biological membranes

The first natural membrane to be examined for asymmetry was that from erythrocytes. Such cells are very convenient for membrane studies since erythrocytes can be easily purified in large quantities and contain only a single membrane. In order to study the two leaflets of such membranes, intact cells were probed (or labelled) in order to establish those components which were in the outer leaflet (i.e. that were accessible). In a second experiment, the erythrocytes were broken by a suitable method (e.g. osmotic shock, sonication) to allow access of the probe to the inner surface as well (Figure 6.12). Control experiments are also necessary to evaluate the total membrane proteins and lipids, since not all of these may be accessible to the probes used.

Various methods have been used to evaluate the lipid composition of inner and outer membrane leaflets. These are listed in Table 6.9. Enzymic digestion is probably the most common method since it is easily carried out and allows analysis of almost all membrane lipids. A drawback with the method, however, is that removal of lipids from the outer leaflet enhances transfer of replacement lipids from the inner leaflet (transbilayer movement). In addition, the products of lipid digestion may be detergents which could themselves alter membrane structure (e.g. lysophospholipids). In some instances, similar methods (i.e. radiolabelling, enzyme digestion, antibody labelling) may be used to analyse membrane proteins and, in addition, if the latter are enzymes, then they can be measured with their substrates.

An important caveat for all experiments to probe membrane sidedness is

(a) Intact erythrocyte

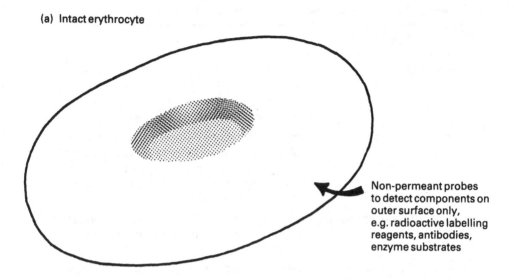

Non-permeant probes
to detect components on
outer surface only,
e.g. radioactive labelling
reagents, antibodies,
enzyme substrates

(b) Erythrocyte ghost

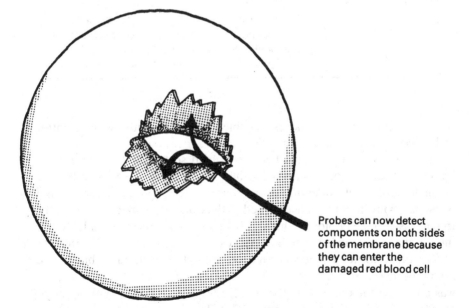

Probes can now detect
components on both sides
of the membrane because
they can enter the
damaged red blood cell

Figure 6.12 Design of an experiment to establish transbilayer composition in erythrocytes (red blood cells).

Table 6.9 Methods used to determine membrane sidedness

Method	Notes
Antibody labelling	Antibodies can be raised to specific proteins or certain lipids. Distribution of antibodies is measured with a radiolabel (I^{125}) tag, a linked enzyme or by immunogold visualization
Enzymic digestion	Specific enzymes are the best to use since large-scale phospholipid or protein digestion can destroy the membrane's structure. Measurement of the decrease in given constituents allows an estimation of their accessibility and, hence, distribution
Chemical (or radio) labelling	Various compounds which react with primary amino groups (phosphatidylethanolamine, phosphatidyl-serine, proteins) most used
Lipid exchange proteins	These allow exchange of specific lipids between the outer leaflet and liposomes in the suspension medium. Thus, the natural membrane composition is retained. Unfortunately, such exchange proteins are few and not widely available
Assay of enzymic function	Only useful for a limited number of proteins
NMR	Signals measured in the absence and presence of paramagnetic ions – these act as chemical shift or broadening agents. Difficult to interpret and not very specific

that the membrane itself is not permeable to the probe, otherwise the latter will, of course, label both sides of the bilayer.

All membranes which have been examined so far show very different lipid (and protein) compositions for their two leaflets. A common feature of mammalian cell membranes is that the amino-phospholipids and phosphatidylinositol are concentrated in the cytosolic leaflet (Figure 6.13). In contrast, sphingolipids are highly concentrated in the external leaflet of the plasma membrane. In the bacterial example also shown in Figure 6.13 phosphatidylinositol is concentrated on the cytoplasmic face, but in the castor bean glyoxysomal membrane, whereas phosphatidylethanolamine was mainly in the outer (cytoplasmic) leaflet, the accessible phosphatidyl-inositol was mainly in the opposite leaflet. This latter example emphasizes two points – first that generalizations should not be made from system to system without firm evidence and, second, that the distribution calculated

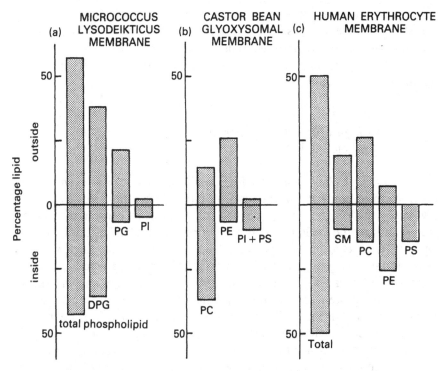

Figure 6.13 Asymmetry of membrane lipids. DPG = diphosphatidylglycerol, PG = phosphatidylglycerol, PI = phosphatidylinositol, PC = phosphatidylcholine, PE = phosphatidylethanolamine, PS = phosphatidylserine, SM = sphingomyelin.

may be biased if a significant amount of the lipid is inaccessible to probes and, therefore, cannot be located.

Nevertheless, lipid asymmetry in membranes raises an important question as to how it arises and also has obvious implications for membrane function. The preservation of asymmetry clearly depends on a very low frequency of exchange from one surface to another (section 6.5.4) or a controlled reciprocal exchange.

An interesting example of the importance of lipid asymmetry has emerged recently. All body cavities in mammalian organisms are lined with epithelia. In most cases (e.g. the gastro-intestinal and urinary tracts) a single layer of cells is involved and these cells have their plasma membrane divided into two parts. An apical portion which borders the lumen is connected to adjacent cells through tight junctions. The remainder of the plasma membrane faces the underlying tissue and is called the basolateral portion. Several studies have shown that the apical portion is significantly less fluid (and, hence, less passively permeable and more mechanically stable) than the basolateral membranes. It turns out that much of rigidity is due to a very high proportion of glycosphingolipids in the outer leaflet. Such lipids are much more highly ordered than phospholipids due to their extensive hydrogen-bonding capacity.

6.5.6 Lateral heterogeneity is probably important in some membranes at least

So far, techniques have not been developed which allow us to examine the distribution of lipids within the plane of most membranes. However, in a few cases the obvious morphological differences in certain regions of membranes can be exploited to allow subfractionation and afterwards lipid analysis. Thus, chloroplast thylakoids can be divided into the single stromal lamellae and the granal stacks or alternatively into appressed and non-appressed membranes (Gounaris *et al.*, 1986). The latter are known to have very different protein compositions and this introduces different charges on to the membrane surfaces. The membrane vesicles deriving from appressed or non-appressed regions can then be separated using two-phase partition systems. Analysis of such membranes showed that, in contrast to the very different protein contents, the lipids are rather similar. The ratio of lipid to protein is, however, very different with non-appressed regions being enriched 3–4 fold in lipid.

In photosynthetic bacteria, where the cytoplasmic membrane becomes infolded to produce photosynthetic membranes (Figure 6.1) the two can be separated. Here, there is a clear difference in lipid composition with the photosynthetic membrane being enriched in phosphatidylglycerol compared to the cytoplasmic membrane. In addition, the carotenoid and bacteriochlorophyll pigments are exclusive to the photosynthetic membrane though this probably reflects their binding to specific proteins rather than a special property of the lipids themselves.

Micro-lateral heterogeneity is present in membranes where particular lipids are found tightly bound to individual proteins. The clearest examples of this are the pigment-protein complexes of photosynthetic organisms. However, certain purified proteins appear to be highly enriched in specific lipids which are very tightly bound and survive solubilization and purification steps in the presence of detergents. Thus, chloroplast ATPase from spinach and the green alga *Dunaliella salina* has been reported to contain highly saturated molecular species of sulphoquinovosyldiacylglycerol. In addition, cytochrome c from mitochondria is a very basic protein which *in vivo* has been suggested to interact with diphosphatidylglycerol. However, in many other cases, what were thought to be specific interactions with proteins turn out to be general properties of lipids which can be replaced by artificial compounds or even detergents while still preserving functional activity. The functions of lipids in providing permeability barriers (sections 6.6 and 8.12), interacting with proteins (section 6.3.5), aiding membrane fusion (section 8.3) and in drug delivery (section 8.5) are discussed later.

6.5.7 Physical examination of membranes reveals their fluid properties

Fluidity in membranes is an easily visualized phenomenon and is a term which is widely used. However, it can be misleading. For example, it is commonly assumed (and stated) that making membrane lipids more saturated or adding cholesterol makes a membrane less fluid. The assumption is that such alterations will reduce the speed of movement of lipids. However, introduction of cholesterol into phosphatidylcholine model membranes has no effect on lateral movement and may actually increase rotational diffusion rates. What cholesterol and increased saturation do is to increase the order in the hydrocarbon matrix and this is what can be measured easily by NMR or ESR order parameters.

Spectroscopic studies have demonstrated that lipid molecules in membranes normally show considerable freedom of movement. Lipids can rotate, bob up and down in the plane of the bilayer, show fluid movement down the hydrocarbon chain and, to a small extent, undergo flip-flop (i.e. movement from one leaflet to the other). Flexing of the hydrocarbon chains occurs fastest with the methyl ends of the chains undergoing the most movement. Rotation of the molecule is also fast (say about every 10^{-7} sec in artificial systems) but flip-flop is much slower (Figure 6.14) and preserves the bilayer asymmetry of natural membranes.

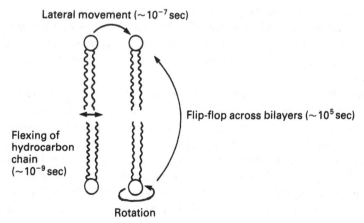

Lateral movement ($\sim 10^{-7}$ sec)

Flip-flop across bilayers ($\sim 10^{5}$ sec)

Flexing of hydrocarbon chain ($\sim 10^{-9}$ sec)

Rotation

Figure 6.14 Lipid movements in membranes.

In addition to lipid movement proteins are also usually able to migrate laterally at rapid rates. This was originally shown following cell fusion where antibody labelling revealed that the proteins from one cell were rapidly found distributed evenly across the entire surface of the fused cell. However, not all membrane proteins exhibit such freedom of movement. For example, the purple patches in the membrane of the photosynthetic bacterium *Halobacterium halobium* have proteins arranged in a regular geometric array where rotational freedom is severely restricted.

6.5.8 Lipid-protein interactions change membrane properties and alter enzyme activities

Some specific examples of the role of lipids in protein function will be given in Chapter 8, but a few general remarks will be made here. Two types of general interaction can be identified. First, those with hydrophobic parts of proteins which penetrate into or through the bilayer. These are interactions with **intrinsic** or **integral** proteins. Secondly, the water-soluble **extrinsic** proteins interact electrostatically with the (negatively) charged groups of the (phospho-) lipid head groups.

Interactions of lipids with a variety of extrinsic proteins have been studied. The proteins which have been studied (e.g. cytochrome c, spectrin) are basic molecules which interact with negatively charged lipids. Even if no negatively charged lipid is tightly bound to these proteins, they induce a time-averaged enrichment of negatively charged lipids in their vicinity. However, this has not been shown to lead to the formation of crystalline domains via local fluidity changes. Indeed there may be an increase in fluidity due to a disturbance of the hydrophobic interactions of the acyl chains in the bilayer. In addition divalent cations compete with basic proteins for the negatively charged head groups. In fact both calcium and basic proteins are able to induce the formation of hexagonal II phases in micelles of diphosphatidylglycerol. This property may explain the ability of cytochrome c to induce non-bilayer structures in diphosphatidylglycerol-containing systems and also to translocate rapidly across bilayers such as the inner mitochondrial membrane.

To remove intrinsic proteins from a membrane it is necessary to solubilize the molecules using detergents which can break the hydrophobic interactions with lipids. The lipids adjacent to the proteins experience a different environment from the bulk lipid of the membrane and this has led to the concept of boundary lipids. ESR (electron spin-resonance) studies showed that lipids residing at the lipid-protein interface exhibited increased order parameters (restricted motion) in the acyl chain region. Recent experiments, however, paint a different picture. First, the lipid-protein interactions in general appear to be relatively non-specific with most isolated integral proteins capable of being reactivated by a whole host of membrane lipids. Secondly, the evidence for a long-lived boundary of lipid is less convincing when further experiments are considered. When NMR (nuclear magnetic resonance) data are used, boundary-bulk lipid exchange appears rapid. This reflects the time scale of NMR ($\sim 10^{-6}$ sec) versus ESR ($\sim 10^{-10}$ sec) measurements.

Although the general concept of long-lived boundary lipids surrounding intrinsic proteins does not appear to be viable at present, there are well documented cases where particular lipids are associated with particular proteins. For example, the purple membranes of *Halobacterium halobium* apart from having a rigidly defined protein structure also contain a unique lipid composition. The lipids appear to be interdigitated between the protein rods and have a composition quite distinct from the rest of the membrane.

6.6 LIPIDS AS COMPONENTS OF THE SURFACE LAYERS OF DIFFERENT ORGANISMS

6.6.1 Cutin, suberin and waxes – the surface coverings of plants

Stems and leaves of plants are covered with a layer of fatty material. The structural component is an hydroxy fatty acid polymer – cutin. Underground parts and wound surfaces are covered in another type of lipid-derived polymeric material – suberin. Both these polymers are associated with or embedded in a complex mixture of lipids imprecisely called waxes. (Strictly speaking a wax is an ester between a long chain fatty acid and a fatty alcohol: section 4.7.)

The main components of plant waxes are listed in Table 6.10. In general, most of the major constituents are non-polar molecules with long hydrocarbon chains. The pathway of synthesis determines whether the final products have odd- or even-numbered carbon chains.

Whereas cutin contains dihydroxy fatty acids as major components, suberin has ω-hydroxy and dicarboxylic fatty acids. Phenolic constituents are present in both (Table 6.11). They are especially prevalent in suberin where they are believed to have an antimicrobial function. The exact composition of these polymers varies with the age and development of a given tissue.

Cell walls of fungi have significant amounts of aliphatic hydrocarbons which are presumed to act as a desiccation barrier – similar to higher plants. In contrast, bacteria do not usually contain large quantities of waxy

Table 6.10 The major components of plant waxes

Compound	Structure	Occurrence
n-Alkanes	$CH_3(CH_2)_nCH_3$	Most plants; major components usually 29C or 31C
iso-Alkanes	$CH_3CH(CH_3)R$	Not as widespread as *n*-alkanes; usually 27C, 29C, 31C and 33C
Ketones	$R_1 \cdot CO \cdot R_2$	Not as common as alkanes; usually 29C and 31C
Secondary alcohols	$R_1 \cdot CH(OH)R_2$	As common as ketones; usually 29C and 31C alcohols
	$CH_3CH(OH)R$	Uncommon in cuticular waxes; more common in suberin waxes; odd-chain length, 9–15C
β-Diketones	$R_1 \cdot COCH_2CO \cdot R_2$	Usually minor, but in some species (e.g. barley) may be major components; mainly 29C, 31C and 33C
Primary alcohols	$R \cdot CH_2OH$	Most plants; even chains predominate, usually 26C and 28C
Acids	$R \cdot COOH$	Very common; even chains predominate, usually 24–28C

Abbreviations: R, R_1 and R_2 are alkyl chains.

Table 6.11 The main compositional differences between cutin and suberin

Monomer	Cutin	Suberin
Dicarboxylic acids	Minor	Major
In-chain-substituted acids	Major	Minor (sometimes substantial)
Phenolics	Low	High
Very long chain (20–26C) acids	Rare and minor	Common and substantial
Very long chain alcohols	Rare and minor	Common and substantial

Source: Kolattukudy, P. E. (1980), in *Biochemistry of Plants*, Vol. 4, P. K. Stumpf and E. E. Conn (eds), 591, New York: Academic Press.

materials in their surface layers. However, mycobacteria contain a number of cell wall lipids, some of which have novel structures. Some related species of bacteria (such as nocardiae or corynebacteria) also contain large amounts of lipid – mainly in the cell wall.

6.6.2 Mycobacteria contain specialized cell wall lipids

The mycobacterial cell wall contains three components. First, there is a skeleton consisting of arabinogalactan mycolate covalently linked through a phosphodiester bond to peptidoglycan. In addition, peptides which can be removed by proteolysis and free lipids are important components.

The skeleton is a branched polymer of D-arabinose and D-galactose in a 5:2 ratio. Every tenth arabinose has a mycolic acid esterified to the 5'-hydroxyl. These mycolic acids are 60–90 carbon fatty acids which are 2-branched or 3-hydroxylated (Figure 6.15). They may also contain other substituted groups such as cyclopropane rings or methoxy groups. The acids are often named after the type of bacteria in which they are found (e.g. nocardomycolic acid from *Nocardia* spp.).

The extractable free lipids are a mixture of cord factors, mycosides, sulpholipids and wax D. Together they represent about 25–30% of the mycobacterial cell wall. Cord factors are esters of the disaccharide, trehalose, with two mycolic acids (Figure 6.16). Cord factor is so called because it is found in the waxy capsular material of virulent strains of tubercle and related bacteria. The factor causes the bacteria to string together in a long chain or cord. The compound is highly toxic and is somehow intimately associated with the virulence of the organism. It interacts strongly with host cell membranes thus impairing their function. Cord factor probably acts as a hapten and binds to albumin in plasma, thus forming an antigen. Like extracts of mycobacterial cell walls, cord factor acts as an immunostimulant. Killed mycobacteria (or their cell walls) are suspended in Freund's adjuvant which is commonly injected with antigens to increase the titre of antibody.

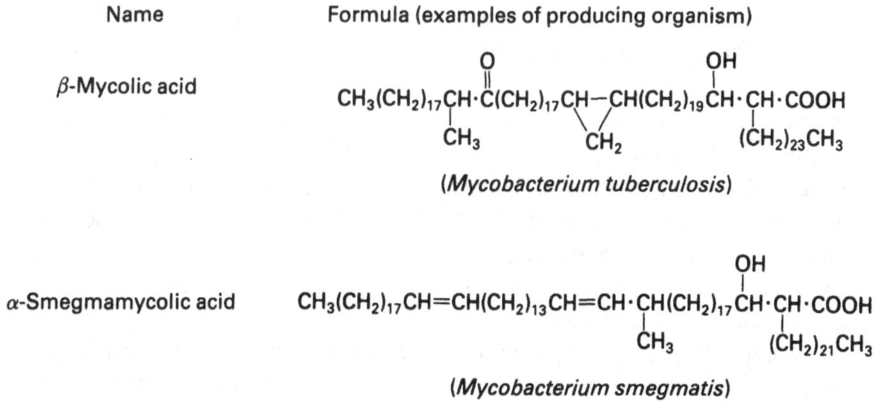

Name	Formula (examples of producing organism)
β-Mycolic acid	$CH_3(CH_2)_{17}CH \cdot C(CH_2)_{17}CH-CH(CH_2)_{19}CH \cdot CH \cdot COOH$ (with O, CH_3, CH_2, $(CH_2)_{23}CH_3$, OH substituents)
	(*Mycobacterium tuberculosis*)
α-Smegmamycolic acid	$CH_3(CH_2)_{17}CH=CH(CH_2)_{13}CH=CH \cdot CH(CH_2)_{17}CH \cdot CH \cdot COOH$ (with CH_3, $(CH_2)_{21}CH_3$, OH substituents)
	(*Mycobacterium smegmatis*)

Figure 6.15 Structure of two mycolic acids.

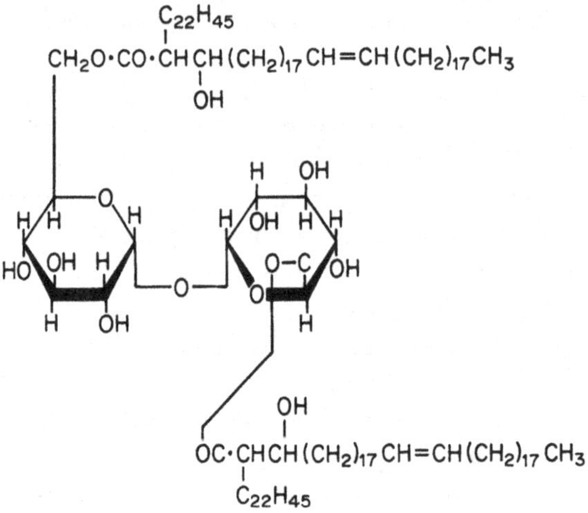
C22H45
CH2O·CO·CHCH(CH2)17CH=CH(CH2)17CH3
OH
H OH
H H H
H OH H H
HO OH H O–C OH
O O
H OH H
OH
OC·CHCH(CH2)17CH=CH(CH2)17CH3
C22H45

Figure 6.16 Cord factor.

The sulpholipids (Figure 6.17) consist of trehalose which is sulphated at the 2-position and acylated at several positions on both sugar residues. Instead of mycolic acids, the sulpholipids have a mixture of palmitic acid and very long chain (31–46 C) fatty acids with up to 10 methyl branches – known as phthioceranic acids.

Mycosides (Figure 6.17) are again characteristic of *Mycobacteria.* The basic structure for mycosides A and B is a long chain, highly branched, hydroxylated hydrocarbon terminated in a phenol group. Acyl groups are esterified to hydroxy groups of the hydrocarbon chain. In contrast, mycoside C is a glycolipid peptide.

6.6.3 Lipopolysaccharide forms a major part of the cell envelope of gram-negative bacteria

In gram-negative bacteria, the wall is far more complex than for gram positives and contains glycopeptide, lipopolysaccharide, phospholipid and protein. Also, there is not such a clear distinction between the wall and the membrane as in gram positives. Up to 20% of the wall contents may be lipids but only a proportion of these are readily extractable by conventional solvent methods. That is because of the covalent nature of the lipopolysaccharide linkages.

The cytosol of gram-negative bacteria is surrounded by a complex cell envelope consisting of at least three layers (Figure 6.18). The cytoplasmic membrane is composed of the familiar phospholipid bilayer with integral

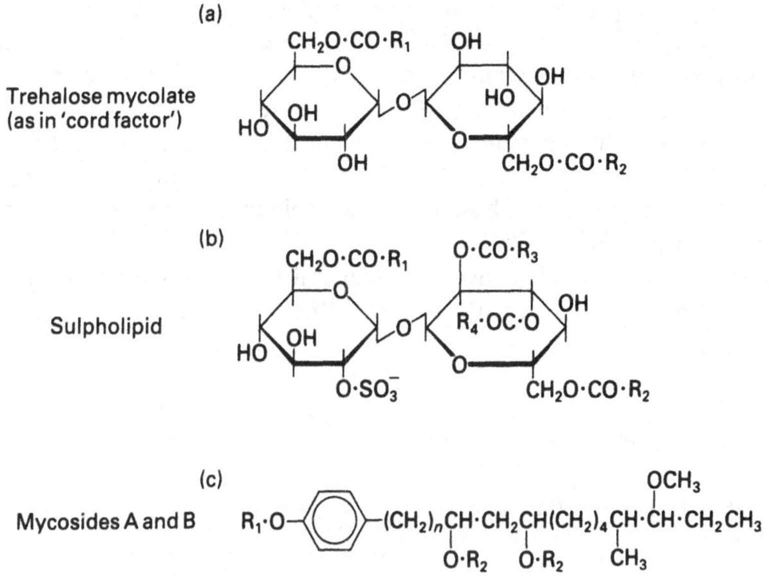

(a)

Trehalose mycolate
(as in 'cord factor')

$CH_2O \cdot CO \cdot R_1$ OH

OH

HO

HO   OH

OH $CH_2O \cdot CO \cdot R_2$

(b)

Sulpholipid

$CH_2O \cdot CO \cdot R_1$ $O \cdot CO \cdot R_3$

OH

HO $R_4 \cdot OC \cdot O$

HO   OH

$O \cdot SO_3^-$ $CH_2O \cdot CO \cdot R_2$

(c)

Mycosides A and B $R_1 \cdot O-\bigcirc-(CH_2)_n CH \cdot CH_2 CH(CH_2)_4 CH \cdot CH \cdot CH_2 CH_3$

$O \cdot R_2$ $O \cdot R_2$ CH_3 OCH_3

Figure 6.17 Some mycobacterial lipids. Structures of mycobacterial lipids: (a) trehalose mycolate, where R_1 and R_2 are mycolic acids; (b) sulpholipid, where R_1–R_4 are palmitic acid or very long chain branched fatty acids; (c) mycosides A and B, where R_1 is a mono- or trisaccharide and R_2 is a 12–18C saturated fatty acid or a mycocerosic acid (e.g. $CH_3(CH_2)_{21}(CH(CH_3)CH_2)_2CH(CH_3)COOH$).

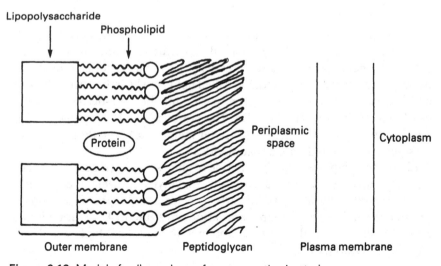

Figure 6.18 Model of cell envelope of gram-negative bacteria.

and peripheral proteins. Then comes a peptidoglycan layer which is separated from the outer membrane by a periplasmic space. The outer membrane is extremely asymmetric but contains lipopolysaccharide and enterobacterial common antigen in its outer leaflet. Phospholipid molecules are missing from the outer membrane in many gram-negative bacteria but are present in some species. In those bacteria which contain phospholipids in the outer membrane, phosphatidylethanolamine is, by far, the most common constituent ($\sim 85\%$).

The lipopolysaccharide is involved in several aspects of pathogenicity. It is a complex polymer in four parts (Figure 6.19). Outermost is a carbohydrate

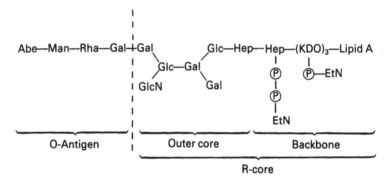

Figure 6.19 Generalized structure of lipopolysaccharide from a gram-negative bacterium.

chain of variable length (called the O-antigen) which is attached to a core polysaccharide. The core polysaccharide is divided into the outer core and the backbone. These two structures vary between bacteria. Finally the backbone is attached to a glycolipid called lipid A. The link between lipid A and the rest of the molecule is usually via a number of 3-deoxy-D-manno-octulosonic acid (KDO) molecules. The presence of KDO is often used as a marker for lipopolysaccharide (or outer membrane) even though it is not present in all bacterial lipopolysaccharides.

Lipid A is composed of a disaccharide of glucosamines (Figure 6.20). The amino groups are substituted with 3-hydroxymyristate while the hydroxyl groups contain saturated (12–16 carbon) acids and 3-myristoxymyristate. Phosphate and KDO groups are also substituted. Unsaturated and cyclopropane fatty acids which are common in other lipid types are absent from lipopolysaccharide.

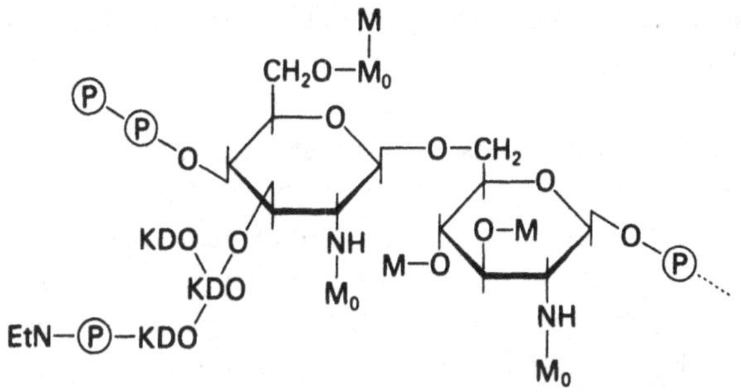

KDO = deoxy-D-mannooctulosonic acid, P = phosphate, EtN = ethanolamine, M = myristate and M_0 = β-hydroxy-myristate.

Figure 6.20 Lipid A.

Studies on the synthesis of lipopolysaccharide have made extensive use of bacterial mutants which are deficient in one or more of the necessary reactions. Rothfield, Horecker and their colleagues in the USA initiated such experiments. It was discovered that formation took place in several stages.

Lipid A, which anchors lipopolysaccharide in the membrane, is made first. Hydroxy acids are added first to the disaccharide, followed by KDO and then saturated fatty acids. The hydroxy fatty acids come from acyl-CoA substrates whereas CMP-KDO is the source of the second addition units. After the addition of saturated fatty acids, sugars are added from nucleotide diphosphate derivatives. Various deficient mutants which lack either glucosyl- or galactosyl-transferase have been isolated. These reactions build one half of the molecule. Another lipid, phosphatidylethanolamine, has been suggested to be intimately involved in the binding of the transferase enzymes to the lipopolysaccharide acceptor.

The O-antigen is made in three stages. The oligosaccharide units are transferred from nucleotide diphosphate carriers to a galactose attached to another lipid carrier; a 55 carbon polyisoprenoid molecule. The oligosaccharide units are then polymerized and lipid carriers are released in the process. Finally, the complete O-antigen is transferred to the R-core, with the release of the final isoprenoid carrier.

Lipopolysaccharide synthesis occurs on the inner membrane from which the molecule must be transferred to the outer face of the outer membrane. This occurs at specific membrane sites generally where the two membranes adhere to each other.

6.6.4 Gram-positive bacteria have a completely different surface structure

The cell walls and membranes of most gram-positive bacteria contain a series of highly anionic polymers. Most important among these is teichoic acid, which is a polymer of glucose 1-phosphate or ribitol phosphate. Membrane teichoic acids are based on glucose 1-phosphate and a proportion may be linked to glycolipids to give lipoteichoic acids. The details of these structures and the amount of substitution of the teichoic acids vary with bacterial species.

Apart from the glycolipid linked to teichoic acid, the only significant lipid in most gram-positive bacteria is that in the cytoplasmic (protoplast) membrane. Notable exceptions are the mycobacteria whose special lipids have been dealt with above.

6.6.5 Lipids of skin – the mammalian surface layer

The skin of mammals contains large amounts of lipids which act as a protective barrier. The terpenoid squalene is a major component along with more usual mammalian lipids. In addition, the sebaceous glands secrete significant quantities of lipids which serve to maintain the skin in a water-proof and flexible condition.

Some remarks on the particular nature of skin lipids are made in section 8.12 where aspects of skin diseases are discussed.

6.7 SUMMARY

A major characteristic of eukaryotic cells is the presence of cellular organelles. These organelles are bound by one or more membranes which serve as a barrier to the free passage of molecules – thus maintaining the unique composition and environment of individual organelles. Not only do individual organelles have a characteristic complement of enzymes (and, therefore, metabolic pathways) they also have membranes made up of a special blend of lipid molecules. In some cases, such as the diphosphatidylglycerol of the inner mitochondrial membrane, these lipid distributions are so specific that they can be used to aid identification of subcellular fractions. In some instances, organelles can be induced or caused to undergo functional development or morphological change. By the use of such systems, the role of lipids in membrane biogenesis and function can be studied.

Most cellular membranes are composed predominantly of proteins and phosphoglycerides. Therefore, phosphoglycerides are significant components of prokaryotes and eukaryotes. In animals and the non-photosynthetic cells of higher plants phosphatidylcholine is the main phosphoglyceride. This molecule is not, however, a major component in bacteria. Phosphatidylethanolamine is a major component in most organisms while phosphatidylglycerol is common in prokaryotes and is the dominant (in many cases exclusive) phosphoglyceride of the photosynthetic membranes of oxygen-evolving species. As well as these molecules, phosphatidylinositol is quantitatively important forming a smaller but significant percentage of the membrane lipids in many cases. Phosphorylated derivatives of phosphatidylinositol are found in the plasma membranes of animals and plants, while diphosphatidylglycerol is exclusively located in the inner mitochondrial membrane.

In contrast to the quantitative importance of the phosphoglycerides, phosphonolipids (which contain a stable carbon–phosphorus linkage) have only been found in significant quantities in a few organisms such as sea urchins.

Glycosylglycerides are the major lipids of the photosynthetic membranes of higher plants, algae and cyanobacteria. Because of the preponderance of such membranes in the world the main glycosylglyceride, mono-galactosyldiacylglycerol, is actually the most prevalent lipid. The other glycosylglycerides found in plants and algae are digalactosyldiacylglycerol and the plant sulpholipid, sulphoquinovosyldiacylglycerol. Different lipids containing various glycosidic residues are found in microorganisms such as *Pneumococcus* or *Mycoplasma* while the galactosylglycerides are minor constituents of mammalian brain.

The glycerolipids are distributed in individual membranes at characteristic ratios. There is also good evidence that biological membranes are asymmetric with regard to their lipid as well as their protein distribution. There may also be some lateral heterogeneity with particular regions of the membrane being enriched with certain lipids.

In contrast to the glycerolipids, sphingolipids are based on a sphingosine backbone. Sphingolipids are only found in any quantity in animals where they tend to be concentrated at the cell surface – in keeping with their function there. Sphingolipids have an acyl residue attached through the amino group of sphingosine while carbohydrates (or phosphorylcholine in the case of sphingomyelin) are linked via the alcohol residue. Depending on the nature of the sugar residues, ceramides, sulphatide and gangliosides are formed. Each cell contains a characteristic pattern of such molecules.

Sterols are based on the cyclopentanoperhydrophenathrene structure which consists of a fused 4-ring system. Three of the rings are six-membered while ring D has five carbons. A side-chain is attached to ring D and sterols contain an hydroxyl function at C-3 of ring A. Cholesterol is, by far, the most common sterol in animals while different sterols occur in other organisms. Thus, ergosterol is important in yeasts while stigmasterol and β-sitosterol are common in plants.

Important steroids include the (pro-)vitamins D, bile acids, and various steroid hormones such as testosterone, oestrogens, progesterones and glucocorticoids. Sterols may also be derivitized to give sterol esters, sterol glycosides and acylated sterol glycosides.

The basic structure of membranes is usually envisaged as the fluid-mosaic model of Singer and Nicholson. In this, amphiphilic lipids are arranged in a bimolecular layer with the hydrophobic moieties in the centre of the membrane and the hydrophilic head-groups at the two surfaces. Stability is achieved by various ionic or hydrophobic interactions between the lipids and the membrane proteins. The latter can be embedded within the membrane (intrinsic proteins) or attached to the surface (extrinsic proteins). Many membrane proteins are large enough to actually span the membrane.

By a combination of enzyme digestion, chemical or antibody labelling or physical techniques such as NMR, it has been shown that all membranes thus examined are clearly asymmetric with regard to their lipids as well as their proteins. This asymmetry is maintained because, although lipids show rapid rotational and translational movement, they only move from one leaflet to the other at very slow speeds.

For a membrane to function effectively, the acyl chains of its lipids must be above their gel–liquid transition temperature. Organisms take care to maintain membrane fluidity by a variety of methods. Membrane fluidity contributes to important properties such as the operation of transport proteins and to the rapid diffusion of hydrophobic electron carriers like quinones.

The surface layers of different organisms often have lipids as significant components. The surface coverings of plants consist of waxes and cutin on the aerial parts while suberin covers the underground tissues or wound surfaces. Very long chain lipids are characteristic of these surface coverings. Dependent on whether the surface layer is wax, cutin or suberin and also, to some extent, the plant tissue in question then the individual lipid composition will vary. Hydrocarbons, very long chain acids and alcohols, their esters and long chain aldehydes and ketones are all found with hydroxylated fatty acids prominent in suberin.

In contrast to plants, microorganisms have rather specialized and, in some cases, unique surface coverings. Mycobacteria contain large amounts of cell wall lipids such as mycosides while gram-negative bacteria contain lipopolysaccharide. The latter is composed of four parts. On the outside of the cell is a polysaccharide of variable structure often called the O-antigen. This is attached to a core polysaccharide in two parts – the backbone of which is linked to lipid A. In contrast, gram-positive bacteria have a completely different surface structure. Lipids are less prevalent although a proportion of the membrane teichoic acids are linked to a glycolipid, such as diacyldiglycosylglycerol, to give lipoteichoic acid.

The skin of mammals contains large amounts of lipids which act as a protective barrier. The terpenoid squalene is a major component along with more usual mammalian lipids. In addition, the sebaceous glands secrete significant quantities of lipids which serve to maintain the skin in a water-proof and flexible condition.

REFERENCES

General – lipids

Gunstone, F.D., Harwood, J.L. and Padley, F.B. (eds) (1986) *The Lipid Handbook*, Chapman and Hall, London.

Harwood, J.L. and Russell, N.J. (1984) *Lipids in Plants and Microbes*, George Allen and Unwin, Hemel Hempstead.

Ratledge, C. and Wilkinson, S.G. (eds) (1988) *Microbial Lipids, Vol. 1*, Academic Press, London.

Stumpf, P.K. and Conn, E.E. (eds) (1980, 1987) *The Biochemistry of Plants, Vol. 4* and *Vol. 9*, Academic Press, New York.

Vance, D.E. and Vance, J.E. (eds) (1985) *Biochemistry of Lipids and Membranes*, Benjamin/Cummings, Menlo Park, CA, USA.

General – membranes

Finian, J.B. and Michell, R.H. (eds) (1981) *Membrane Structure*, Elsevier, Amsterdam.

Gounaris, K., Barber, J. and Harwood, J.L. (1986) The thylakoid membranes of higher plants. *Biochem. J.*, **237**, 313–326.

Harwood, J.L. (1988) Trans-bilayer lipid interactions. *Trends Biochem. Sci.*, **14**, 2–4.

Hoffman, J.F. and Giebisch, G. (eds) (1970-) *Current Topics in Membrane and Transport*, annual series, Academic Press, New York.

Houslay, M.D. and Stanley, K.K. (1982) *Dynamics of Biological Membranes*, Wiley.

Israelachvili, J.N. (1978) The packing of lipids and proteins in membranes, in *Light Transducing Membranes : structure, function and evolution* (ed. D.W. Deamer), Academic Press, New York, pp. 91–107.

Kates, M. and Kuksis, A. (1980) *Membrane Fluidity, Biophysical techniques and cellular recognition*, Humana Press, Clifton, N.J.

McIlhaney, R.N. (1982) The effects of membrane lipids on transport and enzyme activities. *Curr. Top. Memb. Transport*, **17**, 317–380.

Manson, L.A. (1971–) *Biomembranes*, series. Plenum Press, New York.

Prebble, J.N. (1981) *The Membranes of Mitochondria and Chloroplasts*, Longman Group, Harlow, Essex.

Quinn, P.J. (1981) The fluidity of cell membranes and its regulation. *Prog. Biophys. Molec. Biol.*, **38**, 1–104.

Robertson, R.N. (1983) *The Lively Membranes*, U.P., Cambridge.

Glycerolipids

Ansell, G.B. and Hawthorne, J.N. (eds) (1982) *Phospholipids*, Elsevier, Amsterdam.

Joyard, J. and Douce, R. (1987) Galactolipid synthesis, in *Biochemistry of Plants* (eds P.K. Stumpf and E.E. Conn), *Vol. 9*, Academic Press, New York, pp. 215–274.

Harwood, J.L. (1980) Plant acyl lipids: structure, distribution and analysis, in *Biochemistry of Plants* (eds P.K. Stumpf and E.E. Conn), *Vol. 4*, Academic Press, New York, pp. 1–55.

Harwood, J.L. (1980) Sulpholipids, in *Biochemistry of Plants* (eds P.K. Stumpf and E.E. Conn), *Vol. 4*, Academic Press, New York, pp. 301–320.

Mangold, H.K. and Paltauf, F. (1983) *Ether Lipids: biochemical and biomedical aspects*, Academic Press, New York.

Sphingolipids

Kanfer, J.N. and Hakomori, S-I. (1983) *Sphingolipid Biochemistry*, Plenum Press, New York.

Leeden, R.W. and Yu, R.K. (1982) Gangliosides : structure, isolation and analysis, in *Methods of Enzymology* (ed. V. Ginsberg), *Vol. 83*, Academic Press, New York, pp. 139–219.

Makita, A., Handa, S., Taketonii and Nagai, Y. (1982) *New Vistas in Glycolipid Research*, Plenum, New York, pp. 55–69.

Sweeley, C.C. (1985) Sphingolipids, in *Biochemistry of Lipids and Membranes* (eds D.E. Vance and J.E. Vance), Benjamin/Cummings, Menlo Park, CA, USA, pp. 361–403.

Sterols

Bloch, K. (1983) Sterol structure and membrane function. *Crit. Rev. Biochem.,* **17**, 47–92.

Goodwin, T.W. (1980) Biosynthesis of sterols, in *The Biochemistry of Plants* (eds P.K. Stumpf and E.E. Conn), *Vol. 4,* Academic Press, New York, pp. 485–508.

Yeagle, P.L. (1985) Cholesterol and the cell membrane, *Biochim. Biophys. Acta,* **822**, 267–287.

Membrane structure

Anholt, R. (1981) Reconstitution of acetylcholine receptors in model membranes. *Trends Biochem. Sci.,* **6**, 288–291.

Carruthers, A. and Melchior, D.L. (1986) How bilayer lipids affect membrane protein activity. *Trends Biochem. Sci.,* **11**, 331–335.

Cullis, P.R., *et al.* (1983) Structural properties of lipids and their functional roles in biological membranes, in *Membrane Fluidity in Biology* (ed. R.C. Aloia), *Vol.1,* Academic Press, New York, pp. 39–81.

Jain, M.K. and Wagner, R.C. (1980) *Introduction to Biological Membranes,* Wiley, New York.

Jost, P.C. and Griffith, O.H. (eds) (1982) *Lipid-Protein Interaction, Vols 1* and 2, Wiley, New York.

Op den Kamp, J.A.F. (1979) Lipid asymmetry in membranes. *Annual Rev. Biochem.,* **48**, 47–71.

de Kruiff, B. *et al.* (1984) Lipid polymorphism and membrane function, in *Enzymes of Biological Membranes* (ed. A. Martinosi), Plenum, New York, pp. 131–204.

Singer, S.J. and Nicholson, G.L. (1972) The fluid mosaic model of the structure of cell membranes. *Science,* **175**, 720–731.

Storch, J. and Kleinfeld, A.M. (1985) The lipid structure of biological membranes. *Trends Biochem. Sci.,* **10**, 418–421.

Van Meer, G. (1988) How epithelia grease their microvilli. *Trends Biochem. Sci.,* **13**, 242–243.

Surface layers

Goldfine, H. (1982) Lipids of prokaryotes-structure and distribution. *Curr. Top. Memb. Transport,* **17**, 1–43.

Kolattukudy, P.E. (ed.) (1976) *Chemistry and Biochemistry of Natural Waxes,* Am. Elsevier, New York.

Kolattukudy, P.E. (1980) Cutin, suberin and waxes, in *The Biochemistry of Plants* (eds P.K. Stumpf and E.E. Conn), *Vol. 4,* Academic Press, New York, pp. 571–645.

Marks, R. (ed.) (1979) *Investigative Techniques in Dermatology*, Blackwell, Oxford.

Ratledge, C. and Stanford, J. (1982) *The Biology of the Mycobacteria, Vol. 1*, Academic Press, London.

Walton, T.J. (1990) in *Methods in Plant Biochemistry* (eds J.L. Harwood and J. Boyer), *Vol. 4*, Academic Press, London, in press.

7 *Metabolism of structural lipids*

7.1 PHOSPHOGLYCERIDE BIOSYNTHESIS

The first phospholipid to be discovered was phosphatidylcholine, which was demonstrated by Gobley (1847) in egg yolk. The lipid was named lecithin originally after the Greek lekithos (egg yolk). Interest in the chemistry of phospholipids began with the extensive investigations of Thudicum, who isolated and analysed lipids from many animal tissues, particularly the brain, and published his results in *A Treatise on the Chemical Constitution of the Brain* (1884). It began to be appreciated later that the difficulties involved in handling these substances and obtaining a pure product were enormous. Another factor tending to discourage research into phospholipids was the very prevalent but erroneous idea that they were metabolically inert and that once laid down during the initial growth of the tissue their turnover was very slight and hence a purely structural significance was ascribed to them.

7.1.1 Tracer studies revolutionized concepts about phospholipids

The myth that phospholipids were slowly turning-over structural molecules was exploded by the Danish chemist Hevesy who in 1935 demonstrated that the radioactive isotope of phosphorus (^{32}P) could be rapidly incorporated as inorganic orthophosphate into tissue phospholipids. By this time it was also known that the stable isotopes (^{2}H) and (^{15}N) were incorporated into proteins and fats. These studies gave rise to two important concepts. First, molecules (including lipids) in living cells were subject to turnover and were continuously replaced (or parts of them were replaced) by a combination of synthesis and breakdown. Secondly, tracer methods showed the presence of metabolic pools. These are circulating mixtures of chemical substances, in partial or total equilibrium with similar substances derived by release from tissues or absorbed from the diet, which the organism can use for the synthesis of new cell constituents.

That tissues are able to synthesize their own phospholipids could also be inferred from the fact that most of their constituents (choline is one exception) are not essential dietary requirements. However, in spite of the impetus given to biosynthetic studies by the tracer work in the 1930s and 1940s, the major treatise on phospholipids in the early 1950s (Wittcoff's *Phosphatides*) could give virtually no information about their biosynthesis.

7.1.2 Formation of the parent compound, phosphatidate, is demonstrated

The different parts of phosphoglycerides – fatty acids, phosphate and base – are capable of turning over independently. Thus, to study phospholipid synthesis we must first learn the origin of each constituent and then how they are welded together. Eighteen years after Hevesy's demonstration of the rapid rate of phospholipid turnover, came the first real understanding as to how complete phospholipids are built up.

Two American biochemists, Kornberg and Pricer, found that a cell-free enzyme preparation from liver would activate fatty acids by forming their coenzyme A esters. They then went on to demonstrate that these activated fatty acids could be used by an acyl transferase to esterify 3-*sn*-glycerophosphate forming 1,2-diacyl-3-*sn*-glycerophosphate (phosphatidate). We now know that there are two distinct acyl-CoA : glycerolphosphate-*O*-acyltransferases, specific for the *sn*-1 and *sn*-2 positions (section 4.4.1). Furthermore, organisms containing the Type II fatty acid synthetases (section 3.2.2) which form acyl-ACP products use these as substrates for the acyltransferases. Such organisms are bacteria like *E. coli*, cyanobacteria, algae and higher plants. In fact, plant cells produce phosphatidate within their plastids using acyl-ACPs but also contain acyl-CoA: glycerolphosphate acyltransferases on the endoplasmic reticulum for the synthesis of extra-choloroplast lipids such as phosphatidylcholine or triacylglycerol.

Phosphatidate is the parent molecule for all glycerophospholipids and was at first thought not to be a normal constituent of tissue lipids. Later studies have shown it to be widely distributed but in minute amounts. Thus glycerophosphate is one of the building units for phospholipid biosynthesis. It is mainly derived from the glycolytic pathway by reduction of dihydroxyacetone phosphate, though other methods are used to various extents by different organisms or tissues. Likewise, phosphatidate can also be produced by direct phosphorylation of diacylglycerol using diacylglycerol kinase.

7.1.3 A novel cofactor for phospholipid synthesis was found by accident

After the problem of phosphatidic acid biosynthesis had been solved, interest then grew in the pathways to more complex lipids. What was the origin of the group 'X' attached to the phosphate? The first significant finding in connection with phosphatidylcholine biosynthesis was made when Kornberg and Pricer demonstrated that the molecule phosphorylcholine was incorporated intact into the lipid. This they did by incubating phosphorylcholine, labelled both with ^{32}P and ^{14}C in a known proportion, with an active liver preparation and finding that the ratio of ^{32}P to ^{14}C radioactivities remained the same in phosphatidylcholine. Choline may come via a number of pathways having their origins in protein metabolism. For instance the amino acid serine may be decarboxylated to ethanolamine, which may then be methylated to form choline. Phosphorylcholine arises by phosphorylation of choline with ATP in the presence of the enzyme choline kinase. The story of how the base cytidine, familiar to nucleic acid chemists, was found to be involved in complex lipid formation is a good example of how some major advances in science are stumbled upon by accident, although the subsequent exploitation of this finding by the American biochemist, Kennedy, typifies careful scientific investigation at its best. Kennedy proved that cytidine triphosphate (CTP) which was present as a small contaminant in a sample of ATP was the essential cofactor involved in the incorporation of phosphorylcholine into the lipid and later isolated the active form of phosphorylcholine, namely cytidine diphospho-choline (CDP-choline). The adenine analogue had no reactivity.

7.1.4 Further studies have revealed important features of the enzymes concerned in the CDP-base pathways

Some preparations of choline kinase have activity towards ethanolamine and the activities co-purify and cannot be separated. However, in most tissues separate choline and ethanolamine kinases have been found and purified. As with typical kinases, Mg-ATP is the cosubstrate. The other two enzymes involved in the sequence to phosphatidylcholine (Figure 7.1) are a cytidylyltransferase and 1,2-diacylglycerol: choline phosphotransferase. Whereas choline kinase is a soluble enzyme, the cytidylyltransferase is found both in the cytosol and bound to the endoplasmic reticulum. Since the cytidylyltransferase appears to be the rate-limiting enzyme for phosphatidylcholine synthesis in both plants and animals, its regulation is particularly important. There is good evidence that in animal tissues this is

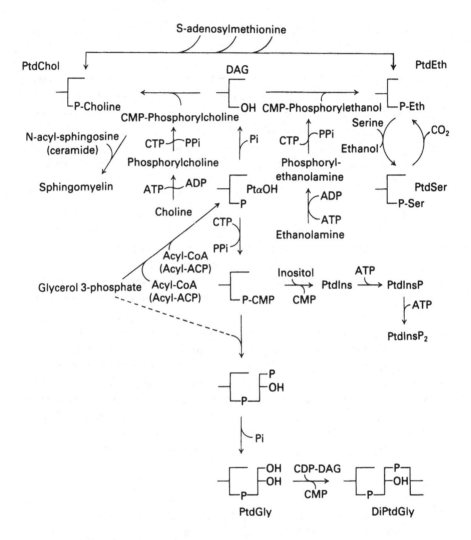

Figure 7.1 The biosynthesis of some important phospholipids.
PtdChol = Phosphatidylcholine; PtdEth = Phosphatidylethanolamine;
PtdSer = Phosphatidylserine; PtdGly = Phosphatidylglycerol;
DiPtdGly = Diphosphatidylglycerol [cardiolipin];
DAG = diacylglycerol; PtdIns = Phosphatidylinositol;
PtdInsP = Phosphatidylinositol 4-phosphate;
PtdInsP$_2$ = Phosphatidylinositol 4,5-*bis*phosphate
PtdOH = Phosphatidic acid

brought about by changing the proportion bound to the endoplasmic reticulum. The membrane-bound enzyme seems to be more active than the soluble and the Canadian biochemist, Vance, and co-workers have shown that binding can be increased 'by the presence of acidic phosphoglycerides, by non-esterified fatty acids and by protein phosphorylation. The CDP-choline produced by cytidylyltransferase is utilized by the final enzyme which releases CMP as a co-product. CTP can be regenerated from CMP using ATP as the phosphate donor. The back reaction for the final step occurs to a significant extent so that, in some tissues, significant amounts of diacylglycerol can be formed from phosphatidylcholine.

The diacylglycerol needed for phosphatidylcholine synthesis is generated from phosphatidic acid by phosphatidate phosphohydrolase. This is a very important enzyme in plants and animals since the diacylglycerol is used for the formation of the major zwitterionic phospholipids, phosphatidylcholine and phosphatidylethanolamine, as well as storage triacylglycerol (section 4.4). The high turnover of phosphatidate in almost every tissue which is exposed to [^{32}P]orthophosphate is a testimony to the rapidity with which it is continually formed and then broken down to yield diacylglycerol. Phosphatidate phosphohydrolase is found in soluble and membrane-bound fractions. In addition, like the cytidylyltransferase referred to above, the proportion of soluble activity can be changed under certain conditions. The soluble activity has to be assayed with membranous phosphatidate substrate while the membrane-bound enzyme is measured with soluble dispersions of phosphatidate. Moreover, Mg^{2+}-dependent and -independent enzymes can often be demonstrated in the same tissue. All these factors make an assessment of the physiological importance of the different forms of phosphatidate phosphohydrolase a matter of some controversy (sections 4.6.2 and 7.2.6).

Phosphatidylethanolamine can also be formed by an analogous series of reactions to phosphatidylcholine; that is by the successive actions of ethanolamine kinase, ethanolamine phosphate cytidylyltransferase and ethanolamine phosphotransferase (Figure 7.1). As in the synthesis of phosphatidylcholine, the activity of the cytidyltransferase appears to be rate limiting for phosphatidylethanolamine synthesis and also the final enzyme, the ethanolamine phosphotransferase, catalyses a freely reversible reaction.

In addition to being involved in the production of phosphatidylcholine, CDP-choline is also used for the formation of the sphingosine-containing phospholipid, sphingomyelin (Figure 7.1). The major pathway for the formation of sphingomyelin transfers phosphorylcholine from CDP-choline to a ceramide (section 7.4.5 and Figure 7.14).

7.1.5 Phosphatidylcholine and phosphatidylethanolamine can be made by other pathways

Both phosphatidylcholine and phosphatidylethanolamine can be made by pathways that are alternative to the CDP-base routes. Indeed, five methods for synthesizing phosphatidylcholine have been described (Table 7.1). In the second method, the primary amino group of phosphatidylethanolamine is methylated three times using S-adenosylmethionine as a donor.

S-adenosylmethionine

In animal and plant tissues this pathway is of much less importance than the one using CDP-choline – in mammals only the liver uses this method to any extent (20–40% of the total phosphatidylcholine synthesized) – but it is the method used in those bacteria which contain phosphatidylcholine. The other three pathways generally have low activity but may be important in special circumstances. For example, the acylation of monoacylphosphatidylcholine is used for the incorporation of specific fatty acids into phosphatidylcholine and may be the main method by which arachidonic acid is acylated at the *sn*-2 position (section 3.4.9). The transacylation (method 5) reaction takes place between two molecules of 1-monoacylphosphatidylcholine and, because saturated fatty acids are esterified at

Table 7.1 Mechanisms for synthesizing phosphatidylcholine

1. Use of CDP-choline intermediate.
2. Step-wise methylation of phosphatidylethanolamine.
3. Ca^{2+}-dependent base exchange reaction.
4. Acylation of monoacylphosphatidylcholine.
5. Transacylation between two molecules of monoacylphosphatidylcholine.

the *sn*-1 position in mammals, this reaction provides one mechanism for generating the disaturated-phosphatidylcholines important in lung function (section 8.13).

Phosphatidylethanolamine can be formed in four ways (Table 7.2). Again, the CDP-ethanolamine pathway is the most active pathway in animals and plants. However, the ethanolamine which is needed for this pathway derives from the decarboxylation of serine and that reaction can only, apparently, take place on a lipid substrate (i.e. phosphatidylserine) in animal tissues. Thus the decarboxylation of phosphatidylserine (method 2) assumes importance not only as a source of phosphatidylethanolamine directly but also to provide ethanolamine. In addition, ethanolamine can be liberated by the catabolism of sphingolipids. The decarboxylation of phosphatidylserine is the method by which bacteria produce phosphatidylethanolamine. The other two pathways (methods 3 and 4) are of minor overall significance but may be used for specialist purposes, e.g. method 4 is used for the formation of molecular species which are poorly formed by the CDP-base pathway.

Table 7.2 Mechanisms for synthesizing phosphatidylethanolamine

1. Use of CDP-ethanolamine intermediate.
2. Decarboxylation of phosphatidylserine.
3. Ca^{2+}-dependent base exchange.
4. Acylation of monoacylphosphatidylethanolamine.

7.1.6 Other phospholipids must be made in significant quantities by certain tissues

Phosphatidylserine has already been mentioned in its connection as a precursor for phosphatidylethanolamine. It is a quantitively minor lipid in animal tissues where it appears to be formed by Ca^{2+}-dependent base-exchange on pre-formed microsomal (probably endoplasmic reticulum) lipids. The enzyme is probably the same as that responsible for the exchange of choline and ethanolamine as well.

$$\text{Phosphatidyl-X} + \text{serine} \underset{\longleftarrow}{\overset{Ca^{2+}}{\rightleftharpoons}} \text{phosphatidylserine} + \text{X} \qquad (7.1)$$

The enzyme has now been purified by Kanfer and, although it was first thought to be a phospholipase, the purified base-exchange enzyme had no such phospholipase activity.

In bacteria, it is well established that phosphatidylserine is synthesized by a reaction involving CDP-diacylglycerol:

$$\text{L-serine} + \text{CDP-diacylglycerol} \rightleftharpoons \text{phosphatidylserine} + \text{CMP} \qquad (7.2)$$

Although this reaction has been detected at low activity in plants and protozoa it does not appear to be present in animals.

Three main types of polyglycerophospholipids have been found in Nature – phosphatidylglycerol, diphosphatidylglycerol and bis(monoacylglycero)-phosphate. In spite of the fact that phosphatidylglycerol is probably the most abundant phospholipid in Nature its generally low levels in animals meant that it was not until 1958 that Benson discovered the lipid (in the alga *Scenedesmus*). CDP-diacylglycerol is used for its synthesis and the pathway occurs in two stages with a phosphatidylglycerophosphate intermediate (Figure 7.1). Because the latter never accumulates, its dephosphorylation is obviously very rapid. In plants and animals the enzymes have a dual subcellular localization: mitochondria and endoplasmic reticulum. In addition, leaves contain significant activity in the chloroplast envelope and can, therefore, generate the phospholipid needed for thylakoids within the organelle. Although phosphatidylglycerophosphate synthetase is a membrane-bound enzyme, it has now been solubilized and purified from such bacteria as *E. coli* and *B. licheniformis*. The phosphatase is also normally membrane-bound but in certain exceptional cases may be soluble. The best example of this was McMurray's discovery that in BHK-21 cell mitochondria phosphatidylglycerophosphate is the main product (92%) as compared with normal liver mitochondria where phosphatidylglycerol is the main product (91%) while the intermediate represented only 7%. Addition of cell supernatant to BHK-21 mitochondria resulted in conversion of phosphatidylglycerophosphate into phosphatidylglycerol and further products.

Diphosphatidylglycerol can be synthesized from phosphatidylglycerol by two mechanisms (Figure 7.2). These pathways have been demonstrated by

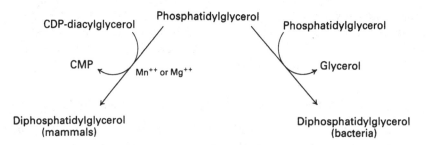

Figure 7.2 The formation of diphosphatidylglycerol (cardiolipin).

appropriate isotope experiments and the mammalian enzyme responsible has been solubilized from mitochondrial membranes.

Phosphatidylglycerol is also the main precursor of bis(monoacylglycero)-phosphate although diphosphatidylglycerol can also function thus. Acyl-phosphatidylglycerol is also found in small amounts in bacteria. Both these derivatives of phosphatidylglycerol have their acyl groups donated from other phospholipids rather than from acyl-CoA.

7.1.7 Comparison of eukaryotic phospholipid synthesis with that in *E. coli*

In order to emphasize the main differences in glycerophospholipid formation in *E. coli* and mammals the major pathways are indicated in Figure 7.3.

7.1.8 The inositol lipids have particularly high turnovers

Although phosphatidylinositol is a significant constituent of most eukaryotic membranes, the phosphorylated derivatives (phosphatidylinositol 4-phosphate; phosphatidylinositol 4,5-*bis*phosphate) are very minor constituents of mammalian cells and even rarer in plant cells. However, ^{32}P-orthophosphate labelling experiments show that these phospholipids

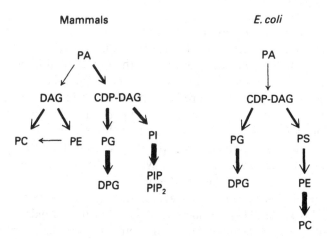

Figure 7.3 Relative importance of pathways for phospholipid formation. Thickness of lines indicate relative importance. For lipid abbreviations see Figure 7.1.

have a very rapid turnover. Phosphatidylinositol itself is made from CDP-diacylglycerol (Figure 7.1) while kinases transfer phosphate from ATP to form the phosphorylated derivatives. The metabolism of these lipids is discussed more fully in connection with their function in the inositide cycle (section 8.8).

7.1.9 Plasmalogen synthesis

Although the pathway for the formation of ethanolamine (and choline) plasmalogens uses CDP-base intermediates, there are several points of difference with those for the synthesis of diacylphospholipids. First, the ethanolamine (choline) phosphotransferase enzyme uses a plasmalogenic diacylglycerol (1-alkenyl-2-acyl-*sn*-glycerol). Hajra and his group showed that the starting point for synthesis was dihydroxyacetone phosphate. This compound was acylated, the acyl group at *sn*-1 substituted by a long chain alcohol and then the keto group at *sn*-2 reduced. The newly created hydroxyl at *sn*-2 could then be acylated and the phosphate group removed to produce a plasmalogenic diacylglycerol (Figure 7.4).

It is interesting to note that the acyldihydroxyacetone phosphate at the start of the pathway can also be reduced to form lysophosphatidate thus providing another source of phosphatidate (section 4.4.1(b)) and a link between acyl- and alkyl-glycerolipid formation.

Further work on the pathway, which particularly involved the laboratories of Snyder in the USA and Paltauf in Austria, demonstrated that the 1-alkyl, 2-acyl-*sn*-glycerol was used by phosphotransferase enzymes to give saturated ether products.

The desaturation of positions 1 and 2 of the alkyl chain to form an alkenyl chain is catalysed by cell free extracts of intestinal epithelial cells, tumour cells and brain. The enzyme is present in the microsomal fraction but the reaction is stimulated by a high molecular weight, heat-labile factor in the soluble cytosol. The fact that a reduced pyridine nucleotide and molecular oxygen are absolute requirements and that the reaction is inhibited by cyanide but not by CO, strongly suggest that this enzyme is very similar to the fatty acyl-CoA desaturase described in section 3.2.4. This provides another interesting example of an enzyme catalysing a modification of a hydrocarbon chain in the intact lipid molecule.

However, choline plasmalogens, at least in mammals, do not appear to be made by this route. There is some evidence that these lipids, which have exceptionally rapid turnover rates in some tissues, are formed from ethanolamine plasmalogens by base exchange.

One of the most exciting discoveries in the ether lipid field has been the discovery of certain acetylated forms of alkylglycerolipids (originally described as platelet activating factor) with potent biological activities. Most important of these is 1-alkyl-2-acetyl-*sn*-glycero-3-phosphocholine whose metabolism and function is detailed next.

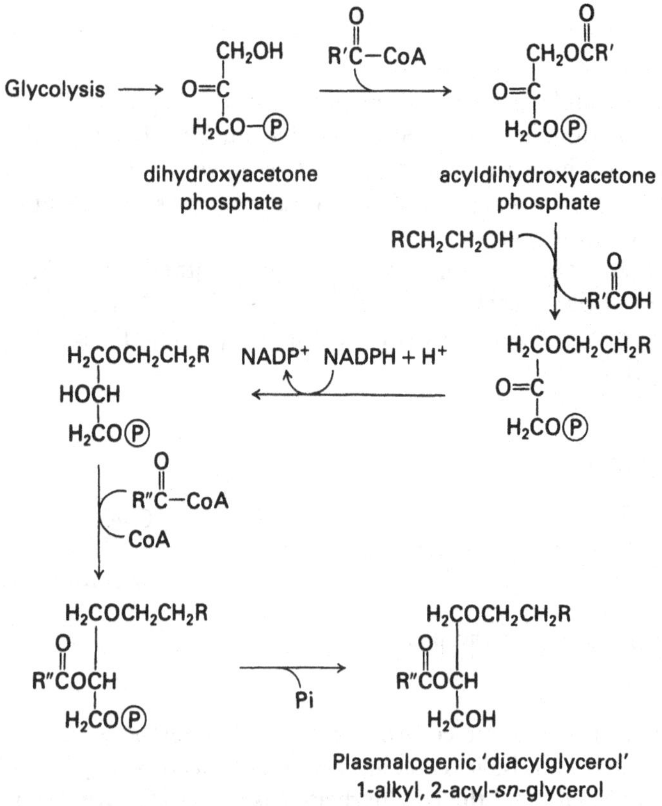

Figure 7.4 Generation of substrate for plasmalogen synthesis.

7.1.10 Platelet activating factor: a biologically active phosphoglyceride

The action of platelet activating factor was first observed when a 'fluid phase mediator' was released from leukocytes to cause platelets to release vasoactive amines. It was identified as 1-*O*-alkyl-2-acetyl-*sn*-glycero-3-phosphocholine by Hanahan's group in Texas in 1980:

$$
\begin{array}{l}
\quad\quad\ \ \overset{\displaystyle O}{\underset{\displaystyle \|}{}}\quad\ CH_2O(CH_2)_nCH_3 \\
CH_3CO-\underset{|}{C}H \quad \overset{\displaystyle O}{\underset{\displaystyle \|}{}} \\
\quad\quad\quad CH_2O\overset{}{P}OCH_2CH_2\overset{+}{N}(CH_3)_3 \\
\quad\quad\quad\quad\ \underset{O^-}{|}
\end{array}
$$

The compound has several unique features: (1) it is the first well documented example of a biologically active phosphoglyceride, (2) it has an ether link at position 1 (fatty acyl derivatives have 1/300th the activity) and (3) an acetyl moiety is present at the 2-position (butyryl substitution lowers activity some 1000-fold).

Platelet activating factor appears to be synthesized mainly by the acetylation of the deacylated (2-lyso) derivatives of choline plasmalogen; the latter being formed by the action of phospholipase A_2 on the plasmalogen:

$$
\begin{array}{l}
\quad\quad CH_2O(CH_2)_nCH_3 \\
HO-\underset{|}{C}H \quad\quad \overset{\displaystyle O}{\underset{\displaystyle \|}{}} \\
\quad\quad CH_2O-\overset{}{P}-OCH_2CH_2\overset{+}{N}(CH_3)_3 \\
\quad\quad\quad\quad \underset{O^-}{|}
\end{array}
\quad + \quad
CH_3\overset{\displaystyle O}{\overset{\displaystyle \|}{C}}-S-CoA \ \rightleftharpoons
$$

Platelet activating factor + CoA—SH

1-*O*-alkyl-2-lyso-glycero-3-phosphocholine

Although there is some controversy as to the metabolic fate of platelet activating factor, it is believed generally that the acetyl group is removed by an acetyl hydrolase and the lyso-derivative is re-acylated (preferentially by arachidonate). The resultant arachidonoyl-containing choline plasmalogen is considered to be a major source of arachidonic acid released during agonist attack on cells (section 3.4.9).

The factor is bound to platelets through a single class of specific receptors (about 250 per cell). Binding sites have also been isolated from other responsive cells.

Platelet activating factor is a very potent molecule having biological activity at 5×10^{-11} M concentrations. Some of its effects are listed in Table 7.3 where it will be seen that this recently discovered molecule has wide-ranging effects in many tissues.

Table 7.3 Effects of platelet activating factors in different cells/tissues

Cell/tissue	Effect	Implications
Platelet	Degranulation, aggregation	Amine release, coronary thrombosis
Neutrophil	Chemotaxis, aggregation, superoxide generation	Antibacterial activity
Alveolar macrophage	Respiratory bursts, superoxide generation	Antibacterial activity
Liver	Inositide turnover, glycogenolysis stimulation	Overall control of activity
Exocrine secretory glands	Similar effects to acetylcholine	Overall control of activity
Leukemic cells	Specific cytotoxicity towards the cells	Treatment?
Vascular permeability	Mimics acute and chronic inflammation	Pathogenesis of psoriasis?
Lungs	Increases airway and pulmonary edema; decreases compliance	Asthma and other respiratory conditions

7.2 DEGRADATION OF PHOSPHOLIPIDS

Phospholipases are classified according to the positions of their attack on the substrate molecule:

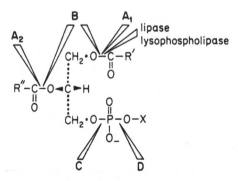

Figure 7.5 Sites of action of phospholipases A, B, C and D.

Phosphopholipase B enzymes differ from phospholipase A_1 and A_2 in that they can hydrolyse acyl groups at both positions, but are relatively rare. All true phospholipases share the same general property of having relatively low activity against monomeric soluble phospholipids but become fully active against aggregated structures, such as phospholipid solutions above their critical micellar concentration or membrane phospholipids in bilayer or hexagonal structures. They play an obvious role as digestive enzymes whether it be in the digestive secretions of mammals (section 5.3.1) or in bacterial secretions. In addition, they are associated with regulatory events in membranes such as the arachidonate cascade (section 3.4).

7.2.1 General features of phospholipase reactions

As mentioned above, phospholipases are distinguished from general esterases by the fact that they interact with interfaces in order to function. The difference in reaction velocity with substrate concentration for these two types of enzymes is illustrated in Figure 7.6. Whereas esterases show classical Michaelis–Menton kinetics, the phospholipases show a sudden increase in activity as the substrate (phospholipid) concentration reaches the critical micellar concentration (CMC) and the molecules tend to form aggregates or micelles with the polar ends in the aqueous environment

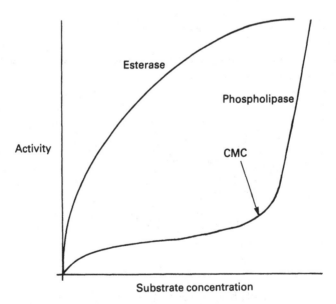

Figure 7.6 Enzyme kinetics for esterases and phospholipases.

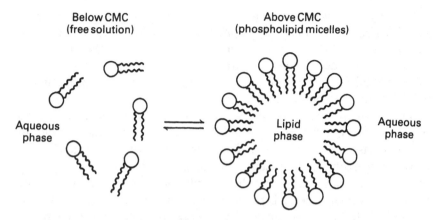

Below CMC
(free solution)

Above CMC
(phospholipid micelles)

Aqueous
phase

Lipid
phase

Aqueous
phase

Figure 7.7 Aggregation of a phospholipid at its critical micelle concentration.

(Figure 7.7). Because of the physical nature of the substrate, enzyme activity is dependent both on the (hydrophobic) interaction with the aggregate and on the formation of a catalytic Michaelis complex. Four factors have been suggested to be responsible for the increased rate of hydrolysis at interfaces:

1. the very high local substrate concentration;
2. substrate orientation at the interface;
3. increased rates of diffusion of products from the enzyme;
4. increased enzyme activity due to conformational changes on binding to the interface.

For a typical phospholipid in solution, formation of a micelle increases its effective concentration by at least three orders of magnitude – depending on the CMC. Factor (3) is particularly important for phospholipases A and B where both products from a typical membrane (natural) phospholipid are hydrophobic. In the case of phospholipases C and D, where one product is water-soluble, action at an interface is still useful because the water-soluble product can move into the aqueous environment while the other product can diffuse into the hydrophobic phase. Similar physicochemical arguments explain why many phospholipases show increased activity in the presence of organic solvents. Thus, when diethylether is used with phospholipase A assays it is thought that accumulating fatty acids are more readily removed. In addition, the solvent may allow more ready access of enzyme to the hydrocarbon chains of the phospholipid and, by reducing micellar size, increases the effective surface area for reaction. Finally, conformational changes (factor 4) have been shown to take place when some digestive phospholipases interact with substrate and/or Ca^{2+}. Kinetic studies indicate that such a conformation change is needed for maximal activity.

The British biochemist Dawson and also Dutch workers including van Deenen, de Haas and Slotboom have studied how the nature of the aggregated lipid influences markedly the activity of phospholipases. Such activity seems to depend on four parameters – surface charge, the molecular packing within the aggregate, the polymorphism of the aggregate and the fluidity of the phospholipid's acyl chains. The surface charge can be quite different from the bulk pH and is influenced by ionic amphipaths as well as ions in the aqueous environment. This explains why, for example, the phospholipase B from *Penicillium notatum* will not attack pure phosphatidylcholine but is active in the presence of activators such as phosphatidylinositol which give phosphatidylcholine micelles a net negative charge. Molecular packing is particularly important for the acylhydrolases but is also relevant to the phosphodiesterases (phospholipase C). Changes in activity with this parameter have been referred to in the example of diethylether activation above. Polymorphic states can include micelles, bilayer structures and hexagonal arrays (sections 6.5 and 8.4). There is some evidence that certain phospholipases attack preferentially hexagonal structures and since these are believed to be formed transiently at sites of membrane fusion (section 8.3), the phospholipase may help to remove fusative lipid and help re-establish the normal membranous bilayer. In addition, it has been well established that gel-phase lipids are attacked preferentially by phospholipases from several sources.

7.2.2 Phospholipase A activity is used to remove fatty acids from intact phospholipids

Early experiments on phospholipase A used an enzyme from snake venom which released one mole of fatty acid from a diacyl phospholipid. The original studies showed quite clearly that unsaturated fatty acids were released from egg or liver phosphatidylcholines and so it was quite natural to assume that the enzyme was specific for unsaturated acids. When fully saturated or fully unsaturated lipids were the substrates, however, one mole of fatty acid was still released from one mole of phospholipid and it was realized that the enzyme must be specific for one or other of the two positions. Early chemical analyses of the products of phospholipase A hydrolysis seemed to indicate that the enzyme released the fatty acid from the 1-position. When it was discovered that the enzyme also liberates fatty acid from plasmalogen (where it is known quite definitely to be located at the 2-position) the problem was reinvestigated more rigorously.

The final, unequivocal proof of the 2-position specificity of this enzyme demonstrates one of the ways in which first-class organic chemists can advance knowledge of biosynthetic mechanisms in the lipid field, just as the work of chemists such as the American Khorana and the Israeli Katchalski have done for nucleic acid and protein biosynthesis respectively. The Dutch chemists de Haas and van Deenen prepared 'mixed acid' phosphatidyl-cholines and ethanolamines by elegant new synthetic techniques so that the 1- and 2-position fatty acids were specified exactly. Hydrolysis of these lipids with phospholipase A could then demonstrate a 2-position specificity directly. Studies with synthetic analogues of the substrate were able to pinpoint the precise structural and stereochemical features necessary for the enzyme to hydrolyse a fatty ester bond. The lipids of the 2,3-diacyl-1-*sn*-glycerol phosphoryl-X series were not hydrolysed, but from 1,3-diacyl-2-*sn*-glycerophospholipids only the 1-position fatty acid was hydrolysed. The enzyme seems to require the presence of only one fatty ester linkage adjacent to the alcohol–phosphate bond and the carbon atom to which this fatty acid is attached must have a precise stereochemical configuration.

Thus, the original work established that the enzyme from *Naja naja* (Indian cobra) venom was a phospholipase A_2. The situation then turned full circle by the discovery of another enzyme in rat liver which was a phospholipase A_1.

Phospholipases A_1 are widely distributed in Nature. In addition, some triacylglycerol lipases will also hydrolyse the *sn*-1 position of phospholipids. Usually, phospholipases A_1 have a broad specificity and act well on lysophospholipids. The first example to be purified was from *E. coli*, which actually has two separate enzymes – a detergent-resistant enzyme in the outer membrane and a detergent-sensitive enzyme in the cytoplasmic membrane and soluble fractions.

In animals, phospholipase A_1 is present in lysosomes and has an appropriately low pH optimum (about 4.0). It does not need Ca^{2+} for activity but Ca^{2+} and charged amphipaths influence hydrolysis by altering the surface charge on the substrate micelle or membrane.

Also present in animals are two enzymes which, while preferentially hydrolysing triacylglycerols, will also hydrolyse the *sn*-1 position of phospholipids. These are the extrahepatic lipoprotein lipase (section 5.3.5(g)) and the hepatic lipase.

Phospholipases A_2 are ubiquitous in Nature, but most emphasis has been placed on studies with the snake venom and pancreatic enzymes, since these are easily purified in large amounts.

Pancreatic phospholipases are synthesized as zymogens that are activated by the cleavage of a heptapeptide by trypsin. Cleavage of this peptide exposes an hydrophobic sequence which then allows interaction of the enzyme with phospholipid substrates. The enzyme is very stable and its seven disulphide bonds no doubt play a key role here. Chemical modification and NMR studies have shown clearly that the catalytic and binding sites are distinct in both the pancreatic and snake venom phospholipases A_2. In confirmation, it is known that the zymogen form of the pancreatic enzyme is also active even though its binding site is masked. Ca^{2+} is absolutely required for activity and seems to interact with both the phosphate and carbamyl groups of the ester undergoing hydrolysis.

The sequences of pancreatic phospholipase A_2 and the enzymes from the venoms of Elapids (cobras, kraits and mambas) are closely related in contrast to the enzymes from vipers (e.g. *Crotalus* (rattlesnakes)). Interestingly, the former enzymes seem to act as monomers while the latter functions as a dimer.

Phospholipase A_2 enzymes also have other important metabolic functions in addition to the overall destruction of phospholipids as catalysed by digestive pancreatic or venom enzymes. An enzyme in mitochondrial membranes seems to be intimately connected with the energy state of this organelle. Thus, the phospholipase is inactive in fully coupled mitochondria and only becomes active when ATP and respiratory control drop to low levels. Also, the widespread distribution of phospholipases A_2 allows many tissues to perform retailoring of the molecular species of membrane lipids by the Lands mechanism. In this process, named after Lands, the American biochemist who first described it, cleavage of the acyl group from the *sn*-2 position yields a lysophospholipid which can be re-acylated with a new fatty acid from acyl-CoA (Figure 7.8).

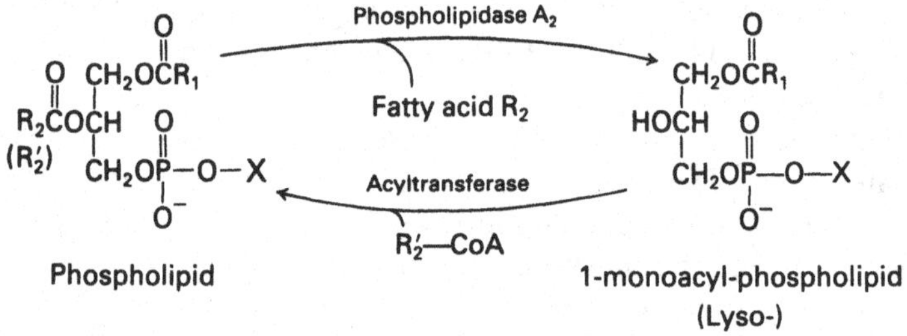

Figure 7.8 The 'Lands Mechanism' for substitutions of the acyl species at the *sn*-2 position of phospholipids.

Finally, phospholipase A_2 plays a key role in the release of arachidonate from certain phospholipids to initiate the arachidonate cascade (section 3.4).

7.2.3 Phospholipase B

Only one phospholipase B has been highly purified – that from *Penicillium notatum*. This enzyme will, of course, also hydrolyse lysophospholipids and, indeed, under most conditions has much greater activity towards the latter. In the presence of detergents (Triton X-100 is usually used) it shows roughly equal activity towards diacyl- and monoacylphospholipids. The enzyme appears to have two distinct binding sites and the intermediate lyso-phospholipid must move from one to the other during hydrolysis.

In contrast to phospholipase B, lysophospholipases are widely distributed and are found in microorganisms, bee venoms and mammalian tissues. Many of them have limited activity towards intact phospholipids. In beef liver, two distinct lysophospholipases have been found – localized in the mitochondrial and microsomal fractions.

The function(s) of lysophospholipases is unclear. It has been suggested that in mammalian tissues they help to prevent the build-up of lytic lysophospholipids.

7.2.4 Phospholipases C and D remove water-soluble moieties

In accordance with the observations showing the independent turnover of phosphate and base moieties, other enzymes named phospholipases C and D are known to hydrolyse each of the two phosphate links (Figure 7.5). Phospholipases C have been traditionally associated with secretions from certain pathological bacteria. Thus, much of the damage caused by *Clostridium welcheii* when giving rise to gas gangrene, is caused by phospholipase C in its toxins. The enzymes from *Bacillus cereus* are the best studied – three different proteins having been isolated, one having a broad specificity, one being a sphingomyelinase and one attacking phosphatidylinositol specifically.

In mammalian tissues, phospholipases C which attack inositol lipids are very important since they initiate the phosphatidylinositol cycle (section 8.8). In addition, broad-specificity phospholipases C are located in lysosomes.

Plants are the main source for phospholipase D although an enzyme catabolizing ether lipids has been found in brain and other tissues. The plant enzymes have been purified from several tissues such as cabbage or carrots. In addition to catalysing hydrolysis of phospholipids the enzyme will cause phosphatidate exchange. This allows the formation of new phospholipids in the presence of an appropriate alcohol. For example:

$$\text{Phosphatidylcholine + glycerol} \rightleftharpoons \text{phosphatidylglycerol + choline} \quad (7.3)$$

Such **transphosphatidylation** reactions were first noticed because phosphatidylmethanol was formed when plant tissues were extracted with methanol. However, transphosphatidylation does not seem to be a physiological function for plant phospholipase D because the phosphatidylglycerol produced in the above reaction is a racemic mixture rather than having the two glycerols in the opposite configurations as occurs naturally (Table 6.3).

7.2.5 Phospholipids may also be catabolized by non-specific enzymes

Not all phospholipid hydrolysis is catalysed by phospholipases. The action of two triacylglycerol lipases in this regard has already been mentioned. In addition, acyl hydrolases are present, particularly in plant leaves. These enzymes have activity towards a variety of lipids, including partial glycerides, glycosylglycerides and phosphoglycerides. The leaf acyl hydrolases have very high activities and they are rather resistant to denaturation. Failure to inactivate the enzymes leads to spoilage of vegetables during freezing.

7.2.6 The hydrolytic enzyme phosphatidate phosphohydrolase is more important in synthesis than in degradation of phospholipids

Phosphatidate lies at a branch-point in glycerolipid metabolism (Figure 7.9). It can be hydrolysed by phospholipases A to glycerolphosphate, converted to CDP-diacylglycerol for acidic phospholipid synthesis or attacked by phosphatidate phosphohydrolase to yield diacylglycerol. The latter is needed for the formation of the important membrane phospholipids, phosphatidylcholine and phosphatidylethanolamine and for triacylglycerol biosynthesis. In terms of the overall rates of carbon flux for lipid synthesis, phosphatidate phosphohydrolase is thought to be a key enzyme in many tissues including plants and animals. Enzyme activity has been reported to occur in many different membrane fractions and in the cytosol. However, the picture is probably complicated by difficulties in assaying the enzyme specifically. Thus many assays follow the release of phosphate (or water-soluble phosphates) from phosphatidate and such compounds can also be formed by alkaline phosphatase (and phospholipase A) action. Phosphatidate phosphohydrolase usually needs Mg^{2+} for activity but, instead of this cation being needed for specific substrate binding, it seems to be required to adjust the surface charge and packing arrangement of membrane phosphatidate. Control of phosphatidate hydrolase activity is discussed in section 4.6.2.

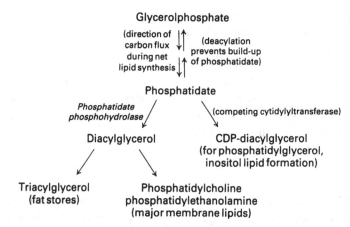

Figure 7.9 The key position of phosphatidate in glycerolipid metabolism.

7.3 METABOLISM OF GLYCOSYLGLYCERIDES

7.3.1 Biosynthesis of galactosylglycerides takes place in chloroplast envelopes

Since the galactosylglycerides are confined (almost exclusively) to chloroplasts, it would seem natural to seek an active enzyme preparation from these organelles. Firstly, it was shown that carefully prepared isolated chloroplasts could incorporate ^{14}C-galactose into the lipids. The nature of the substrates involved was not clear until Ongun and Mudd in California showed that an acetone powder of spinach chloroplasts would catalyse the incorporation of galactose from UDP-galactose into monogalactosyl-diacylglycerol if the acetone-extracted lipids were added back. The acceptor proved to be diacylglycerol or, for the synthesis of digalactosyldiacyl-glycerol, a monogalactosyldiacylglycerol acceptor was needed. Further work on the enzymes involved was carried out by Joyard and Douce in Grenoble who showed that the reactions were confined to the chloroplast envelope – hence the need for carefully-prepared chloroplasts.

The reactions are, therefore:

1,2-Diacylglycerol + UDP-galactose ⇌

$$\text{Monogalactosyldiacylglycerol + UDP} \qquad (7.4)$$

Monogalactosyldiacylglycerol + UDP-galactose ⇌

$$\text{Digalactosyldiacylglycerol + UDP} \qquad (7.5)$$

The UDP-galactose is produced in the cytosol. In contrast, the diacylglycerol is synthesized by a phosphohydrolase localized in the envelope. The two galactosyltransferases have slight differences in their enzymatic characteristics and, of course, result in the formation of β- and α-glycosidic bonds.

More recently, in Wintermans' laboratory in Holland, it was noticed that isolated chloroplast envelopes were capable of forming digalactosyldiacyl-glycerol from labelled monogalactosyldiacylglycerol by a reaction that did not require UDP-galactose. Further examination revealed that inter-lipid galactosyltransfer was involved thus:

2 Monogalactosyldiacylglycerol $\rightleftharpoons$

$$\text{Digalactosyldiacylglycerol} + \text{diacylglycerol} \qquad (7.6)$$

This enzyme is also capable of generating higher homologues by further transfer of a galactose from monogalactosyldiacylglycerol. The trigalactosyl- and tetragalactosyldiacylglycerols thus formed are detected in small quantities in many isolated chloroplasts. It is, however, unclear at present whether such higher homologues are actually present *in vivo* or are artefacts of the isolation and analytical processes. The relative rate of formation of digalactosyldiacylglycerol by the two pathways is also an area of controversy.

One difference in the fatty acyl compositions of galactosylglycerides which has been noticed repeatedly in the analysis of plants is the presence of hexadecatrienoate (16:3) in monogalactosyldiacylglycerol (but not digalactosyldiacylglycerol) from certain plants. The presence of 16:3 in some plants seems to be related to the provision of palmitate at the *sn*-2 position of monogalactosyldiacylglycerol where it acts as a substrate for fatty acid desaturases. In such plants the diacylglycerol for galactolipid biosynthesis is generated by phosphatidate phosphohydrolase within the chloroplast. In contrast, the source of diacylglycerol for galactolipid biosynthesis in other plants comes from outside the chloroplast and does not contain 16-carbon acids at the *sn*-2 position. A glance at Table 7.4 will also reveal that even in plants, like spinach, which do contain hexadecatrienoate in their monoglactosyldiacylglycerol, little is present in the digalactosyl-derivative. It is presumed that this is due to substrate specificity of the second galactosyltransferase.

7.3.2 Catabolism of glycosylglycerides

Enzymes are present in higher plant tissues which rapidly degrade glycosyl-glycerides. The initial attack is by an acyl hydrolase (section 7.2.5) which removes acyl groups from both positions and has very high activity in some tissues such as runner bean leaves or potato tubers. Indeed, homogenization of the latter at a suitable pH in aqueous media results in the complete

Table 7.4 Fatty acid composition of galactosylacylglycerols from *E. gracilis* cells or spinach chloroplasts

Lipid		Fatty acid (% total)						
E. gracilis		16 : 0	16 : 3	16 : 4	18 : 1	18 : 2	18 : 3	Other
	MGDG	6	–	32	9	6	41	6
	DGDG	17	–	7	19	12	26	19
Spinach								
	MGDG	trace	25	–	1	2	72	trace
	DGDG	3	5	–	2	2	87	1

MGDG = monogalactosyldiacylglycerol; DGDG = digalactosyldiacylglycerol.

breakdown of all membrane lipids within a minute! The activity of acyl hydrolases (and other lipid degradative enzymes) in many vegetables even at low temperatures makes it necessary to blanch (boil) such products before storage in a deep-freeze.

Several acyl hydrolases, with slightly different specificities, have been purified from various plant tissues. The enzymes from runner bean leaves are remarkably stable to heating (10% activity lost after 30 min at 70°C) and solvents and can be conveniently purified using hydrophobic chromatography. Further breakdown of the galactosylglycerides occurs by the action of α- and β-galactosidases.

7.3.3 Relatively little is known of the metabolism of the plant sulpholipid

The plant sulpholipid, sulphoquinovosyldiacylglycerol, is also rapidly broken down by the acyl hydrolases mentioned above. It is not known if plants can metabolize the sulphoquinovose moiety itself although it is known to be catabolized by soil microorganisms.

It is the biosynthesis of sulpholipid which is such a mystery. In fact it has been remarked that there are more ideas about the possible synthetic pathways than there are experimental facts! Several experiments have been conducted to see if there was evidence for the sulphoquinovose moiety being produced by a sulphoglycolytic pathway starting from adenosine phosphosulphate (APS) or phosphoadenosine phosphosulphate (PAPS) (Figure 7.10) with glycolytic enzymes catalysing some of the reactions. In some tissues the results are in favour of this mechanism but as many contra-indications have also been found. Regardless of how the sulphoquinovose is

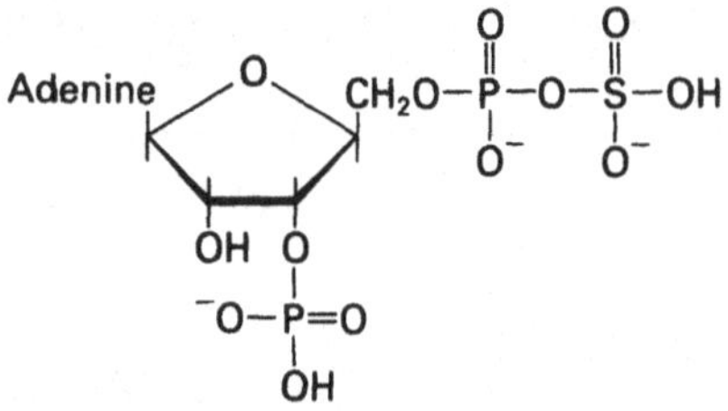

3'-phosphoadenosine-5'-phosphosulphate

Figure 7.10 Phosphoadenosine phosphosulphate: the donor of sulphate groups.

produced, it has been shown to be transferred via a UDP-derivative to diacylglycerol in the final stage. All the enzymes necessary for sulpholipid synthesis seem to be present in chloroplasts.

7.4 METABOLISM OF SPHINGOLIPIDS

7.4.1 Biosynthesis of the sphingosine base

The pathway by which sphingosine is formed was elucidated first by the German biochemist Stoffel in Köln. He demonstrated the pathway by isolating intermediates and thus disproving a previously accepted alternative scheme. The first intermediate is 3-ketodihydrosphingosine, which is formed by the condensation of palmitoyl-CoA with serine in the presence of pyridoxal phosphate. NADPH is then used for the reduction of the intermediate to dihydrosphingosine:

$$\text{Palmitoyl-CoA} + \text{serine} + \text{(pyridoxal phosphate)} \rightarrow \text{3-ketodihydrosphingosine}$$

$$(7.7)$$

$$\text{3-ketodihydrosphingosine} + \text{NADPH} \rightarrow \text{dihydrosphingosine} \quad (7.8)$$

Conversion of dihydrosphingosine to sphingosine uses a flavoprotein enzyme – possibly analogous to the acyl-CoA dehydrogenase of β-oxidation (3.3.1).

7.4.2 Formation of the simplest sphingolipid, ceramide

Synthesis of ceramides was first demonstrated with rat brain microsomal preparations. Acyl-CoAs are used and free fatty acids are inactive. Several *N*-acyltransferases (acyl-CoA : sphingosine *N*-acyltransferases) which differ in substrate specificity with regard to the fatty acyl-CoA moiety, have been demonstrated. There is less specificity with regard to the long chain base – C_{18} and C_{20} sphingosines as well as dihydrosphingosines are all good acceptors.

$$\text{Long chain base} + \text{fatty acyl-CoA} \rightleftharpoons \text{ceramide} + \text{CoA} \qquad (7.9)$$

Other possible alternative routes include the reversibility of ceramidase reaction:

$$\text{Ceramide} \rightleftharpoons \text{fatty acid} + \text{sphingosine} \qquad (7.10)$$

However, this enzyme is more likely to be important in catabolism and in some experimental systems yields not sphingosine but a condensation product between long chain fatty acids and ethanolamine.

7.4.3 Cerebroside synthesis

Two pathways for the synthesis of galactosylceramide from sphingosine have been proposed:

1.   Sphingosine + UDP-gal $\rightleftharpoons$ Psychosine + UDP

         Psychosine + fatty acyl-CoA $\rightleftharpoons$ Galactosylceramide + CoA   (7.11)

2.   Ceramide + UDP-gal $\rightleftharpoons$ Galactosylceramide + UDP   (7.12)

In the experiments which led to the proposal of the first pathway, (7.11), the acylation of psychosine occurred by a non-enzymatic reaction and, therefore, the second pathway (i.e. galactosylation of ceramide, (7.12)) is though to be the major pathway. The reaction was first demonstrated by Morrell and Radin and the UDP-galactose : ceramide galactosyltransferase has now been purified from rat brain. It is a membrane-bound enzyme which has to be solubilized with detergent and contains tightly-bound phospholipid, even in its purified form. Interestingly, in view of the concentration of α-hydroxy fatty acids in galactosylceramides (Table 6.7) the enzyme was inactive with ceramides containing non-hydroxy fatty acids.

Glucosylceramide, which is the precursor for gangliosides, is formed in an analogous reaction by a glucosyltransferase which transfers glucose from UDP-glucose to a ceramide containing non-hydroxy fatty acids.

7.4.4 Formation of cerebroside derivatives

(a) Neutral ceramides

A whole host of higher homologues of ceramide are present in different tissues (section 6.2.4). Lactosylceramide and globosides (in addition to glucosylceramide) are the major glycolipids of most extraneural tissues of mammals. UDP-derivatives of sugars are used by the various transferases involved, except for fucosyltransferases which use GDP-fucose (Table 7.5). It will be noted from the structures given in Table 6.7 that the neutral ceramides (as well as the gangliosides, Table 6.8) contain a basic glucosylceramide structure to which further sugar residues are attached.

(b) Sulphated glycolipids

The major sulphated glycolipid in mammalian tissues is galactosylceramide with a sulphate attached to the 3-position of the sugar (Table 6.7). It is commonly known as sulphatide. However, various other sulphated glycolipids (e.g. lactosylceramide sulphate) have also been detected in small amounts in different animal tissues.

Sulphatide is an important constituent of the myelin sheath of nervous tissue and the incorporation of radioactive sulphate into the sulphatide molecule has been used to study myelination. The donor of the sulphate group is the complex nucleotide 3'-phosphoadenosine 5'-phosphosulphate, usually abbreviated in scientific papers to PAPS (Figure 7.10). PAPS itself is produced from ATP in two steps via an adenosine 5'-phosphosulphate (APS) intermediate. The synthesis of sulphatide is catalysed by a sulphotransferase which has been detected in the microsomal fraction from a number of tissues:

$$\text{Galactosylceramide} + \text{PAPS} \rightleftharpoons \text{3-sulpho-galactosylceramide} + \text{PAP} \qquad (7.13)$$

(c) Gangliosides

The biosynthesis of gangliosides proceeds by stepwise addition of monosaccharide units to the growing carbohydrate chain of the glycosphingolipid. Sugars are attached mainly by the use of UDP-derivatives (Table 7.5) but sialic acid residues are added using a cytidine

Table 7.5 Substrates used for the formation of sphingolipids

Group to be added	Substrate used
Glucose	UDP-glucose
Galactose	UDP-galactose
N-Acetylgalactosamine	UDP-N-acetylgalactosamine
Fucose	GDP-fucose
Sialic acid (N-acetyl neuraminic acid)	CMP-NANA (CMP-NeuAc)
Sulphate	PAPS

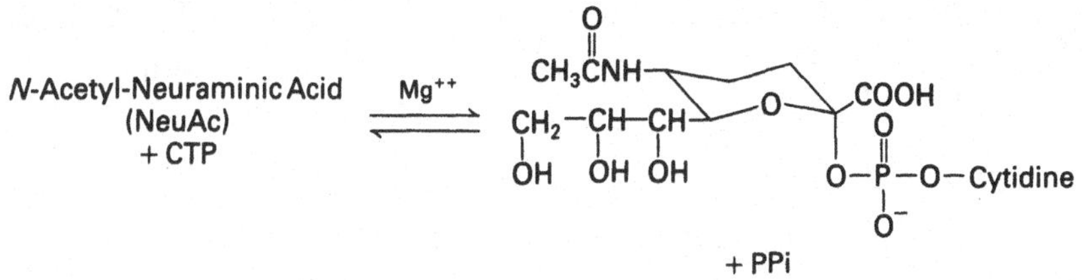

N-Acetyl-Neuraminic Acid (NeuAc) + CTP

$\xrightleftharpoons{\text{Mg}^{++}}$

[structure of CMP-NANA: ring with CH₃CNH (with C=O above), CH₂-CH-CH with OH, OH, OH below, O in ring, COOH, O-P-O-Cytidine with O⁻ groups]

+ PPi

Figure 7.11 Formation of CMP-NANA.

derivative. This compound, CMP-NANA (Figure 7.11), is formed by an enzymic reaction from CTP and is analogous to the cytidine derivatives used in phospholipid biosynthesis.

The picture of ganglioside synthesis was built up largely by the work of Roseman, Brady and their colleagues in the USA. The way in which they worked out the sequence of sugar additions was by testing the specificity of each individual transferase for a specific acceptor. Each enzyme requires a specific substrate as donor and the end product of the last step has much the greatest activity as acceptor. The higher sialylated gangliosides are made by pathway I (Figure 7.12). The major routes for ganglioside synthesis are shown in Figure 7.13.

In recent years, considerable information has accumulated to show that gangliosides, like other glycosphingolipids, have important properties for cells (see also section 8.9 and 8.10). The relative amounts of gangliosides for a given cell are not always constant. Changes occur with the developmental stage of the cell, its environment and whether it is subject to a number of pathological conditions. With neuronal development a general increase in the quantity of total gangliosides and in the proportion of the more highly sialylated compounds is seen. Small shifts in composition also occur after prolonged nerve stimulation or during temperature adaptation in poikilothermic animals or during hibernation in mammals. A number of drug-induced changes (e.g. with opiates) have also been noted.

The changes in ganglioside patterns which occur during cell development and growth imply that these lipids may also play a role in cell contact inhibition. This suggestion is reinforced by several lines of evidence. First, exogenous gangliosides (or antibodies to them) have been shown to control growth and differentiation in a number of systems. Second, oncogenic transformation, which results in a loss of growth regulation mechanisms (such as the cell-to-cell contact-dependent inhibition of growth), is usually paralleled by an irreversible reduction in the levels of the more complex gangliosides (and neutral glycosphingolipids). Third, gangliosides make an important contribution to the characteristic immunoexpression of individual cell types. Fourth, they act as binding sites for a number of important

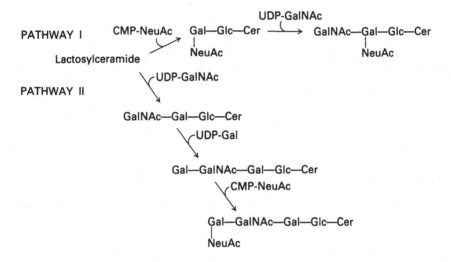

Figure 7.12 Key steps in the initial formation of simple gangliosides.

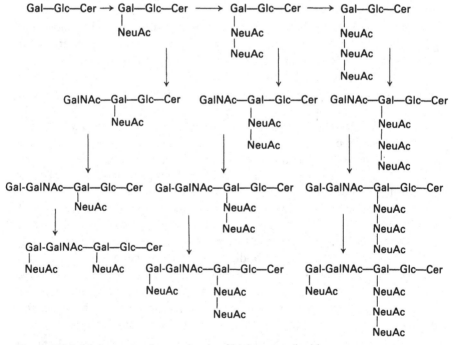

Figure 7.13 Major routes for synthesis of higher gangliosides.

compounds. Originally, several toxins, such as tetanus toxin, botulinum toxin and cholera toxin, were shown to interact tightly with gangliosides but there is now some indication that they may function as receptors for interferon and for certain cell growth and differentiation factors. These functions are discussed further in sections 8.9 and 8.10.

7.4.5 Sphingomyelin is both a sphingolipid and a phospholipid

The formation of sphingomyelin was mentioned in section 7.1.4 but because it is also a sphingolipid a few remarks will be made here also. It can be made by two pathways which differ in the order in which the fatty acyl and X moieties are attached to the sphingosine base (Figure 7.14). Experimental evidence has been obtained which supports both pathways but recent data would indicate that, like cerebroside formation, the main route involves the initial production of a ceramide, followed by the transfer of the phosphocholine group from CDP-choline.

7.4.6 Catabolism of the sphingolipids

Glycosphingolipids are widely distributed in animal tissues and constitute significant components of the human diet. It is not surprising, therefore, that the small intestinal cells contain enzymes for breaking down such lipids into their constituent parts. In addition, turnover of the various endogenous sphingolipids takes place. This turnover is mediated by the lysosomes of phagocytic cells, particularly the histiocytes or macrophages of the

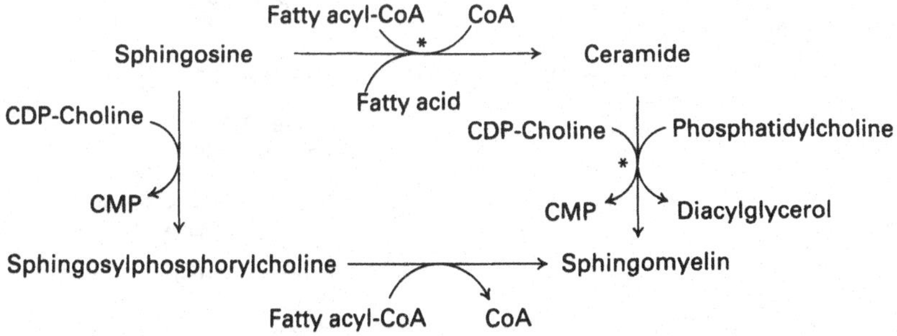

* = Main route.

Figure 7.14 Possible pathways for sphingomyelin formation. N.B. Some workers claim that the conversion of sphingosine to ceramide can use a non-esterified fatty acid instead of acyl-CoA.

reticuloendothelial system located primarily in bone marrow, liver and spleen. The predominant sphingolipid type varies between tissues and, therefore, the important catabolic enzymes will be different. For white or red blood cells, lactosyl-ceramide (cer–glc–gal) and hematoside (cer–glc–gal–NANA) are major components. By contrast the brain composition is dominated by complex gangliosides. During the neonatal period, the turnover of gangliosides is particularly rapid as sphingolipids are broken down and then resynthesized.

Usually each of the catabolic enzymes is specific for a particular chemical bond. Thus, a combination of a number of enzymes is needed to ensure the complete breakdown of a given sphingolipid. Catabolism begins by attack on the terminal hydrophilic portions of the molecules. The enzymes responsible are glucosidases, galactosidases, hexosaminidases,

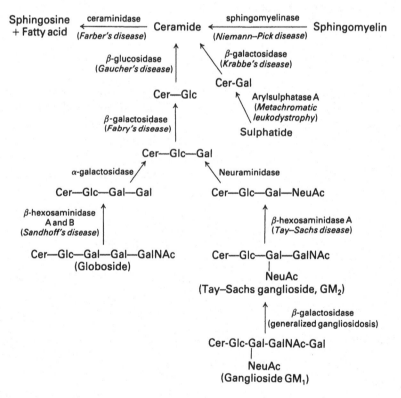

Figure 7.15 Pathways for sphingolipid catabolism showing enzyme deficiencies in lipid storage diseases.

neuraminidases and a sulphatase. As an example of the specificity of such enzymes it has been noted that β-galactosidases have been found which are specific for Cer–glc–gal and for Cer–glc–gal–gal. Another enzyme (from spleen) cleaves the glucose from Cer–glc but is inactive on Cer–gal, whereas an intestinal enzyme is active with both substrates. Brain contains an enzyme which cleaves ceramide (but not cerebroside) to yield free fatty acid and sphingosine. This reaction seems to be freely reversible and so could be used for synthesis (section 7.4.3) in some circumstances. The sulphatases are responsible for the cleavage of the sulphate ester from sulphatides while N-acetyl neuraminic acid is hydrolysed from gangliosides by neuraminidases.

The breakdown of the various sphingolipids usually proceeds smoothly and it has been suggested that the various catabolic enzymes are aligned in an ordered fashion on the lysosomal membrane – thus ensuring more efficient hydrolysis than would be the case for various substrates allowed random access to enzymes freely admixed within the organelle. Sometimes one of the breakdown enzymes is missing or has very low activity. When this happens there is a build-up of one of the intermediate lipids. Such accumulation can impair tissue function and gives rise to a **lipidosis**. These diseases are discussed in section 8.14. The pathways for sphingolipid breakdown are shown in Figure 7.15.

7.5 CHOLESTEROL BIOSYNTHESIS

As mentioned in section 6.4, cholesterol is a major and important consituent of animal membranes. Different sterols may be important in other organisms. Sterols are, however, a relatively small group of a very large class of biogenetically related substances – the polyisoprenoids (or terpenoids). These compounds are all derived from a common precursor 3-methylbut-3-enyl(isopentenyl)diphosphateisoprene. Polyisoprenoids can be open-chain, partly cyclized or fully cyclized substances. They contain a basic structure of a branched–chain C_5 (isoprenoid) unit.

For a membrane constituent, cholesterol has a long scientific history. In 1816, Chevreul coined the term cholesterine (from the Greek 'chole' meaning bile and 'stereos' meaning solid) for an alcohol-soluble substance which could be isolated from bile stones and in 1843 Vogel was one who identified it in several normal animal tissues as well as atheromatous lesions. After various advances in knowledge of its chemistry, the structure of cholesterol was finally solved, principally by Wieland and Windaus around 1932.

7.5.1 Acetyl-CoA is the starting material for terpenoid as well as fatty acid synthesis

The biosynthesis of cholesterol is covered well in standard biochemistry textbooks and only the more important features will be outlined here.

The precursor pool is mammalian cells is the cytosolic acetyl-CoA. This acetyl-CoA may be derived, for example, from β-oxidation of fatty acids by mitochondria or microbodies (section 3.3.1). The acetyl-CoA pool is in rapid equilibrium with intracellular and extracellular acetate which allows radiolabelled acetate to be used conveniently to measure cholesterol synthesis in tissues.

The first two steps involve condensation reactions catalysed by a thiolase and hydroxymethylglutaryl-CoA(HMG-CoA) synthetase. Both enzymes are soluble and the first reaction is driven to completion by rapid removal of acetoacetyl-CoA by the second step (Figure 7.16).

HMG-CoA synthetase has been studied in considerable detail and the reaction mechanism defined. It shows a very high degree of specificity with regard to the stereochemistry of the acetoacetyl-CoA substrate and the condensation proceeds by inversion of the configuration of the hydrogen atoms of acetyl-CoA. In addition to cytosolic HMG-CoA synthetase, a second synthetase is found in mitochondria. Not only has this been shown to be a different protein, the HMG it forms has a different function. Whereas HMG in the cytosol is destined for mevalonate formation, that in mitochondria is broken down by HMG-CoA lyase to yield acetyl-CoA and acetoacetate (section 3.3.1).

The next stage in mevalonate formation, involves two reductions of HMG-CoA by HMG-CoA reductase using NADPH. The enzyme is membrane-bound and is widely distributed in many tissues. It is a highly regulated enzyme with a short half-life ($T_{1/2}$ about 3 hours) and its activity is often considered to control the overall rate of cholesterol biosynthesis. Even in a normal diurnal cycle the activity of the reductase will vary about tenfold.

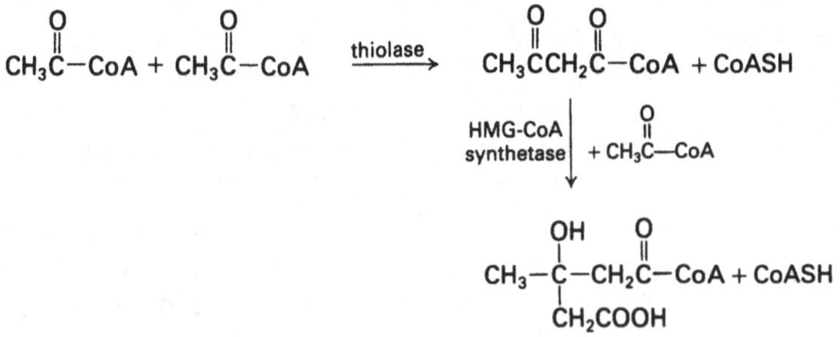

$$CH_3\overset{O}{\underset{||}{C}}-CoA + CH_3\overset{O}{\underset{||}{C}}-CoA \xrightarrow{\text{thiolase}} CH_3\overset{O}{\underset{||}{C}}CH_2\overset{O}{\underset{||}{C}}-CoA + CoASH$$

$$\text{HMG-CoA synthetase} \downarrow + CH_3\overset{O}{\underset{||}{C}}-CoA$$

$$CH_3-\underset{\underset{CH_2COOH}{|}}{\overset{OH}{\underset{|}{C}}}-CH_2\overset{O}{\underset{||}{C}}-CoA + CoASH$$

Figure 7.16 Formation of hydroxymethylglutaryl-CoA (HMG-CoA).

Two types of regulations are important. The diurnal variations in reductase activity are brought about by alterations in the amounts of the enzyme protein and, also, the reductase-specific mRNA. An increase in HMG-CoA reductase appears to be due to an increase in its synthesis while decreases are caused by a cessation of its formation. Several compounds such as oxygenated sterols (but not cholesterol itself) seem to prevent transcription of the reductase gene. Mevalonate (or products of its metabolism), in addition to inhibiting synthesis of the reductase, also appears to enhance its degradation.

HMG-CoA reductase is also regulated by a reversible phosphorylation/ dephosphorylation cycle (Figure 7.17). The phosphorylated reductase is inactive and the amounts of the phosphorylated enzyme can be shown to be increased when its activity is decreased by mevalonate or glucagon. The protein kinase which inactivates HMG-CoA reductase is itself subject to phosphorylation. Interestingly, both ATP and ADP are needed. Apparently ADP binds to a different site on the reductase kinase and acts as an allosteric effector. The phosphatases which activate HMG-CoA reductase are highly sensitive to NaF. Both the kinase and phosphatases are present in both microsomal and cytoplasmic fractions.

Because of the involvement of cholesterol in the aetiology of arterio-vascular disease (section 5.5.3) considerable efforts have been made to develop suitable pharmaceutical agents which could reduce its formation. Two interesting compounds which inhibit HMG-CoA reductase are natural antibiotics isolated from the moulds *Penicillium* spp. and *Aspergillus terreus*

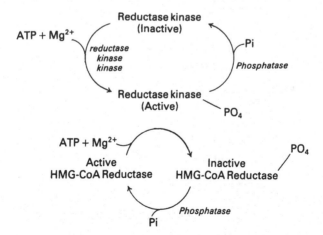

Figure 7.17 Regulation of HMG-CoA reductase activity by phosphorylation-dephosphorylation.

and named compactin and mevinolin, respectively. Very recently, tablets containing mevinolin or its derivatives have been marketed by drug companies as cholesterol-lowering agents.

7.5.2 Further metabolism generates the isoprene unit

Once mevalonate has been formed it is sequentially phosphorylated by two separate kinases yielding mevalonate 5-diphosphate. A third ATP-consuming reaction involving a decarboxylase then generates the universal isoprene unit, isopentenyl pyrophosphate (Figure 7.18). The function of the ATP in this reaction appears to be to act as an acceptor for the leaving OH group in the dehydration part of the reaction.

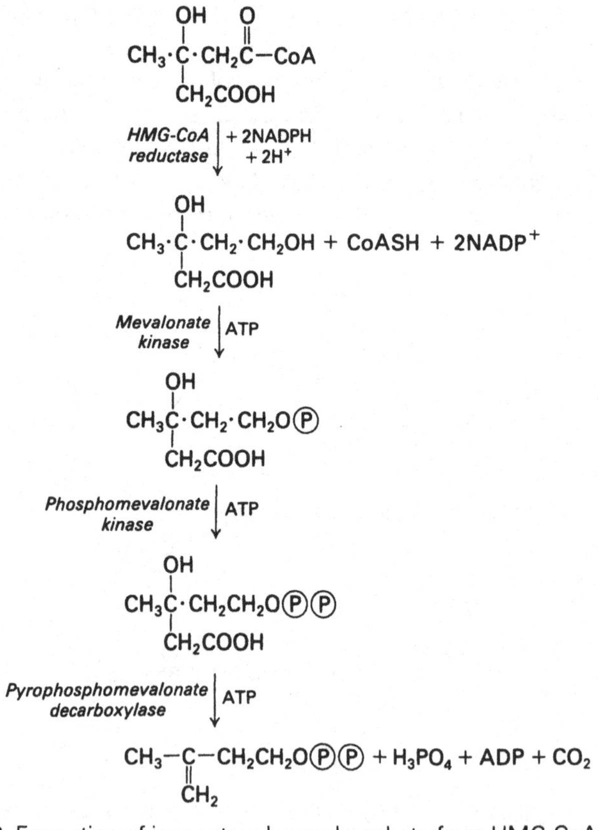

$$
\begin{array}{c}
\overset{\displaystyle OH}{\underset{\displaystyle CH_2COOH}{CH_3 \cdot \overset{|}{\underset{|}{C}} \cdot CH_2 \overset{\displaystyle O}{\overset{||}{C}} - CoA}}
\end{array}
$$

HMG-CoA reductase | + 2NADPH + 2H$^+$

$$
\begin{array}{c}
\overset{\displaystyle OH}{\underset{\displaystyle CH_2COOH}{CH_3 \cdot \overset{|}{\underset{|}{C}} \cdot CH_2 \cdot CH_2OH}} + CoASH + 2NADP^+
\end{array}
$$

Mevalonate kinase | ATP

$$
\begin{array}{c}
\overset{\displaystyle OH}{\underset{\displaystyle CH_2COOH}{CH_3 \overset{|}{\underset{|}{C}} \cdot CH_2 \cdot CH_2O\,\textcircled{P}}}
\end{array}
$$

Phosphomevalonate kinase | ATP

$$
\begin{array}{c}
\overset{\displaystyle OH}{\underset{\displaystyle CH_2COOH}{CH_3 \overset{|}{\underset{|}{C}} \cdot CH_2CH_2O\,\textcircled{P}\textcircled{P}}}
\end{array}
$$

Pyrophosphomevalonate decarboxylase | ATP

$$
\begin{array}{c}
\underset{\displaystyle CH_2}{CH_3 - \overset{||}{C} - CH_2CH_2O\,\textcircled{P}\textcircled{P}} + H_3PO_4 + ADP + CO_2
\end{array}
$$

Figure 7.18 Formation of isopentenyl pyrophosphate from HMG-CoA.

7.5.3 Higher terpenoids are formed by a series of condensations

Isopentenyl pyrophosphate is potentially a bifunctional molecule. Its terminal vinyl group gives a nucleophilic character whereas when it isomerizes to 3,3-dimethylallyl diphosphate, the latter is electrophilic. Thus, longer-chain polyprenyls are formed by a favourable condensation of isopentenyl pyrophosphate first with dimethylallyl pyrophosphate and later with other allylic diphosphates. The initial interconversion of isopentenyl diphosphate and dimethylallyl diphosphate is promoted by an isomerase. The successive condensations yield the C_{10} compound geranyl diphosphate and then the C_{15} farnesyl diphosphate. The two molecules of farnesyl diphosphate condense to form presqualene pyrophosphate which is reduced by NADPH to give the C_{30} open chain terpenoid squalene. The condensation reactions with IPP are a rather novel method of C–C bond formation since in the formation of other types of natural products (peptides, sugars, fatty acids, etc.) the reactions involve Claisen- or aldol-type condensations.

7.5.4 Sterol synthesis requires cyclization

Formation of sterols from squalene involves cyclization. First a microsomal mixed-function oxidase (squalene epoxidase) forms squalene-2,3-oxide in the presence of NADPH, FAD and O_2 (there is no requirement for cytochrome P_{450} in this reaction). The cyclization of the oxide to lanosterol then takes place by a concerted reaction without the formation of any stable intermediates. This conversion, which has been described as the most complex known enzyme-catalysed reaction, depends on a cyclase with a molecular mass of only 90 kDa. In plants and algae squalene-2,3-epoxide is cyclized to cycloartenol which is the precursor of stigmasterol whereas lanosterol is the precursor of cholesterol and ergosterol (Figure 7.19).

The conversion of lanosterol to cholesterol involves a 19-step reaction sequence catalysed by microsomal enzymes. The exact order of the reactions has not been delineated and, indeed, there may be more than one pathway. The main features of the transformation are the removal of three methyl groups, reduction of the 24(25)-double bond and isomerization of the 8(9)-double bond to position 5 in cholesterol.

7.5.5 Cholesterol is an important metabolic intermediate

Cholesterol is not only an important membrane constituent in animals but also plays a vital role as a metabolic intermediate (Figure 7.20). It acts

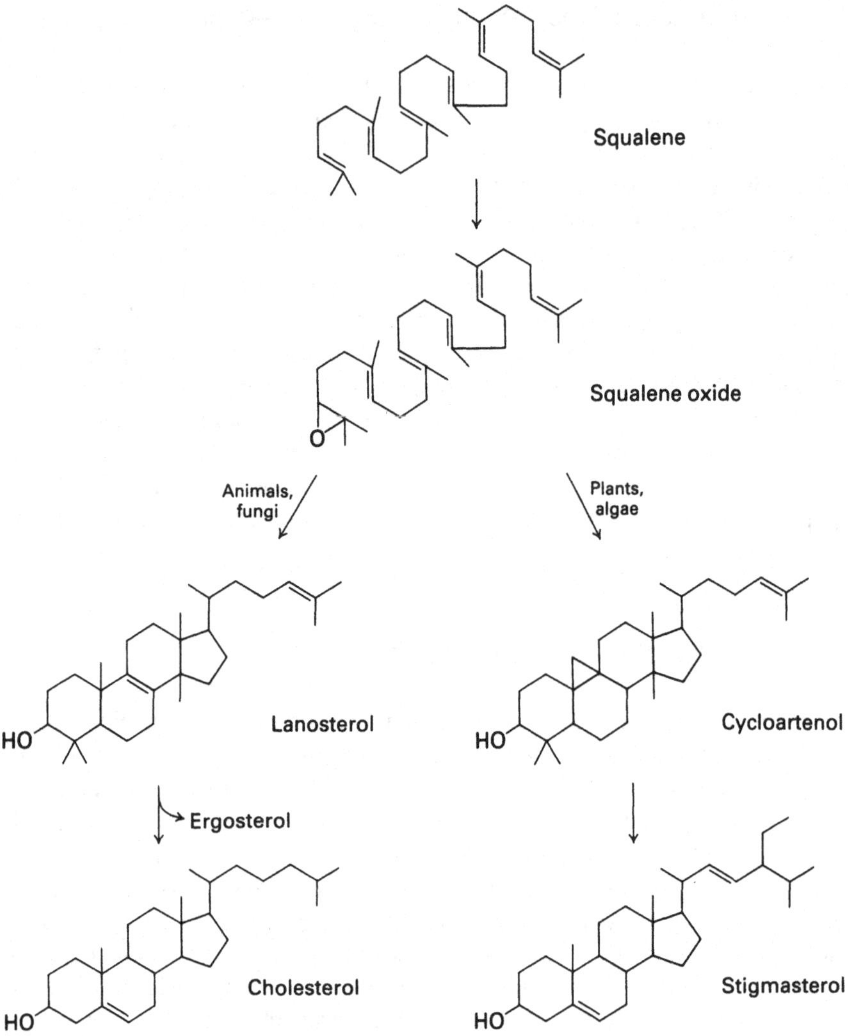

Figure 7.19 Conversion of squalene into sterols.

as a precursor for the steroid hormones (glucocorticoids, aldosterone, oestrogens, progesterones, androgens), for the bile acids (and their salts) and can also be esterified. Even if cholesterol was not required continuously because of membrane turnover, a considerable amount would be needed for bile production – even though greater than 80% of the bile salts are absorbed from the large intestine and reutilized. Thus, under physiological conditions cholesterol is needed mainly in cell division (growth or replacement of

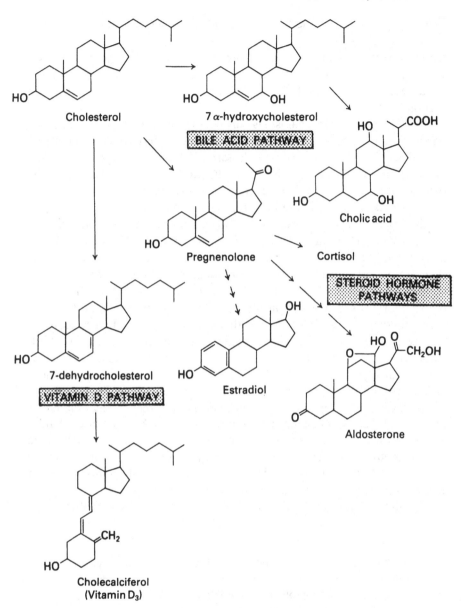

Figure 7.20 Metabolism of cholesterol to bile acids, calciferols and steroid hormones.

desquamated cells), in the replacement of cholesterol metabolized to steroid hormones in the adrenals and other endocrine glands or catabolized to bile acids by the liver.

7.5.6 Cholesterol levels are normally carefully controlled

Cholesterol balance in cells is maintained by a number of factors (Table 7.6). In fact, it has been found that uptake of lipoproteins may influence cholesterol synthesis itself via the LDL receptor mechanism described in section 5.3.5(g).

The LDL receptor mechanism is only one way in which cells may regulate cholesterol formation. In particular oxidosterols derived from squalene-2,3 : 22,23-dioxide or from the cholesterol of LDL (via oxysterol binding protein) have been shown to inhibit HMG-CoA reductase effectively. The oxidosterols are present at high levels in LDL and may well account for some of the effects described in the LDL receptor mechanism.

Although sterols are present in most mammalian body tissues, the proportion of sterol ester to free sterol varies markedly. For example, blood plasma, especially that of Man, is rich in sterols and like most plasma lipids they are almost entirely found as components of the lipoproteins; about 60–80% of this sterol is esterified. In the adrenals, too, where cholesterol is an important precursor of the steroid hormones, over 80% of the sterol is esterified. However, in brain and other nervous tissues, where cholesterol is a majot component of myelin, virtually no cholesterol esters are present. Cholesterol esters are formed by the action of a microsomal acyl-CoA: cholesterol acyltransferase (ACAT) which is present in most cells. Under normal conditions the enzyme is considered rate-limiting for cholesterol esterification. It is regulated by progesterone and may be modulated by phosphorylation/dephosphorylation like HMG-CoA reductase. Under conditions where cells take up a large amount of cholesterol, such as via LDL receptors, ACAT is induced. The enzyme is particularly important in intestine and is relatively low in liver where the lipoproteins made for secretion into the serum contain little if any cholesterol ester.

In contrast, cholesterol esters are formed in blood by another acyl-

Table 7.6 Factors influencing cellular cholesterol concentrations

1. Uptake of intact lipoproteins via receptors.
2. Uptake of free cholesterol from lipoproteins by lipid transfer.
3. Cholesterol synthesis.
4. Cholesterol metabolism (e.g. to hormones).
5. Efflux of cholesterol.
6. Esterification of cholesterol by acyl-CoA : cholesterol acyltransferase.
7. Breakdown of cholesterol esters by neutral cholesterol esterase.

transferase, the lecithin cholesterol acyl transferase (LCAT) as described in section 5.3.5. Most of the cholesterol that accumulates in arterial plaques during the development of atherosclerosis is in the esterified from. An understanding of cholesterol transport and cholesteryl ester metabolism are crucial for the understanding of this disease, discussed in section 5.3.5.

7.6 SUMMARY

Glycerolipids are based on the trihydric alcohol glycerol. Phosphoglycerides are synthesized in one of three basic pathways. Either a CDP-base derivative reacts with diacylglycerol to produce the phospholipid or CDP-diacylglycerol can be used as an intermediate. The third type of pathway involves the conversion of one phospholipid into another. As a generalization, phosphatidylcholine and phosphatidylethanolamine are made by the CDP-base pathway in eukaryotes while the CDP-diacylglycerol pathway is used for acidic phospholipids such as phosphatidylglycerol and phosphatidylinositol. In both these pathways the hydrophobic part of the phospholipid ultimately derives from the acylation of glycerol 3-phosphate. The resultant phosphatidic acid can then be dephosphorylated to yield diacylglycerol or transferred to a cytidine nucleotide to produce CDP-diacylglycerol.

The plasmalogen derivatives of phospholipids are made by the same basic pathway as their diacyl- analogues except that a 1-ether, 2-acyl-glycerol substrate is used instead of diacylglycerol.

Inositol lipids have unusually high turnovers. Phosphatidylinositol can be successively phosphorylated to form phosphatidylinositol 4-phosphate and phosphatidylinositol 4,5 *bis*-phosphate. The latter is a particularly important molecule in the regulation of cellular metabolism and differentiation. Another important lipid for biological regulation is platelet activating factor. This molecule is formed by the deacylation of choline-plasmalogen at the *sn*-2 position followed by acetylation of the free hydroxyl.

Phospholipases, responsible for the degradation of phospholipids, often have unique characteristics as enzymes. They usually operate best at the surface of immiscible solvents such as with lipid micelles. The micellar nature of their substrates makes the application of classical enzyme kinetics difficult. Many phospholipases (and lipases) are extremely stable molecules which exhibit activity in organic solvents and at remarkably high temperatures.

Dependent on the position of attack, phospholipases are classified as A, B, C or D. Phospholipases A remove a fatty acid from phospholipids and are subdivided as A_1 or A_2 according to which acyl group is hydrolysed.

They are important not only in lipid degradation but also in the turnover of acyl groups and in the release of fatty acids for particular purposes such as eicosanoid production. Phospholipase B removes the remaining fatty acid from a monoacyl (lyso) phospholipid while phospholipase C action gives rise to diacylglycerol and a phosphate-base moiety. A phosphatidylinositol 4,5-*bis*-phosphate-specific phospholipase C is responsible for the generation of second messengers from this lipid. Phospholipase D removes the base moiety from a phospholipid to yield phosphatidate. Phospholipase D enzymes are very active in many plant tissues and can give rise to analytical artefacts if their activity is not carefully controlled.

The hydrolytic enzyme phosphatidate phosphohydrolase plays its main role in biosynthetic pathways. Thus, the diacylglycerol which it forms is used for the production of major membrane lipids such as phosphatidylcholine, on the one hand, or storage triacyglycerols, on the other.

In photosynthetic membranes, the major lipid constituents are glycosylglycerides. Monogalactosyldiacylglycerol is formed by the transfer of galactose from UDP-galactose to diacylglycerol. Digalactosyldiacylglycerol can then be synthesized either by transfer of a second galactose from UDP-galactose to monogalactosyldiacylglycerol or by inter-lipid transfer between two molecules of monogalactosyldiacylglycerol. The formation of the third major glycosylglyceride, the plant sulpholipid (sulphoquinovosyldiacylglycerol) remains unclear although it is known that isolated chloroplasts contain all the necessary synthetic enzymes.

Sphingolipids are based on the sphingosine base which is formed from palmitoyl-CoA and serine. The amino group of sphingosine can be acylated to give ceramides while the alcohol group can be attached to phosphorylcholine (from CDP-choline) to yield sphingomyelin or glycosylated to various degrees to form cerebrosides, neutral ceramides and gangliosides. Substrate-specific enzymes are used for the transfer of individual sugars during these syntheses. Usually UDP-sugars are the source of the sugar moiety although N-acetylneuraminic acid is transferred from its CMP-derivative.

Sphingolipids are broken down by substrate-specific enzymes. Sialic acid residues are removed by neuraminidase, galactose by galactosidases and glucose by glucosidases etc. Almost all of these enzymes are found in lysosomes and their absence gives rise to the accumulation of the respective substrate sphingolipid in tissues. This gives rise to various disease states known as lipidoses (section 8.14).

Cholesterol (and other sterols) is derived from acetyl-CoA. By a series of reactions the 5C-isoprene unit is formed and this can then self-condense to give a series of 10C, 15C, 30C etc. isoprenoid molecules. Reduction of hydroxymethylglutaryl-CoA is the key regulatory step in the overall process. To form sterols from the open-chain isoprenes requires cyclization and various other modifications are also needed to yield the final cholesterol molecule. Cholesterol itself is an important metabolic intermediate – being converted to cholesterol esters, to bile acids, to cholecalciferol (and vitamin D) or to various steroid hormones by different tissues. The synthesis of cholesterol and the regulation of its plasma circulating levels or conversion to other compounds is normally carefully controlled.

REFERENCES

General

Harwood, J.L. (1989) Lipid metabolism in plants. *Crit. Rev. in Plant Sciences*, **8**, 1–44.

Harwood, J.L. and Russell, N.J. (1984) *Lipids in Plants and Microbes*, George Allen and Unwin, Hemel Hempstead.

Hawthorne, J.N. and Ansell, G.B. (eds) (1982) *Phospholipids*, Elsevier, Amsterdam.

Mead, J.F., Alfin-Slater, R.B., Howton, D.R. and Popjak, G. (1986) *Lipids: Chemistry, Biochemistry and Nutrition*, Plenum Press, New York.

Vance, D.E. and Vance, J.E. (eds) (1985) *Biochemistry of Lipids and Membranes*, Benjamin/Cummings, Menlo Park, CA.

Phosphoglyceride synthesis

Ansell, G.B. and Spanner, S. (1982) Phosphatidylserine, phosphatidylethanolamine and phosphatidylcholine, in *Phospholipids* (eds J.N. Hawthorne and G.B. Ansell), Elsevier, Amsterdam, pp. 1–50.

Fisher, S.K., van Rooijen, L.A.A. and Agranoff, B.W. (1984) Renewed interest in the polyphosphoinositides. *Trends Biochem. Sci.*, **9**, 53–56.

Hanahan, D.J. (1986) Platelet activating factor: a biologically active phosphoglyceride. *Annual Rev. Biochem.*, **55**, 483–509.

Hawthorne, J.N. (1982) Inositol phospholipids, in *Phospholipids* (eds J.N. Hawthorne and G.B. Ansell), Elsevier, Amsterdam, pp. 263–278.

Hori, T. and Nozawa, Y. (1982) Phosphonolipids, in *Phospholipids* (eds J.N. Hawthorne and G.B. Ansell), Elsevier, Amsterdam, pp. 95–128.

Hostetler, K.Y. (1982) Polyglycerophospholipids: phosphatidylglycerol, diphosphatidylglycerol and bis(monoacylglycero) phosphate in *Phospholipids* (eds J.N. Hawthorne and G.B. Ansell), Elsevier, Amsterdam, pp. 215–261.

Klein, R.A. and Schmitz, B. (eds) (1986) *Topics in Lipid Research* (Chapters on platelet activating factor, eicosanoids, glycolipids and membrane structure and function), *Royal Society of Chemistry*.

Moore, T.S. (1982) Phospholipid biosynthesis. *Annual Rev. Plant Physiol.*, **33**, 235–259.

Raetz, C.R.H. (1986) Molecular genetics of membrane phospholipid synthesis. *Annual Rev Genet.*, **20**, 253–295.

Mangold, H.K. and Paltauf, F. (eds) (1983) *Ether Lipids: biochemical and biomedical aspects*, Academic Press, New York.

Snyder, F. (1985) Metabolism, regulation and function of ether-linked glycerolipids, in *Biochemistry of Lipids and Membranes* (eds D.E. Vance and J.E. Vance), Benjamin/Cummings, Menlo Park, CA, pp. 271–298.

Snyder, F. (1982) *Ether Lipids: chemistry and biology*, Academic Press, New York.

Vance, D.E. (1985) Phospholipid metabolism in eucaryotes, in *Biochemistry of Lipids and Membranes* (eds D.E. Vance and J.E. Vance), Benjamin/Cummings, Menlo Park, CA, pp. 242–270.

Vance, D.E. and Pelech, S.L. (1984) Enzyme translocation in the regulation of phosphatidylcholine synthesis. *Trends Biochem. Sci.*, **9**, 17–20.

Phospholipid degradation

Brindley, D.N. (ed.) (1988) *Phosphatidate Phosphohydrolase, Vols 1 and 2*, CRC Press, Boca Raton, Fl.

Dennis, E.A. (1983) Phospholipases, in *The Enzymes*, 3rd edn (ed. P.D. Boyer), *Vol. 16*, Academic Press, pp. 307–353.

Galliard, T. (1980) Degradation of acyl lipids: hydrolytic and oxidative enzymes in *The Biochemistry of Plants* (eds P.K. Stumpf and E.E. Conn), *Vol. 4*, Academic Press, New York, pp. 85–116.

Waite, M. (1985) Phospholipases, in *Biochemistry of Lipids and Membranes* (eds D.E. Vance and J.E. Vance), Benjamin/Cummings, Menlo Park, CA, pp. 299–324.

Glycosylglyceride metabolism

Harwood, J.L. (1980) Sulfolipids, in *The Biochemistry of Plants* (eds P.K. Stumpf and E.E. Conn), *Vol. 4*, Academic Press, New York, pp. 301–320.

Joyard, J. and Douce, R. (1987) Galactolipid synthesis, in *The Biochemistry of Plants*, (eds P.K. Stumpf and E.E. Conn), *Vol. 9*, Academic Press, New York, pp. 215–274.

Mudd, J.B. and Kleppinger-Sparace, K.F. (1987) Sulfolipids, in *The Biochemistry of Plants* (eds P.K. Stumpf and E.E. Conn), *Vol. 9*, Academic Press, New York, pp. 275–290.

Metabolism of sphingolipids

Barenholz, Y. and Gatt, S. (1982) Sphingomyelin: metabolism, chemical synthesis and physical properties, in *Phospholipids* (eds J.N. Hawthorne and G.B. Ansell), Elsevier, Amsterdam, pp. 129–177.

Kanfer, J.N. and Hakomori, S-I. (1983) *Sphingolipid Biochemistry*, Plenum Press, New York.

Kishimoto, Y. (1983) Sphingolipid formation, in *The Enzymes*, 3rd edn (ed. P.D. Boyer), *Vol. 16*, Academic Press, New York, pp. 358–408.

Makita, A., Handa, S., Taketomi, T. and Nagai, Y. (eds) (1982) *New Vistas in Glycolipid Research*, Plenum Press, New York.

Sweeley, C.C. (1985) Sphingolipids, in *Biochemistry of Lipids and Membranes* (eds D.E. Vance and J.E. Vance), Benjamin/Cummings, Menlo Park, CA, pp. 361–403.

Cholesterol biosynthesis

Brown, M.S. and Goldstein, J.L. (1986) A receptor-mediated pathway for cholesterol homeostasis, *Science, 232*, 34–47.

Endo, A. (1981) Biological and pharmacological activity of inhibitors of 3-hydroxy-3-methylglutaryl coenzyme A reductase. *Trends Biochem. Sci., 6*, 10–13.

Fielding, C.J. and Fielding, P.E. (1985) Metabolism of cholesterol and lipoproteins, in *Biochemistry of Lipids and Membranes* (eds D.E. Vance and J.E. Vance), Benjamin/Cummings, Menlo Park, CA, pp. 404–474.

Myant, N.B. (1981) *The Biology of Cholesterol and Related Steroids*, Heinemann, London.

Porter, J.R. and Spurgeon, S.L. (1981) *Biosynthesis of Isoprenoid Compounds, Vol. 1*, Wiley, New York.

Schroepfer, G.J. (1982) Sterol synthesis. *Annual Rev. Biochem. 52*, 411–439.

8 *Lipid functions*

8.1 GENERAL FUNCTIONS OF STRUCTURAL LIPIDS

Detailed discussion of membrane properties and functions is beyond the scope of this book – numerous detailed reviews being available. Nevertheless, a few remarks centred on the role of lipids can be made. In Chapter 6 we described current ideas for membrane structure and the concept that, although lipids are capable of adopting non-bilayer structures when isolated, by-and-large they participate in bilayers *in vivo*.

Under normal conditions the membrane bilayer is in a fluid state. Membrane proteins can migrate within the plane of the membrane with diffusion coefficients of about 10^{-10} cm^2 sec^{-1} while lipids diffuse with coefficients of about 10^{-8} cm^2 sec^{-1}. Overall behaviour might be considered, therefore, in thermodynamic terms. But generalized deductions relating fluidity of the membrane to enzyme activity are difficult to make for several reasons. For example, motion in a given lipid molecule may include rapid rotations but slow lateral movement. Also, increased disorder in a bilayer may not correlate with increased translational motion. Moreover, all membranes so far examined have shown transbilayer asymmetry while there is evidence in several cases for at least small areas of concentration of certain lipids, i.e. micro-lateral heterogeneity. These sorts of consideration complicate the interpretation of experiments designed to show how the bilayer lipids affect membrane enzyme activities at a molecular level.

Enzymologists frequently study how reaction rates, substrate binding etc. vary under a variety of experimental conditions. Parameters which may be changed include pH, ionic concentration and the addition of water-soluble inhibitors. Thus, the membrane protein is only probed in regions where it is exposed to bulk water or at the interfacial region where such water meets the bilayer. This often only represents a small part of the whole membrane protein.

Membrane proteins which clearly have an important part spanning the bilayer are those concerned with transport. These molecules must have at least two portions – (a) a domain which allows the movement of the substrate vectorially across the membrane. The domain could be a channel or a series of gates which are opened successively as the substrate moves; (b) a binding site for the substrate which confers selectivity and also, when occupied, initiates transport. There may also be regulatory domains and each of the above three could be modulated by lipids in the bilayer.

The passive transport system in erythrocytes which is used for sugars has been studied in detail by Melchior and Carruthers. The protein is particularly useful because it can be purified and reconstituted into lipid vesicles of different composition and the exact number of reconstituted transporters determined through the binding of radiolabelled cytochalasin B – a specific inhibitor. Furthermore, the kinetic properties, complete amino acid sequence and arrangement of the protein in erythrocyte membranes are known. Comparison of transport rates with the number of reconstituted transporters, allowed estimation of the turnover number (Tn), and allowed comparison between different lipids used for the reconstitution (Table 8.1). It was found that several factors influenced transport rates. In order of importance these were: lipid head group > lipid acyl chain length and saturation index > lipid backbone ≫ bilayer fluidity. Negatively charged phosphoglycerides such as phosphatidylserine were the most active for reconstitution and suggested that the normal tertiary structure of the translocating domain might need a high surface potential. In fact, for the best lipids, such as phosphatidylserine, the fluidity of the bilayer had little

Table 8.1 Factors influencing the activity of the erythrocyte sugar-transporter in reconstituted lipid vesicles

Factor and effect	Experimental result
Nature of the phospholipid head group (alters surface charge/potential)	Vesicles composed of single phospholipids had the following order of activity PS > PA > PG ≫ PC
Acyl chain length (important to be correct for size of transmembrane protein portions)	For saturated acyl chains: C18 > C20 > C16 > C14 For C18 chains: stearoyl > elaidoyl > oleoyl
Bilayer fluidity (affects immediate environment of hydrophobic domains)	Relatively little change in activity on going from gel to crystalline phases except for micelles containing PC
Cholesterol inclusion (affects lipid lateral mobility)	Lipid packing alterations correlate with transporter activity changes

effect on transport rate. By contrast, for a lipid giving slower turnover numbers, such as phosphatidylcholine, transporter Tn increased markedly when the bilayer changed from the crystalline to the gel state. Thus, the nature of the lipid head group(s) influenced the degree to which bilayer fluidity altered transport activity.

The length of the acyl chains (and hence membrane thickness) was also important – not surprisingly because the sugar transporter is known to contain 12 membrane-spanning hydrophobic domains. In effect the thickness of the membrane has to be right for the correct arrangement of the hydrophobic domains.

Of course, natural membranes contain not just a single lipid species but a whole host of different molecular types. Thus, experiments to determine the effect of lipids on membrane proteins *in vivo* are extremely difficult to interpret. In animal membranes, cholesterol is frequently present in addition to phosphoglycerides. Addition of cholesterol to a lipid bilayer induces complex alterations in lipid packing. McConnell and colleagues at Stanford showed that these alterations in lipid packing were associated with altered lateral mobility within the bilayer. In experiments with the sugar transporter referred to above, changes in transporter activity correlated closely with cholesterol-induced alterations in lipid packing. They did not, however, apparently result from cholesterol-induced changes in bilayer lipid order. Experiments with other systems (such as the Ca^{2+}-Mg^{2+}-ATPase from endoplasmic reticulum) have also shown no consistent correlation in bilayer fluidity and enzyme activity. Once more, the length of the acyl chains seemed a more important parameter.

8.2 MODIFICATIONS OF MEMBRANE LIPIDS DURING TEMPERATURE ADAPTATION

We have already seen that membrane lipids can exist in the gel- or liquid-crystalline phases, depending on the prevailing temperature. Whereas isolated lipid samples show a sharp gel-liquid phase transition, naturally occurring lipids show broad non-cooperative transitions due to the heterogeneity of their acyl chains. For a series of molecular species of, say, phosphatidylcholine, the temperature (Tc) for the gel-liquid phase transition depends on the fatty acid composition (with *cis* unsaturated fatty acids lowering Tc) and even on the position of acyl esterification (Table 8.2).

It used to be thought that the membranes of natural organisms could not function when their lipids were in the gel-crystalline phase and that changes in growth or other physiological parameters seen in prokaryotes on lowering temperature could be explained simply by proposing that phase transitions had taken place. However, it is now clear that while a correlation can

Table 8.2 Effect of fatty acyl groups on the temperature (*Tc*) of the gel– to liquid–crystalline phase transition of phosphatidylcholines

Lipid species	Tc (°)
(a) Saturated	
Dimyristoyl–PC (14 : 0, 14 : 0–PC)	24
Dipalmitoyl–PC (16 : 0, 16 : 0–PC)	41
Distearoyl–PC (18 : 0, 18 : 0–PC)	55
(b) Unsaturated	
Palmitoyl, oleoyl–PC (16 : 0, 18 : 1–PC)	− 1
Dioleoyl–PC (18 : 1, 18 : 1–PC)	− 19
Stearoyl, oleoyl–PC (18 : 0, 18 : 1–PC)	6
Stearoyl, linoleoyl–PC (18 : 0, 18 : 2–PC)	− 16
Stearoyl, linolenoyl–PC (18 : 0, 18 : 3–PC)	− 13
Stearoyl, arachidonyl–PC (18 : 0, 20 : 4–PC)	− 13

sometimes be seen between, for example, enzyme activity changes and the temperature for membrane lipid phase changes, this relationship is by no means simple nor the correlation universal. Nevertheless, as a general rule it can be stated that once significant areas of a natural membrane are in the gel state then function is significantly impaired. It is known, for example, that as temperatures are lowered phase separation of certain lipid species takes place and membrane proteins are squeezed out of such areas leading to the formation of protein-free areas which can be seen by electron microscopy. Studies with various fatty acid-requiring mutants of *E. coli* have shown that if less than half of the lipids of the cytoplasmic membrane are fluid then it becomes non-functional and the bacterium dies.

Thus, poikilotherms (organisms which cannot regulate their own temperatures) which grow at low temperatures have a pattern of membrane lipids which have lower phase transition temperatures than of organisms existing under hotter conditions. Furthermore, during changes in environmental temperature, an alteration in the composition of membrane lipids is seen. Although an increase in fatty acyl unsaturation at lower temperatures is the best known, it is not, by any means, the only change which can take place.

The process of temperature adaptation has been studied in some detail in cyanobacteria by Murata's group and in the protozoan *Tetrahymena* and the alga *Dunaliella* by Thompson in Texas. A number of lipid changes have been observed by these workers and others (Table 8.3) but they occur on a different time scale. A sudden change in growth temperature could

Table 8.3 Changes in membrane lipid composition caused by lowered growth temperatures

1. Re-tailoring of molecular species	Quick change involving swapping acyl chains on the glycerol backbone
2. Increase in fatty acyl unsaturation	Relatively quick for organisms using the aerobic pathway of desaturation; involves *in situ* modification of lipids
3. Chain shortening	Relies on the lower melting temperatures of shorter acyl chains; relatively uncommon to see shortening of existing acyl chains; same net effect more commonly seen when *de novo* synthesis produces shorter chain products
4. Alteration of fatty acid type	Some microorganisms may alter amounts of branch chain acids – this will also modify fluidity
5. Lipid classes altered	Changes in membrane lipid classes often seen; where phospholipid interconversion is easy (e.g. PE → PC) then this could be adaptive; PC has a lower *Tc* than PE for the same molecular species
6. Lipid to protein ratio changed	Usually the ratio is increased significantly at lower temperatures – due to increased net lipid synthesis

immediately alter membrane fluidity with harmful consequences for the organism concerned. Murata has proposed that an emergency response involves re-tailoring (altering the position of fatty acid acylation) of molecular species. Such re-tailoring has been shown *in vitro* to lead to significant differences in the *Tc* of the lipid concerned. This is because the glycerol backbone of a membrane lipid is not exactly parallel to the surface of the bilayer and, hence, two acyl chains of equal length penetrate the hydrophobic membrane region to a different extent (Figure 8.1). After the emergency response has occurred, other changes in lipid composition can take place on a longer time scale. For organisms operating the aerobic pathway for desaturation an increase in *cis* unsaturation is a relatively quick and easy modification to make. In contrast, bacteria such as *E. coli* could only increase their unsaturation by *de novo* synthesis of more palmitoleate and vaccenate (Figure 3.10). Such microorganisms have been known, however, in some cases to lower the *Tc* of any given lipid by shortening the

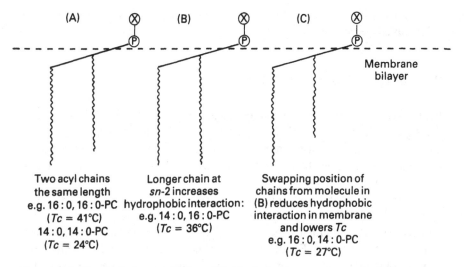

Figure 8.1 The basis for re-tailoring of lipids during temperature adaptation.

acyl chain. The slowest changes to take place are those involving membrane lipid proportions or the ratio of lipid to protein in the membrane. These alterations might include, for example, an increase in lipid to protein ratio in the plasma membranes of cold-stressed plant roots (Table 8.3).

An interesting application of membrane lipid biochemistry to a natural phenomenon has been made by Murata in Okasaki. He has examined the process of frost sensititivity in plants. It is well known that, whereas some plants can survive freezing temperatures, many others drop their leaves and/ or are killed by the first frosts of autumn. It seemed logical to look for differences in the membrane compositions of these two groups of plants to explain the difference. A superficial examination showed that both types of leaves contained the same classes of lipids with the usual exceptionally high content of linolenic acid. Murata noticed, however, that the phosphatidylglycerol fractions from frost-sensitive as opposed to frost-resistant plants were markedly different. Thus, frost-sensitive species had a much higher proportion (40–80%) of the 'disaturated' species which had Tc's of about 40°C. (For this purpose the unusual *trans*-3-hexadecenoic acid which is present in phosphatidylglycerol and which has a similar melting point to palmitic acid is classed as saturated.) In contrast, frost-resistant leaves only had about 15% of the disaturated species. Murata, therefore, proposed that the basis for frost sensitivity lay in the phase separation of a significant quantity of phosphatidylglycerol during exposure to low temperatures – this leading to membrane damage, leakage of intra-organellar contents and cell death. The experiment also emphasized how important it is to examine lipid composition in detail rather than merely taking average values.

8.3 LIPIDS AND MEMBRANE FUSION

Fusion is a very common membrane-mediated event. It occurs, for example, during fertilization, cell division, exo- or endocytosis and intracellular membrane transport. Just a short consideration of the overall event of fusion will make it obvious that during the process the normal lipid bilayer structure must disappear.

First of all, biochemists studied the process in model systems, employing lipid vesicles. It has been recognized for some time that such vesicles can be induced to fuse when incubated in the region of their gel to liquid-crystalline transition temperature. Furthermore, it was also noted that Ca^{2+} addition would induce fusion in phosphatidylserine-containing systems; possibly by causing local crystalline points for fusion. The role of Ca^{2+} in many (but not all) natural fusion events led to a great deal of interest in this phenomenon.

An important clue as to the mechanism of lipid membrane fusion has come from the work of de Kruijff and Cullis. They proposed that fusion proceeded because of the ability of lipids to undergo polymorphism (i.e. to adopt different structures). Three types of observation support the hypothesis. First, **fusogens** (such as monoacylglycerols) induce H_{II} (inverted micellar structures; Figure 6.11) phase structures, consistent with a role for non-bilayer structures in fusion. Second, promotion of fusion of, for example, phosphatidylserine-containing systems by Ca^{2+} is accompanied by H_{II} structures. Third, a number of factors such as pH changes or elevated temperatures that cause H_{II} formation also promote fusion of lipid vesicles.

A diagrammatic representation of the mechanism of fusion is shown in Figure 8.2.

Extension of the Cullis/de Kruijff hypothesis to natural membrane events has been difficult to prove. Nevertheless, confirmatory evidence has been obtained with a number of systems such as the release of chromaffin granule contents during stimulation of the adrenal medulla. The exocytosis accompanying this event seems to depend on the ability of Ca^{2+} to promote non-bilayer structures.

In the foregoing section we have suggested ways in which cells could initiate membrane fusion through allowing areas of hexagonal II phase. Conversely, there may be situations where synthesis of large amounts of hexagonal II-forming lipids could destabilize the normal bilayer structure. Thus, when the anaerobic bacterium *Clostridium butyricum* is grown in the absence of biotin (when it is reliant on exogenous fatty acids) its membrane

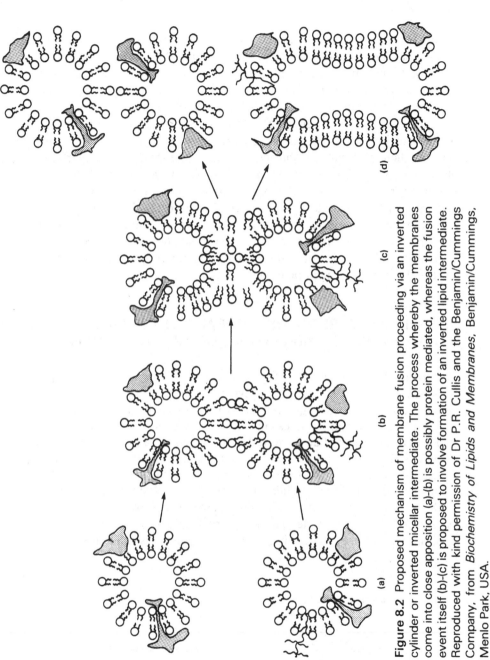

Figure 8.2 Proposed mechanism of membrane fusion proceeding via an inverted cylinder or inverted micellar intermediate. The process whereby the membranes come into close apposition (a)-(b) is possibly protein mediated, whereas the fusion event itself (b)-(c) is proposed to involve formation of an inverted lipid intermediate. Reproduced with kind permission of Dr P.R. Cullis and the Benjamin/Cummings Company, from *Biochemistry of Lipids and Membranes*, Benjamin/Cummings, Menlo Park, USA.

lipid composition can be made more unsaturated by supplying *cis*-monounsaturated acids rather than saturated ones. Since phosphatidylethanolamine and its plasmalogen are major membrane lipids in this bacterium, and their unsaturated species form the hexagonal II phase (Figure 6.11) an increase in unsaturation might destabilize the membrane. *C. butyricum* reacts by reducing the proportion of phosphatidylethanolamine and increasing that of its glycerol acetal. Since the latter forms bilayers then this change stabilizes the membrane structure. Similarly, in *Acholeplasma laidlawii* where monoglucosyldiacylglycerol and diglucosyldiacylglycerol are major membrane lipids, the former is hexagonal II-forming while the latter gives bilayers (Figure 6.11). Under conditions such as increased temperature or unsaturation which would tend to favour the formation of non-bilayer structures, the monoglucosyl-lipid is converted to diglucosyldiacylglycerol. This change in lipid proportions would be expected to preserve membrane bilayer stability.

8.4 LIPIDS AND PROTEINS INTERACT IN ORDER TO DETERMINE MEMBRANE STRUCTURE AND SHAPE

As discussed in section 6.5 (Figure 6.11) the shape of complex lipid molecules can give rise not only to polymorphism but also to curved regions of bilayers. Israelachvili suggested that, for regions of high curvature (such as at the ends of thylakoid stacks in chloroplasts) cone-shaped lipids would be needed in one leaflet and inverted cone-shaped lipids in the other leaflet (Figure 8.3). This is an attractive idea but examination of the trans-bilayer distribution of chloroplast lipids by various techniques has shown that the enrichment of monogalactosyldiacylglycerol (inverted cone) is the opposite of that required, i.e. it is more abundant in the outer leaflet. Nevertheless, because it is entirely possible that a small proportion of certain lipids may be concentrated laterally in certain membrane regions, it is still plausible that

Curved membrane

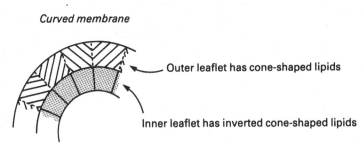

Outer leaflet has cone-shaped lipids

Inner leaflet has inverted cone-shaped lipids

Figure 8.3 How lipids could cause bilayer curves.

lipids alone may play a significant role in determining natural membrane shape. Experiments with whole membrane fractions would not pick up such a specific distribution, of course, but would merely provide an averaging of trans-bilayer composition.

Much more likely is the cooperation of membrane lipids and proteins in controlling membrane shape and properties. For example, membranes containing very large integral protein complexes often contain high amounts of lipids which would form H_{II} phases when purified. Thus, chloroplast thylakoids contain around 50% of the non-bilayer forming mono-galactosyldiacylglycerol while the photosynthetic membrane of purple non-sulphur bacteria is enriched in unsaturated phosphatidylethanolamine (also H_{II}-forming). It has been suggested that these inverted cone-shaped lipids may help to package the integral proteins and ensure a good bilayer structure which must be impermeable to ions and other molecules.

A situation where proteins and lipids appear to interact in determining membrane properties is in the erythrocyte. This simple cell can have its membrane composition altered in a non-destructive fashion by the use of phospholipid exchange proteins. The Dutch biochemists op den Kamp and van Deenen and their colleagues introduced specific phosphatidylcholine molecules into the outer leaflet by such means and showed that, depending on the molecular species substituted, the cells became more or less fragile and displayed susceptibility to osmotic shock. They could lose their flexibility and show an altered ion permeability. A surprising finding was that substitution with certain species resulted in a changed cell shape (Figure 8.4). These results were explained by assuming that replacement of the natural slightly cone-shaped phosphatidylcholine molecules by the cylindrical dipalmitoyl-species would cause the membrane to become convex while enrichment with the cone-shaped dilinoleoyl-species would alter the shape to concave (Figure 8.4).

Sickle cell anaemia is characterized by an amino-acid substitution of the β-chain of haemoglobin. The erythrocytes have a sickled shape and have changes in their membrane characteristics. Studies with such cells including experiments on repeated sickling/unsickling suggest that sickling results from a local uncoupling of lipid bilayer and the cytoskeleton. This leads amongst other things to a very much increased trans-bilayer movement (flip-flop) of lipids. In ageement with this suggestion it has been noted that membranes which do not contain a strictly organized cytoskeleton such as the endoplasmic reticulum of rat liver or the plasma membrane of gram-positive bacteria also show relatively fast bilayer lipid movement.

Other experiments with erythrocytes also confirm the interaction of protein and lipids. In patients with hereditary pyropoikilocytosis the spectrin dimers cannot associate into the normal tetramers and this gives rise to enhanced thermosensitivity of the cytoskeleton. The cells also have an

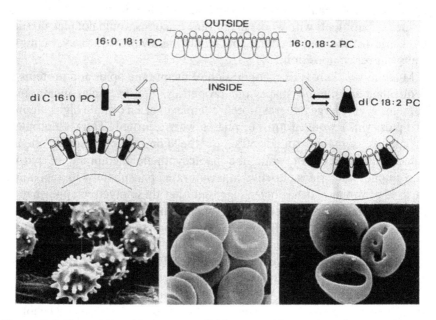

Figure 8.4 The relationship between lipid composition and erythrocyte shape. Reproduced with kind permission of Professor L.L.M. van Deenen and Elsevier Trends Journals, from *Trends in Biochemical Sciences* (1985), p. 322, Figure 3.

abnormal cell shape, increased osmotic fragility and an instability of the lipid bilayer manifesting itself in the temperature-sensitive nature of flip-flop movement. Also, in mutants which lack one or more of the membrane-spanning proteins which anchor the cytoskeleton in the bilayer, trans-bilayer mobility of lipids was enhanced. These observations emphasize that proteins play a role in determining lipid properties in membranes as well as *vice versa*.

8.5 LIPOSOMES AND DRUG DELIVERY SYSTEMS

In Chapter 6 we described how many membrane lipids, when isolated, could form natural bilayers in aqueous systems. To prevent any contact of the hydrophobic acyl chains with the aqueous medium, such bilayers close to form vesicles. Usually, under the experimental conditions used to rehydrate lipid samples and with the commonly used membrane lipids (such as phosphatidylcholine) stable multi-layered structures are formed. These structures, which have been compared to the multi-layered appearance of onions, were termed **liposomes** by Bangham. He realized when studying such systems that, because of their non-toxic nature and impermeability to many solutes, they would make very good agents for drug delivery and other clinical applications. Because liposomes contain alternating layers of bilayer and aqueous space, hydrophobic solutes can be contained within the bilayer and water-soluble materials within the aqueous compartments. Both types of molecule are prevented from leaking by the hydrophilic or hydrophobic layers, respectively.

There are three fundamental problems to the use of liposomes for drug delivery. First, the liposomes must be able to carry a sufficient quantity of trapped drug. Their capacity is usually increased by making liposomes of very large volumes using special procedures such as the reversed-phase evaporation method. Second, once in the blood circulation, serum lipoproteins tend to interact with the liposomes and cause leakage. This leakage can be reduced by using more saturated phospholipids and increased cholesterol proportions in the liposomes. Third, because of the size of liposomes they tend to be taken up by the macrophages of the reticuloendothelial system. This means that the liposomal contents are almost always discharged in the liver or spleen, regardless of which organ site is really the target. Furthermore, the blood-brain barrier precludes treatment of brain tissue by liposomal delivery. Attempts are now being made to introduce specificity into the targeting of liposomes by attachment of antibodies to the vesicle surface.

Despite these problems, the enormous potential for the use of liposomes in medicine ensures a very active research field. Some successes have been recorded already. Thus Fabry's disease (a lipidosis; section 8.14) which affects the liver mainly has been treated by enzyme-replacement therapy while parasites such as *Lieshmania* which reside in macrophages have been destroyed by drugs delivered in liposomes. Macrophages can also be activated by certain activating factors which can be liposome-delivered. Such treatment renders the macrophages remarkably active in recognizing and destroying diseased tissues including transformed cells.

A particularly exciting development of the use of liposomes has been the attempt to utilize the changed patterns of surface glycoproteins and glycolipids in malignant cells (section 8.10). The aim is to incorporate a specific recognition molecule onto the surface of liposomes which carries highly toxic molecules such as the protein ricin. Treatment of patients should result in the specific uptake of liposomes only by cancer cells which would then be destroyed. In theory these techniques would enable very small numbers of cancer cells to be destroyed before they had a chance to form a sizeable malignancy and might be particularly useful for the treatment of secondaries or in following up surgical intervention.

An interesting commercial application of liposomes has been the recent development of a new type of skin preparation. These varieties of moisturizing creams are lotions which are claimed to remove wrinkle lines and are anti-ageing. The main constituents of these preparations are liposomes and the use of such vehicles might be expected to aid in the skin penetration of the other components. Although the scientific basis for some of the claims made by the manufacturers are difficult to understand, the lotions do represent a novel and unexpected use of liposomes!

8.6 MEMBRANE RECEPTORS: THESE ARE PROTEINS EMBEDDED IN THE LIPID BILAYER AND THEIR ACTIVITY IS MODULATED BY THE NATURE OF THOSE LIPIDS

Membrane receptors are proteins designed to recognize specific molecules and bind them in preparation for the initiation of a biological process that takes place in the membrane. The process may be the transport of a molecule (e.g. glucose, LDL) across the membrane or the triggering of chemical message by a hormone or a growth factor. The recognition step demands that the receptor has specificity, like an enzyme for its substrate. Binding is followed frequently, but not always, by internalization in which the membrane in the vicinity of the receptor forms a vesicle that encapsulates the receptor-ligand complex and enters the cell. Once inside, the receptor complex can be degraded by lysosomal enzymes and the receptor proteins can often be recycled to be reinserted into the plasma membrane again. The number of membrane receptors is usually responsive to the availability of the specific ligand. This requires a mechanism for the control of the biosynthesis of the receptor by the endoplasmic reticulum. Receptors are usually glycoproteins. A well worked out example of a receptor protein involved in lipid metabolism is the LDL receptor (section 5.3.5(g), references and Figure 8.5). The receptor is synthesized on the rough endoplasmic reticulum as a precursor of molecular mass 120 kDa. About 30 minutes after its synthesis, the protein is modified to a mature receptor with an apparent mass of 160 kDa as estimated by SDS migration. The increase in molecular mass coincides with extensive modifications of the carbohydrate chains. The precursor molecule contains up to 18 *N*-acetylglucosamine molecules attached in *O*-linkage to serine and threonine residues as well as two high mannose-containing *N*-linked chains. The latter are modified extensively and the *O*-linked chains are elongated by addition of one galactose and two sialic acid residues to each *N*-acetyl-glucosamine. These changes do not represent an increase in sugar content of 40 kDa, so it is presumed that a conformational change in the receptor protein also contributes to the apparent increase in molecular mass. Other receptor proteins (e.g. the transferrin receptor) may have phosphate and

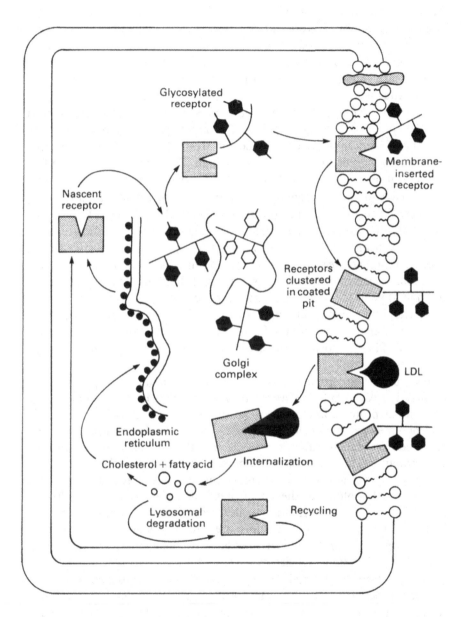

Figure 8.5 The synthesis and migration of LDL receptors.

fatty acyl groups covalently bound into the structure in addition to sugar residues.

The overall process directed by a membrane receptor may be studied by a number of techniques *in vivo*, but to establish its fine structure and the details of receptor-ligand interaction, the functional receptor must be solubilized, removed from the membrane and examined in isolation. Removal of proteins from membranes almost always involves the use of detergents and it is crucial that an appropriate detergent is chosen for a specific task. First, it is essential that the receptor is truly solubilized, not simply attached to a rather large fragment of non-specific membrane. The criteria for defining solubility may be separation from insoluble or sedimentable material by centrifugation under specified conditions or the use of gel filtration. A general rule seems to be that successful solubilization occurs only with detergents that have a high CMC (section 3.1.8) and that concentrations of detergent above the CMC usually lead to loss of function. In the case of the β-adrenergic receptor, however, digitonin is the only detergent that solubilizes this receptor in a form that still binds ligands, but the low CMC of this particular detergent makes its removal difficult. In this case, it has been necessary to use a second detergent (e.g. octyl glucoside or sodium deoxycholate) during the next step, which is the reconstitution of the functional receptor in lipid vesicles. The actual lipids used for reconstitution do not appear to be critical and successful reconstitutions have used dimyristoyl phosphatidylcholine, crude soya bean lecithin or lipid extracts derived from the membranes from which the receptors were originally prepared. Final removal of the detergent is generally accomplished by gel filtration.

The functional properties of receptors in membranes appear to be intimately related to the microenvironment provided by membrane lipids. Membrane proteins are on average associated with 30–40 molecules of phospholipid per molecule of peptide. These annular phospholipids are required for functional activity or for the stabilization of protein conformation. The structural and compositional changes of the lipid bilayer, that provides a fluid matrix for proteins, can induce alterations in functional properties of proteins in the intact membrane, for example by allowing changes in protein conformation or in the diffusion or position of proteins in the membrane. Lipid structures are also necessary for the defined orientation of the proteins once they have been solubilized by detergents and stabilized in lipid vesicles as described above. We may speak about membrane-associated proteins being dependent on the presence of lipids for functional activity and this can best be demonstrated by observing restoration of receptor activity, lost during detergent solubilization, when suitable lipids are added back. A possible specific role for phosphatidylinositol in the insulin receptor is discussed in section 8.7. More subtle modulations of receptor activity by the type of lipids in the bilayer can best be demonstrated using techniques *in vivo* that do not involve gross disruption of existing protein–lipid interactions in the native membrane. Such lipid compositional changes can be achieved conveniently by dietary means or by manipulating the lipid composition of the media in which various cultured cell lines are grown. Alternatively membrane lipid modifications can be made *in vitro*, for example, altering lipid composition by incubation of membranes with different lipids in the presence of exchange proteins.

8.7 INOSITOL LIPIDS PLAY SPECIFIC ROLES IN MEMBRANE PROTEIN ANCHORING

Intrinsic proteins may be anchored in the membrane by several different mechanisms. Of these, the hydrophobic interaction of an amino acid sequence with the interior of the lipid bilayer is the most common. Such interactions include hydrophobic sequences at one end of the protein only as well as polypeptide chains which traverse the membrane (several times). In addition, some proteins may be attached through interactions with lipid. Thus, fatty acid acylation of the NH_2-terminus or the formation of fatty acid thioesters with cysteine residues have been observed. In addition, a number of membrane proteins are believed to be covalently attached to phosphatidylinositol.

The evidence that phosphatidylinositol has a role in protein anchoring comes from two sources: (a) the release of certain membrane proteins following digestion with a phosphatidylinositol-specific phospholipase C and (b) analysis of the anchoring domain of some of these membrane proteins. Examples of proteins believed to be attached thus to membranes include alkaline phosphatase, acetylcholinesterase, 5'-nucleotidase and the variant surface glycoprotein (VSG) from *Trypanosoma brucei*. (The latter is responsible for the ability of this parasitic protozoan to evade the victim's immune defence system and so cause sleeping sickness.)

The anchoring domain for VSG has been determined in some detail and, together with information from other proteins, a generalized structure for the membrane link proposed (Figure 8.6). It will be noted that it is only the diacylglycerol domain of phosphatidylinositol which holds the membrane protein in place. However, such an anchor is at least as strong as a hydrophobic sequence of amino acids spanning the entire membrane.

At present we do not have a definite reason as to why some membrane proteins are attached through phosphatidylinositol. One possibility is that it may provide a means of selectively mobilizing these proteins. Indeed, there is some evidence that endogenous phospholipases may cause the release of VSG and alkaline phosphatase in soluble forms and other proteins such as acetylcholinesterase are also known to be released under certain physiological conditions.

Recently, Cuatrecasas and associates claim to have achieved the partial purification and characterization of two novel mediator substances produced by insulin action. These substances appeared to be complex

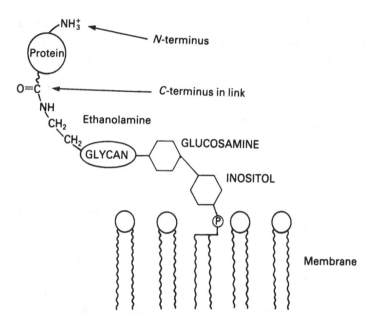

Figure 8.6 Probable structure of the phosphatidylinositol anchor for membrane proteins.

carbohydrate phosphates containing inositol and it has been suggested that insulin causes their release as follows. Binding of insulin to its receptor allows the α-subunit to interact with a specific G protein (G_{Ins}). A phospholipase C type enzyme is activated in consequence and this enzyme is specific for phosphatidylinositol (rather than its phosphorylated derivatives). The enzymic action would generate the water-soluble inositol derivative as well as diacylglycerol and these compounds could have interactions with the changed cAMP concentrations also induced by insulin binding. As confirmation, Cuatrecasas showed that the action of insulin in releasing mediator from plasma membranes could be mimicked by the addition of purified phosphatidylinositol-specific phospholipase C. In addition, other workers have shown that insulin causes an increase in cellular diacylglycerol and stimulates the turnover of phosphatidylinositol in a number of cell types. Certainly it is easy enough to see how diacylglycerol could be released into the cell and have important physiological effects such as via the arachidonic acid cascade (section 3.4). However, it is more difficult to envisage a role for the water-soluble inositolphosphate derivative, given that the protein anchor is in the outer leaflet of the plasma membrane bilayer. There is much current interest in elucidating insulin action so further advances are awaited with interest.

8.8 INOSITOL LIPIDS AND SECOND MESSENGERS

Mention has already been made of the relatively high turnover of phosphatidylinositol and other inositol-containing lipids (section 7.1.8). In fact Mabel and Lowell Hokin first coined the term 'the phosphatidylinositol cycle' nearly 40 years ago. They found that acetylcholine caused a stimulation of labelling of phosphatidate and phosphatidylinositol from ^{32}P-orthophosphate in pigeon pancreas slices. They believed that the high labelling was due to the rapid synthesis and degradation of phosphatidylinositol and subsequent research has confirmed this, although our current picture is much more complicated than originally envisaged.

It is now known that the initial event following binding of the agonist to its receptor (usually within 30 secs) is a rapid hydrolysis of phosphatidylinositol bisphosphate by a specific phospholipase C. This produces inositol 1,4,5-triphosphate and diacylglycerol, both of which have second messenger functions.

First of all, it is necessary to consider the various reactions involved in the metabolism of the inositol lipids. These are shown in Figure 8.7. Phosphatidylinositol is converted into its phosphorylated derivatives by the

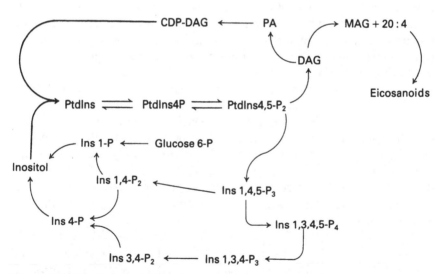

Figure 8.7 The phosphatidylinositol cycle and ancillary reactions.
Abbreviations: CDP-DAG = CDP-diacylglycerol;
PA = phosphatidic acid (phosphatidate);
MAG = monoacylglycerol;
DAG = diacylglycerol; 20:4 = arachidonic acid;
PtdIns = Phosphatidylinositol;
Ins = Inositol

addition of phosphate from ATP through the action of two plasma membrane-associated kinases. Although phosphatase enzymes are also present in cells which can reverse the reactions, they seem to have low activity under physiological conditions *in vivo*. (They do, however, result in the disappearance of virtually all phosphatidylinositol bisphosphate within a minute or so of cell death, in the absence of appropriate precautions.) The specific phospholipase C then cleaves phosphatidylinositol 4,5-bisphosphate to give the water-soluble inositol 1,4,5-triphosphate and diacylglycerol.

Both inositol triphosphate and diacylglycerol can be metabolized by two separate pathways. Either diacylglycerol can be phosphorylated to yield phosphatidate, which then can go on to complete the cycle back to phosphatidylinositol or a lipase will cleave the fatty acid at the *sn*-2 position (often arachidonate). If arachidonic acid is released then it can initiate the arachidonate cascade (section 3.4.2 and 3.4.5). Likewise inositol 1,4,5-triphosphate can either be phosphorylated to give a tetraphosphate (which may itself produce cellular effects) or it can be hydrolysed via intermediates to yield inositol and again complete the cycle.

The hormone-sensitive inositol lipid pool is confined to the plasma membrane where most of the phosphatidylinositol bisphosphate is located. Although the other inositol lipids have been suggested to be used for cell signalling, it is phosphatidylinositol bisphosphate which is by far the most important. Increased hydrolysis of this lipid is caused by a host of stimuli including neurotransmitters, releasing factors, hormones, growth factors, fertilization and light.

The main reactions involved in the generation of the inositol lipid signalling system are shown in Figure 8.8. Binding of an agonist to its receptor can activate a G-protein, which is responsible for signal transduction across the membrane. For inositol lipids, this protein is referred to a Gp-protein (p standing for phospholipid) and there is the intriguing possibility that Gp may resemble or actually be the product of the *ras* oncogene. Expression of the latter appears dramatically to increase the ability of receptors to activate phosphoinositides – a process which is normally rate-limiting.

The Gp-protein causes activation of a specific phospholipase C which generates inositol 1,4,5-triphosphate and diacylglycerol from phosphatidylinositol 4,5-bisphosphate. By analogy with the control of adenylate cyclase, the presence of an inhibitory G-protein has also been suggested for phosphoinositides. Although some data are consistent with its presence, the existence of an inhibitory G-protein has not been proved.

Both the products, inositol 1,4,5-triphosphate and diacylglycerol, function as second messengers (Figure 8.8). In addition, removal of phosphatidylinositol 4,5-bisphosphate from the plasma membrane may,

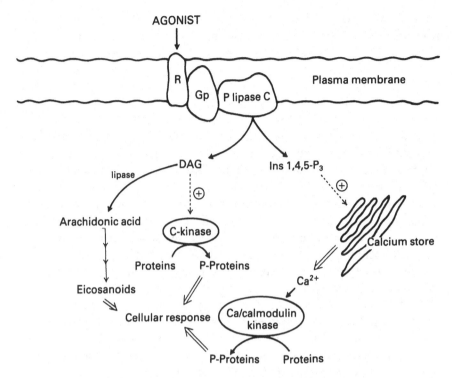

Figure 8.8 The dual signal hypothesis of phosphoinositide hydrolysis.
Abbreviations: Ins 1,4,5-P$_3$ = inositol 1,4,5-triphosphate; DAG = diacylglycerol.

itself, play a role in cell metabolism. Thus, this lipid appears to function in attaching the spectrin system to the plasma membrane. Its breakdown may, therefore, contribute to the remodelling of the cytoskeleton which has been observed following stimulation of secretory cells (section 8.4).

The diacylglycerol which is generated in the membrane may also restructure microfilaments by being involved in attachment of α-actinin to the platelet membrane. However, its main effect is to activate protein kinase C. Interestingly, this activation also needs phosphatidylserine and Ca^{2+} (the latter being mobilized by inositol triphosphate). Active protein kinase C can phosphorylate a large number of proteins *in vitro* and it is thought that at least some of these contribute to the physiological response involving inositol lipids. One of the proteins which has been identified for such a function is the epidermal growth factor (EGF) receptor. However, at the present time it is thought that the diacylglycerol/protein kinase C part of the response mainly plays a modulating role. Thus, it is thought to sensitize calcium-dependent processes, but at the same time has a negative effect on intracellular calcium levels. It may control the latter by reducing calcium efflux from stores and by stimulating calcium pumps that remove calcium from the cytosol. Protein kinase C also appears to stimulate the formation of phosphatidylinositol bisphosphate, thus readying the cell for a new agonist-induced stimulation. Because protein kinase C needs calcium for activity but has various effects on calcium levels it occupies a pivotal position. Thus it can either enhance or inhibit calcium signalling and its net effect will depend on the degree to which these functions are expressed in a given cell. For example, activation of protein kinase C with phorbol esters enhances contractibility in some smooth muscle cells but relaxes others.

Of all the various inositol phosphates, inositol 1,4,5-triphosphate is the only one for which a clear second messenger role has been identified. In particular, it acts to release calcium from the endoplasmic reticulum. To release calcium, inositol 1,4,5-triphosphate acts through a receptor and the resultant rise in cytosolic calcium causes changes in activity in a number of enzymes, including protein kinase C, but notably protein kinases modified through the calcium/calmodulin system. These proteins can then cause a large number of changes in cell metabolism and regulation.

We have already seen how the two products of phosphatidylinositol bisphosphate hydrolysis can interact through their cellular effects. A good example is liver, where the inositol 1,4,5-triphosphate/Ca^{2+} pathway controls the activity of phosphorylase kinase whereas the diacylglycerol/protein kinase C pathway switches off glycogen synthetase.

The principal secondary messenger diacylglycerol can also give rise to additional messengers. As shown in Figure 8.7 it can either be phosphorylated to phosphatidate or hydrolysed by a lipase. Because phosphoinositides usually have a high concentration of arachidonate at their *sn*-2 position, lipase action will cause a significant rise in arachidonic acid levels. The latter can then be converted to the eicosanoids (prostaglandins, thromboxanes and leukotrienes) as described in section 3.4.

Two other aspects of the inositol cycle should also be mentioned. First, hydrolysis of inositol 1,4,5-triphosphate is necessary to complete the cycle. The initial breakdown is very rapid ($T_{1/2} = 4$ sec) but further stages can be altered by various compounds notably lithium which is used in the therapy of manic-depression. Second, some of the other inositol phosphates have been suggested to have messenger functions – notably inositol 1,3,4,5-tetraphosphate which may be utilized in stimulating extracellular calcium entry during fertilization of sea urchin eggs.

Considering the explosion of interest in and knowledge gained about phosphoinositides in the last ten years, it will not be surprising if many other fascinating aspects are uncovered in the next decade.

8.9 ROLE OF LIPIDS IN IMMUNITY: LIPIDS PLAY A ROLE IN IMMUNITY BY INFLUENCING THE PROPERTIES OF MEMBRANES IN CELLS OF THE IMMUNE SYSTEM AND BY ACTING AS PRECURSORS OF REGULATORY SUBSTANCES

Lipids can be involved in immunity in a number of ways, principally through their roles in membrane integrity and in providing precursors for eicosanoid formation. Thus, the proliferation of lymphocytes involves rapid synthesis of new membranes; the interaction of antigens with receptors on the lymphocyte surface is influenced by the physico-chemical properties of the membrane (sections 6.5.4 and 8.1); and the eicosanoids play a key role in inflammatory reactions. Dietary lipids may modulate immune reactions, therefore, either by modifying the composition of cell membranes and membrane surface components or by influencing the nature of the precursor pool for eicosanoids.

In addition to the rather general influence of lipids on membrane properties, certain lipids may also play a more specific role and, in this respect, attention has focussed on the glycosphingolipids. Essentially all the glycosphingolipids are immunologically active, either in haptenic reactivity *in vitro* or in antibody-producing potency. Carbohydrate antigens may be distributed in both glycoproteins and glycolipids. Regardless of the nature of the carbohydrate moiety, however, the lipid moiety of glycosphingolipid antigens markedly strengthens the antigenic potency.

Complex glycosphingolipids that contain L-fucose (fucolipids) have attracted particular interest because they possess antigenic activity of the human ABH and Lewis blood group systems. In man, the ABH locus has been assigned to chromosome 9. The H gene gives rise to a carbohydrate structure that is the immediate precursor of the A and B determinants. Thus, in group O persons, the H-determinant does not undergo further change. The ABH antigens occur in erythrocytes and possibly in other cell types having ABO antigenicity in the form of low molecular weight glycosphingolipids. The classic Lewis system consists of Lewis-a (Le_a) and Lewis-b (Le_b) antigens and is related to the ABH antigens in body fluids. Individuals who have Le_a antigens on their red cells are non-secretors, while in contrast, those with Le_b antigens are secretors of ABH antigens. About 10% of the population lacks both Le_a and Le_b antigens which, interestingly, are not constitutive in erythrocyte membranes but are acquired from blood plasma. Two other systems of glycosphingolipid antigens have also been described, namely the Ii and P systems.

The glycosphingolipid content of plasma membranes and, hence, their antigenicity changes during embryogenesis and differentiation. The embryonic stage-specific substances have been postulated to regulate cell interactions and cell sorting. In particular, antigens belonging to the human blood group ABH, Ii system and the Forssman antigen are thought to be important. Among the glycosphingolipids on the lymphocyte cell surfaces of rats and mice, gangliotetraglycosylceramide has been identified as being immunologically important.

Anti-glycosphingolipid antibodies have been detected in patients with a number of diseases. Several autoimmune diseases (e.g. experimental allergic encephalitis of the central nervous system) can be augmented by repeated immunization with galactosylceramide. Moreover, antibodies against galactosylceramide were able to induce demyelination and to prevent myelination and sulphatide synthesis in tissue cultures of the central nervous system. These observations suggest a role for some glycosphingolipids in the aetiology of autoimmune diseases.

In the 1970s it was found that diets rich in polyunsaturated fatty acids prolonged the survival of skin allografts in mice and were subsequently employed as adjuncts to conventional immunosuppressive therapy to reduce rejection of human kidney grafts. Such diets also appeared to be beneficial in treating patients with multiple sclerosis. It has been suggested that one of the features of this disease is the production by the nervous system of a protein antigen to which the person's own immune system responds, eventually resulting in damage to the myelin membranes. These findings strongly suggest that polyunsaturated fatty acids were acting to suppress the responses of the immune system and that suggestion is reinforced by findings that antibody production following immunization by antigen can be reduced in animals given high fat diets.

Progress in understanding the mechanisms by which lipids or their metabolites were able to influence immunity could only be made by working at the cellular level. Several model systems have been used, these include: (a) Cultured lymphocytes which are stimulated to divide by specific antigens or non-specific mitogens (e.g. phytohaemagglutinin, PHA). Cell division is measured by the incorporation of tritiated thymidine into DNA. Addition of fatty acids to the culture medium has usually been found to suppress the increase in DNA labelling in response to mitogens. The effect has usually, but not invariably, been greater with polyunsaturated fatty acids than monounsaturated or saturated fatty acids, and even more pronounced with prostaglandins (Table 8.4). (b) Macrophage suspensions that migrate in an electric field. Mitogens inhibit this migration and the lipids mentioned above tend to release this inhibition. (c) Host versus graft reaction. In this model, lymphoid cells are taken from a donor animal and grafted into a genetically different host animal which recognizes them as foreign and mounts a rejection. Animals given diets containing a large quantity of polyunsaturated fatty acids are able to suppress this rejection. Fish oil fatty acids in the n-3 series are particularly powerful in suppression. At low dietary intakes, suppression does not occur; indeed there may even be an enhancement of immunity. A frequent finding is that there is a bell-shaped response of various types of immune response to essential fatty acids (Figure 8.9). At levels consistent with deficiency, immunity is poor because membrane integrity is affected. More normal levels are consistent with a fully functioning immune system whereas high levels are suppressive.

Several other tests for humoral immunity (plaque-forming assays, passive haemagglutination) and cellular immunity (skin tests, rosette formation by red blood cells) have been used and the general finding is suppression by polyunsaturated fatty acids or their eicosanoid metabolites. The degree of suppression varies enormously and depends on many factors inherent in the experimental design. The mechanisms by which lipids influence immunity are not yet understood. Changes in membrane phospholipid fatty acid composition occur in response to prolonged consumption of diets containing lipids of different composition or even to acute administration of lipids. Several authors have concluded that lipid unsaturation is crucial to the activation of lymphocytes but measurements really need to be made of the

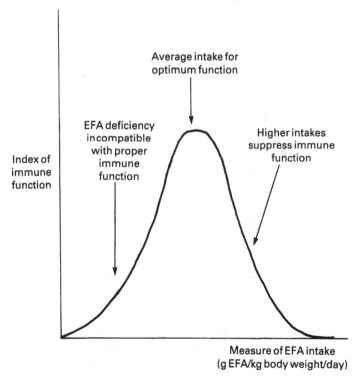

Figure 8.9 Dependence of immunity on essential fatty acids.

Table 8.4 Effect of exogenous fatty acids and prostaglandins on human lymphocyte stimulation *in vitro**

Fatty acid	Dose	Form of addition	Lymphocyte test	Serum in culture	Stimulant	Result
Palmitic stearic oleic linoleic arachidonic	80 µg/ml	Ethanol solution	Lymphocyte transformation [³H]thymidine	10% human AB serum	PHA PPD	All FA inhibited basal incorporation of ³H-thymidine; 18:2, 20:4 inhibited incorporation into cells stimulated by PHA and PPD
Myristic palmitic heptadecanoic stearic oleic linoleic arachidonic	0.2 mM	Albumin complex	Lymphocyte transformation [³H]thymidine	10% fetal calf	PHA	Each fatty acid except 14:0 inhibited when added singly; when UFA and SFA present together, little or no inhibition, 14:0 did not inhibit but relieved inhibition of both UFA and SFA
Oleic linoleic α-linolenic 8, 11, 14–20:3	0.1–100 µg/ml	Sodium salts	Lymphocyte transformation [³H]thymidine [³H]uridine	1. 0 serum 2. lipid-poor albumin 3. 1% fetal calf 4. 10% fetal calf	PHA	20:4, 18:2, 18:1 increased stimulation at concentrations of 0.1–5 µg/ml and decreased stimulation at higher concentrations

Substance	Concentration	Preparation	Test	Serum	Stimulant	Result
Oleic linoleic arachidonic	80 μg/ml	Sonicated suspension in culture medium	Macrophage migration inhibition test	10% fetal calf	Measles antigen parainfluenza virus type I antigen	18 : 2, 20 : 4 (not 18 : 1) decreased inhibition of migration by antigens in subjects that display vigorous response to antigen and increased in patients showing poor response
PGA PGE$_1$ PGE$_2$ PGF$_{1\alpha}$ PGF$_{2\alpha}$	3×10^{-8} M– 1.5×10^{-5} M	Ethanol solution	Lymphocyte transformation [^{3}H]thymidine	20% fetal calf	PHA PWM ConA	PGE inhibited 50% at 10^{-7} M. Other PGs only inhibit at high concentrations. PHA and ConA stimulation blocked more effectively than PWM stimulation
PGE$_1$ PGE$_2$	10^{-8} M– 10^{-7} M	Ethanol solution	Lymphocyte transformation [^{3}H]thymidine	10% fetal calf	PHA	10^{-7} PGE inhibited LT in mixed T cell population 35%. High density T-cells were 60% inhibited, medium density 15% and low density enhanced
PGE$_1$ PGE$_2$	2.75×10^{-8} M– 2.75×10^{-6} M	Ethanol solution	Macrophage migration inhibition test	10% fetal calf	Measles and para-influenza type I virus	Inhibited lymphocyte activity in lymphocytes showing vigorous response to virus and enhanced in those showing weak response

* Each set represents the results of different published experiments. For original literature see references Gurr (1983).
UFA = unsaturated fatty acid; SFA = saturated fatty acid; PHA = phytohaemagglutinin; PPD = purified protein derivative of tubercle bacillus; PWM = pokeweed mitogen; PG = prostaglandin; LT = lymphocyte transformation; ConA = concanavalin A.

membrane environment in the vicinity of the receptor for the antigen or mitogen. Likewise, although eicosanoids are clearly involved, their mode of action is obscure. Finally, there is also evidence for an immunoregulatory function of certain subclasses of low density lipoproteins. Washed lymphocytes from the plasma of patients with hyperlipoproteinaemias responded normally to mitogenic stimulation in presence of plasma from subjects with normal plasma lipoprotein concentrations but all responses were inhibited when the plasma was taken from patients with hyperlipoproteinaemias. An LDL subfraction has been isolated that caused 50% inhibition of lymphocyte transformation by PHA and suppresses humoral immunity *in vivo* but insufficient research has been done to assess the significance of such immunoregulatory lipoproteins in the overall functioning of the body's immune defence system.

Finally, a discussion of the role of lipids in immunity would be incomplete without mentioning that vitamin A has long been known to be associated with the immune system. Vitamin A-deficient animals are prone to infection and protection is restored by the small amounts of the vitamin that are necessary to overcome the deficiency. Larger doses have an adjuvant effect and stimulate the immune system. Extremely large doses, however, are toxic to the cells of the immune system as well as having general toxic effects on other organs. The precise mechanisms are ill-defined but the action of retinoids on cells of the immune system may be considered as a specialized facet of the general process of cellular differentiation. Thus, they affect the early processing of antigens by macrophages, activate B-cells and thereby stimulate antibody production and stimulate T-cell formation in the thymus gland.

8.10 LIPIDS AND CANCER: LIPIDS MAY BE INVOLVED IN THE DEVELOPMENT OF CANCER BECAUSE OF THEIR ROLE IN MEMBRANES OR THROUGH DIETARY EFFECTS INCLUDING INFLUENCES ON THE IMMUNE SYSTEM

Whereas normal cells undergo very carefully regulated growth and division, tumour cells have lost nearly all control and grow autonomously. These differences are conveniently induced with tissue cultures *in vitro* by transforming the cells with oncogenic DNA or RNA viruses. The transformed cells appear to be analogous to tumour cells derived *in vivo*, either spontaneously or by treatment with carcinogenic agents. Typical alterations in transformed cells include a loss of contact inhibition of growth and movement, enhanced sugar transport, increased cell agglutinability by lectins and a lowered requirement for serum factors.

In 1968, the Japanese biochemist, Hakamori, found that upon oncogenic transformation, a dramatic change in the pattern of glycosphingolipid components takes place. Following the characterization of these lipids by workers in Rapport's laboratory, further research showed that there were two major types of change:

1. Deficiency of specific glycosyltransferases leads to a shortening of the carbohydrate chains of the more complex cellular glycosphingolipids.
2. Unusual glycosphingolipids, that are either present in only trace amounts or are undetectable in normal tissues, become major constituents of malignant tissues.

Two examples of changes observed are shown in Table 8.5. Since it is the more complex glycosphingolipids that possess antigenic properties, the loss of these lipids would be expected to modify differentiation and growth in the manner observed. It is sometimes difficult, however, to demonstrate unequivocally whether the glycosphingolipid pattern and metabolism of a tumour *in vivo* are different from those of the parent tissue. This is because of cellular heterogeneity in tissues and the fact that glycosphingolipid patterns are cell-dependent and change with development (section 8.9). For example, analysis of the glycosphingolipid patterns of lung cancer tissues may be complicated by those lipids present in leukocytes that invade these tissues prolifically. Nevertheless, where tumour tissues contain basically a single cell type (e.g. leukaemia), comparisons with normal tissue can be rewarding. The general finding with transformed tissue cultures that there is a decrease in the amount of complex glycosphingolipids and an increase in specific lipids normally absent or present only in small quantities appears to be substantially true for many tumour tissues (Table 8.6).

Table 8.5 Changes in glycosphingolipid contents associated with transformed cells

Cell	Lipid accumulating	Lipid decreasing
Hamster BHK[a]	LacCer	NeuAc-LacCer
Rat LV embryo[b]	NeuAc-LacCer	NeuAc-G_{T1}Cer NeuAc-G_{Q1}Cer $(NeuAc)_2 G_{Q1}$Cer

[a] Transformed by polyoma virus.
[b] Transformed by Rauscher-leukemia virus.

Table 8.6 Increment or appearance of specific glycosphingolipids in cancer

Tumour type	Glycosphingolipid increased
Colonic adenocarcinoma	Fucosylceramide
Burkitt lymphoma	Trihexosylceramide
Melanoma	Sialylated Le$_a$ antigen*
Gastric and colonic adenocarcinoma	X hapten*

* Structures

Sialylated Le$_a$ antigen = NeuAc-Gal-GlcNAc-Gal-Glc-Cer
                                                         |
                                                        Fuc

X hapten = Gal-Gal-GlcNAc-Gal-Glc-Cer
                            |
                           Fuc

It has also been demonstrated that tumour cells cultured *in vitro* often release tumour-associated antigens into the growth medium. Similar antigens can be detected in the blood of patients with advanced cancer. The development of monoclonal antibodies should make possible the detection of (tissue specific) antigens in quite small amounts and methods are now being developed to improve diagnosis.

Antibodies against glycosphingolipids have been used for therapeutic purposes. For example, antibodies against a ceramide trihexoside that accumulates in murine lymphoma tissue caused a suppression of growth of such tissue. This technique has now been developed further and combined with the knowledge that avidin is a very specific inhibitor of biotinyl-proteins (section 3.2.2). An antibody to ceramide trihexoside was coupled to biotin and added to virus-transformed mouse 3T$_3$ cells in culture. Avidin was complexed to the anti-cancer drug, neocarzinostatin, and when this complex was added, it successfully targeted to and killed the transformed cells. These techniques offer hope that it will be possible specifically to destroy malignant cells and so avoid the unwanted side-effects currently experienced by patients undergoing chemotherapy.

Lipids may also be associated with cancer development through their effects in the diet. Knowledge about the involvement of any particular nutrient in the development of a disease often comes from epidemiological studies. The epidemiologist collects statistics about the occurrence of specific diseases in populations and relates them to certain characteristics in the lifestyles or in the biochemistry of persons in those populations. These techniques have brought to light strong positive correlations between the consumption of fat per head of population and the mortality from breast and colon cancer in many different countries in the world. It must be emphasized that a strong correlation between A and B arising from an epidemiological study is in no way indicative that A causes B. The difficulties involved in drawing conclusions from epidemiological data are nicely illustrated by the apparent inverse correlation between cholesterol at low plasma concentrations and cancer that has been debated vigorously in the 1980s.

The biggest problem in the epidemiology of a slow developing disease like cancer is that one does not know whether the correlation existed at the time when the disease was initiated. The low plasma cholesterol might be the end result of a series of metabolic changes brought about by the cancer itself or there might be a direct cause and effect relationship between the low plasma cholesterol concentration and the development of the cancer as suggested by recent clinical findings that low plasma cholesterol values may precede the detection of colon cancer by more than five years. Other types of clinical studies, such as case-control studies and prospective studies, have been unable to confirm the positive correlations between dietary fat and cancer provided by epidemiology.

From the biochemist's point of view, epidemiology is useful in giving clues about the most worthwhile avenues for research. The formation of tumours arising from the action of a carcinogen proceeds through three distinct stages. The first is **initiation**, in which a carcinogen (either from the diet or produced during body metabolism) causes a permanent alteration to the DNA in a particular cell. A typical initiator is a substance like dimethyl-benzanthracene (DMBA). In the next step, **promotion**, the promoter increases the chance of the misinformation, inserted into the genetic code by the initiator, being expressed. Finally, **propagation** involves anything that stimulates cell growth.

Animal experiments have shown that high levels of dietary fat promote the development of mammary tumours initiated by DMBA although some researchers have pointed out that dietary fat does not seem to work through the same mechanisms as other promoters. Dietary fat appears to enhance carcinogenesis by decreasing the latent period for tumour development rather than increasing the final tumour yield. There must be a minimal level of polyunsaturated fatty acids in the diet for promotion to occur but this will only be effective when the total fat content of the diet is high: about 40% of dietary energy. A low fat diet rich in polyunsaturated fatty acids was ineffective in promoting the effect of DMBA as was a high fat diet that contained no polyunsaturated fatty acids. As most human diets contain at least enough essential fatty acids to protect against essential fatty acid deficiency, the main correlation observed is, therefore, between total fat intake and cancer prevalence. The situation is complicated, however, by the finding that a most effective way to limit the occurrence of spontaneous tumours or the growth of induced tumours is to restrict an animal's energy intake. Part, at least, of the effect of dietary fat may be in its contribution to energy intake, but the mechanism by which energy restriction inhibits tumorigenesis remains to be elucidated.

The induction of cancer by oxidized lipids is referred to in section 8.15 but we have no idea of the relative importance of these substances in the sum total of human cancer mobidity. Some authors attach more importance to the processes of lipid peroxidation *in vivo*, discussed briefly in section 8.11. Peroxisomes are responsible for the oxidation of an appreciable proportion of dietary fatty acids, especially those whose structural features (some isomeric and long chain acids for example) limit their oxidation in mitochondria. The removal of each two-carbon unit in peroxisomal oxidation generates one molecule of hydrogen peroxide which can be a carcinogen, a mutagen or a promoter. Some hydrogen peroxide escapes degradation by peroxisomal catalase, thereby contributing to the supply of oxygen free radicals. These can damage DNA and proteins, increasing the potential for tumorigenesis. We also have to consider the possibility that peroxidation can have a protective as well as a promoting role. Malignant tissues have a much lower capacity for peroxidation than healthy tissues. If one of the positive roles of lipid peroxidation is to terminate the life of old and worn out cells (section 8.11), the failure of lipid peroxidation in tumour cells may encourage their survival and multiplication. This illustrates the dangers of regarding any broad class of biological molecules as toxic or beneficial, suppressive or stimulatory. More often than not, the same molecules exhibit the full range of activities at different dose levels and dose-response relationships should always be established early in any investigation.

8.11 FREE RADICALS AND CELLULAR DAMAGE

Polyunsaturated fatty acyl groups (or acids) can become oxidized by chemical or enzymic means (section 3.3). Lipid hydroperoxides (ROOH) are stable intermediates of peroxidation and accumulate in peroxidized lipids. These are capable of reacting with transition metals to generate $RO\cdot$ and $ROO\cdot$ radicals. The latter can then cause structural changes in proteins including cross-linking and fragmentation.

One well studied system is that of the serum low density lipoprotein (LDL; sections 5.3.5(e) and 5.5.3). When the lipids of this lipoprotein are peroxidized, the apoprotein becomes fragmented and there is cross-linking and residue modification. Two products of lipid oxidation, malonaldehyde and 4-hydroxy nonenal, are responsible, the latter reacting with lysines of apoprotein-B by an addition reaction. As a result LDL no longer binds to the fibroblast receptor (section 5.5.3) and has an extended half-life in blood.

This gives an increased chance of uptake by macrophages and consequent production of atherosclerotic foam cells.

During radiolysis it is known that lipid peroxidation takes place and that inactivation of membrane enzymes occurs. This inactivation can be reduced by free radical scavengers and by inhibitors of lipid peroxidation such as vitamin E. The modification of enzymes is believed to proceed in two phases with an initial attack on the proteins by free radicals followed by prolonged further damage due to the continued production of radicals as the lipid ROOH decomposes in the presence of transition metals. Such changes have extremely widespread implications in that denatured proteins may accumulate in cells and affect function or, alternatively, radical generation at particular sites may lead to pathological tissue degeneration.

Organisms have a number of ways in which to reduce the accumulation of toxic lipid hydroperoxides. They may contain tocopherols or carotenoids as well as a variety of enzymes such as glutathione peroxidase which changes the hydroperoxide group to the much less toxic hydroxy moiety. However, for glutathione peroxidase to act it needs a non-esterified fatty acid substrate, the enzyme being inactive towards peroxidized membrane lipids. Interestingly, phospholipase A_2 which can release such peroxidized fatty acids from membranes is much more active towards these acids than for lipids with normal acids. Thus, phospholipase A_2 can play a role in protecting membranes from lipid peroxidation damage. It not only allows glutathione peroxidase to remove the toxic ROOH moiety but also permits regeneration of a normal membrane lipid molecule through reacylation.

8.12 LIPIDS AND SKIN DISEASES

The surface layers of multicellular organisms form an important barrier to the outside world. On the one hand this barrier prevents the escape of important substances such as water, while, on the other, it reduces the entry of harmful materials or organisms. Lipids are key constituents of these surface layers.

The composition of human skin surface lipids varies over the body but in sebaceous gland-enriched areas (such as the face) the secretion, sebum, may represent 95%. When sebum is freshly formed it contains mainly triacylglycerols, wax esters and squalene. However, due to bacterial action, unesterified fatty acids are rapidly released. These acids are both irritant and comedogenic (give rise to 'black-heads') and have been implicated in acne vulgaris – an extremely common complaint of puberty. It is often stated that the severity of acne correlates with the amount of seborrhoea and many treatments are designed primarily to remove epidermal lipids from the skin surface.

More serious than acne, but less prevalent, are the skin diseases atopic eczema and psoriasis. Lipids have been suggested to be involved in both of these complaints although it must be stated that the evidence at present is equivocal. It has been noted in patients with atopic eczema that there is a disturbance in the normal tissue complement of polyunsaturated fatty acids. For convenience blood samples have usually been analysed and these usually exhibit reduced levels of γ-linolenic ($\Delta6,9,12$-octadecatrienoic) acid. γ-Linolenate is an intermediate in the normal conversion of the essential fatty acid, linoleate, to the eicosanoid precursor arachidonate:

$$\Delta9,12\text{-}18:2 \xrightarrow{\text{desaturase}} \Delta6,9,12\text{-}18:3 \xrightarrow{\text{elongase}} \Delta8,11,14\text{-}20:3$$

$$\xrightarrow{\text{desaturase}} \Delta5,8,11,14\text{-}20:4 \tag{8.1}$$

Moreover, symptoms of atopic eczema have been reported to be alleviated by γ-linolenate-rich lipids such as the oil from evening primrose. These results are interpreted to indicate a deficiency of the $\Delta6$-desaturase responsible for linoleate metabolism but they do not explain the beneficial effect of some other treatments.

Psoriasis is a severely debilitating skin disease which affects 2% of the population world-wide. It is responsible for more days spent in hospital than any other skin complaint and present treatments are rudimentary and rather unsatisfactory. Characteristically there is an impairment of normal cell differentiation and development, excessively high levels of some arachidonate-metabolites and a potentiation of the symptoms in patients with lithium therapy. The symptoms have been drawn together recently with the suggestion that defective phosphoinositide metabolism (section 8.8) is involved.

Finally a few remarks should be made about essential fatty acid (EFA) deficiency (section 5.2.2). Linoleic acid is the precursor of the eicosanoids, leukotrienes, prostaglandins, thromboxanes and related compounds (section 3.4). Until recently it was thought that the classic skin symptoms resulting from a deficiency were due to a lack of eicosanoids. However, the specific role of n-6 acids (such as linoleate) as opposed to n-3 acids (such as α-linolenate) in membrane fluidity and their function cannot explain their essentiality. For example, aspirin ingestion which effectively prevents prostanoid formation does not result in the appearance of EFA-deficiency symptoms. It is also generally accepted that at least some cell lines in culture do not have an EFA requirement – thus suggesting that EFAs are not essential for the formation and function of cellular membranes in general.

Recently, several linoleic acid-rich lipids such as acylglucosylceramide, acylceramide and a specific wax ester have been identified in human, pig and

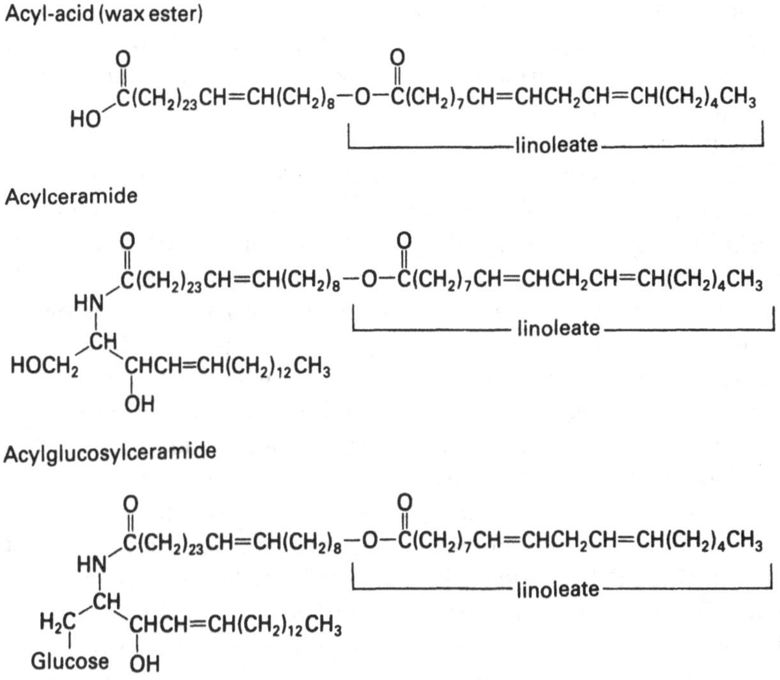

Acyl-acid (wax ester)

$$\underset{HO}{\overset{O}{\underset{\|}{C}}}(CH_2)_{23}CH=CH(CH_2)_8-O-\overset{O}{\overset{\|}{C}}(CH_2)_7CH=CHCH_2CH=CH(CH_2)_4CH_3$$

linoleate

Acylceramide

$$\overset{O}{\overset{\|}{C}}(CH_2)_{23}CH=CH(CH_2)_8-O-\overset{O}{\overset{\|}{C}}(CH_2)_7CH=CHCH_2CH=CH(CH_2)_4CH_3$$

HN
|
CH
HOCH$_2$ CHCH=CH(CH$_2$)$_{12}$CH$_3$
|
OH

linoleate

Acylglucosylceramide

$$\overset{O}{\overset{\|}{C}}(CH_2)_{23}CH=CH(CH_2)_8-O-\overset{O}{\overset{\|}{C}}(CH_2)_7CH=CHCH_2CH=CH(CH_2)_4CH_3$$

HN
|
CH
H$_2$C   CHCH=CH(CH$_2$)$_{12}$CH$_3$
| |
Glucose OH

linoleate

Figure 8.10 Structures of linoleic acid-rich epidermal lipids.

rat epidermis (Figure 8.10). These lipids form part of the intercellular lipid-rich matrix serving as a water permeability barrier. In EFA deficiency the linoleate of these lipids is replaced by oleate, a change which is associated with a loss of the barrier function. Moreover, when EFA deficiency is cured by feeding a mixture of linoleate and arachidonate, only linoleate appears in the epidermal sphingolipids. These results suggest a biochemical mechanism for some of the EFA-deficiency skin symptoms whereby linoleic acid *per se* is an essential part of the human diet.

8.13 PULMONARY SURFACTANT

Every time we breathe out, our lungs are prevented from collapsing by a unique lipoprotein mixture–pulmonary surfactant. This lipoprotein is absorbed at the alveolar air/liquid interface where it lowers the surface tension and, therefore, reduces the contractile force at the surface and the work of lung expansion. Collapse is prevented by the formation of a solid film on the alveolar surface during expiration. A dramatic example of its importance is provided by the disease *acute respiratory distress of the newborn*. In this complaint, premature infants who have not yet begun to synthesize pulmonary surfactant are unable to expand their lungs properly and suffer from various complications which, even with the best modern treatment, results in death for over 25% of affected individuals.

Surfactant has a unique composition. Pulmonary surfactant can be isolated by carefully washing out lungs repeatedly with isotonic saline. Such material, of course, only represents a proportion of the total lung surfactant – that which has already been secreted into the alveolar spaces. The important compositional features of pulmonary surfactant are emphasized in Table 8.7. First, surfactant is a lipid-rich lipoprotein. Moreover, in contrast to other lipoproteins the lipids are dominated by a single class and, indeed, a single molecular species – dipalmitoylphosphatidylcholine. Thus, the lipids of lung surfactant have a composition particularly suited to the rapid formation of stable lamellar structures. The small amounts of unsaturated phosphatidylcholine are believed to help in the rapid spreading of surfactant at body temperatures and phosphatidylglycerol may assist in the morphological dissolution of lamellar bodies in the aqueous sub-phase to provide a constant source for renewal of the surface monolayer.

The proteins of surfactant are also thought to be important in its function. Two types are found generally – the so-called surfactant apoprotein which is a glycoprotein present in several species with varying amounts of glycosylation giving molecular weights in the 28–35 kDa range. This protein is thought to be involved in the control of surfactant secretion from its site of synthesis in the alveolar type II cell. In addition, a low molecular weight protein of about 10 kDa is present. Although the function of this protein has

Table 8.7 Composition of alveolar surfactant

| | (% w/w) | | |
	Human	Rabbit	Rat
Total protein	15	8	10
Total lipid	85	91	89
	(% w/w of total lipid)		
Phosphatidylcholine	74	83	73
(dipalmitoylphosphatidylcholine)	(68)*	(63)*	(82)*
Phosphatidylglycerol	6	3	5
Sphingomyelin	4	2	2
Other phospholipids	7	3	9
Cholesterol (free and esterified)	4	3	3
Other lipids	5	6	8

* % of phosphatidylcholine.

not been demonstrated convincingly, it is believed to help in the surface properties of surfactant.

The type II cells which are present at the alveolar surface are the exclusive site of surfactant synthesis. In these cells the necessary proteins and lipids are made on the endoplasmic reticulum from where they are assembled into lamellar bodies. Surfactant is then released from the latter by exocytosis.

Detailed studies of lipid metabolism in type II cells have been made either by the use of normal cells (isolated from whole lungs by collagenase treatment to separate them carefully from the more abundant type I cells or the underlying tissue) or with tumour cells which have the same morphological characteristics as type II cells. In particular, attention has been focussed on the formation of dipalmitoylphosphatidylcholine. This major component of surfactant is made primarily by the CDP-base pathway (sections 7.1.3 and 7.1.4). However, the operation of this route results in phosphatidylcholine containing a number of molecular species without an especial abundance of the dipalmitoyl species. Two possibilities exist for enrichment of palmitate into phosphatidylcholine (or re-modelling as it has been termed). Either fatty acids can be removed from phosphatidylcholine by phospholipase A attack and then replaced by an acylation with palmitoyl-CoA (method A; Figure 8.11) or two molecules of monopalmitoyl-phosphatidylcholine could react together (method B; Figure 8.11). Careful experiments by the Dutch workers Batenburg and van Golde have shown that method A (i.e. deacylation/reacylation) is the major method used.

The type II cell is important not only for surfactant synthesis but also seems to be involved in its recycling. In much the same way as it would be wasteful for an animal continually to produce bile salts without any reabsorption, so pulmonary surfactant is constantly taken up by alveolar cells. Little is known of this process (and still less of its control) but it appears

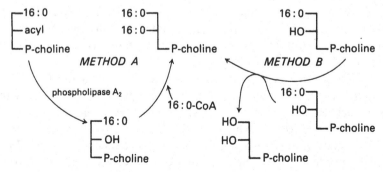

Figure 8.11 Methods for the enrichment of surfactant phosphatidylcholine with palmitate.

that all the major lipid components are recycled. The careful balance of secretion and re-uptake of surfactant ensures that optimal amounts are always available to function in the alveoli.

There would be little point for a young fetus in producing surfactant until near the time for birth and active breathing. Accordingly, surfactant is produced only towards the end of term – in Man after about 8 months. After this time the typical surfactant components can be found, and tested for, in the amniotic fluid. Various tests have been used of which the phosphatidylcholine/sphingomyelin ratio is the most common. A ratio of at least 2 is taken to indicate significant surfactant production.

Respiratory distress syndrome of the newborn is a developmental disorder which is caused by immaturity of the baby's lungs. The lack of normal amounts of surfactant causes morphological alterations (hyaline membranes, atelactasis) and physiological changes (decreased lung compliance, hypoxaemia) and is a major cause of death in premature infants. About 25% of such babies die and some others are left with handicaps. There have been two approaches to the treatment of the disease. The first has been to hasten the development of the fetal lungs by the use of hormones, particularly corticosteroids. Such treatments obviously require considerable prewarning of a possible caesarian delivery or premature natural birth and cannot be used in all cases due to pregnancy complications. The second treatment is to instill lipid mixtures into the baby's lungs in order to reproduce the effects of natural surfactant until such time as the baby can synthesize its own. Some success has been achieved recently by such replacement therapy. However, although artificial surfactant can effectively lower the surface tension at the alveolar surface and thus help breathing it is not effective in ameliorating other complications of respiratory distress. At the present time, a combination of steroid treatment, replacement therapy and careful post-natal care seems to be the best course to take.

Although respiratory distress of the newborn is a very tragic and dramatic disease, that of adults is far more prevalent and has a worse prognosis. In the USA alone it has been estimated that at least 150 000 persons die of the disease each year. Since the problem is not caused by lung immaturity, hormone therapy is of little use but, so far, little attention has been paid to its treatment by replacement therapy.

A number of other diseases or conditions are known to affect surfactant metabolism and, hence, lung function. For example, dust-related industrial diseases such as silicosis lead to a massive accumulation of surfactant (up to 40-times normal amounts) which impairs gas exchange and breathing. In paraquat (a herbicide) poisoning the opposite occurs and the type II cells no longer produce surfactant – leading to particularly painful efforts at breathing by the victim who dies within a few days.

8.14 LIPID STORAGE DISEASES (LIPIDOSES): IF TISSUES LACK A KEY BREAKDOWN ENZYME, LIPIDS ACCUMULATE

Several inborn errors of metabolism exist in which the missing enzyme is one that is involved in the breakdown of a specific lipid molecule. Since the biosynthesis of these lipids is not impaired, the result of the enzyme deficiency is the gradual accumulation of lipids in the tissues. Most of the important diseases of this type are ones that involve structural lipids, frequently glycosphingolipids, of the central nervous system and they are summarized in Table 8.8. The diseases are rare and frequently fatal, which serves to indicate how important it is that the amounts and types of lipids in membranes are strictly controlled to preserve biological function. Many of the lipids involved in these disorders are readily synthesized in the body, so that dietary treatment is ineffective. There is one lipid storage disease, Refsum's disease, however, that can be controlled by strict exclusion of a fatty acid from the diet. This disease is due to a failure to break down by α-oxidation, the branched chain fatty acid, phytanic acid (Figure 8.12), which is formed from phytol, a universal constituent of green plants. In patients, there is a characteristic build-up of phytanic acid in the blood where it may represent 30% of the total fatty acids. A condition of ataxic neuropathy develops and the disease is normally fatal. To survive, the patients must have a low phytol diet.

Several exceedingly rare triacylglycerol storage diseases have been described, resulting from a deficiency of lipase in adipose tissue. There is no evidence to suggest that obesity (section 5.5.2) is a triacylglycerol storage disease caused by a specific lipase defect.

Since an important consequence of the accumulation of sphingolipids in the central nervous system is to produce mental retardation, it is important to be able to detect the defect as early as possible. It is now possible to detect the enzyme deficiency responsible for some of the lipid storage diseases *in utero*. Once the deficiency has been established, a possible treatment is enzyme replacement therapy. Thus, patients with Fabry's disease (Table 8.8) can be infused with normal plasma to provide active enzyme (ceramide

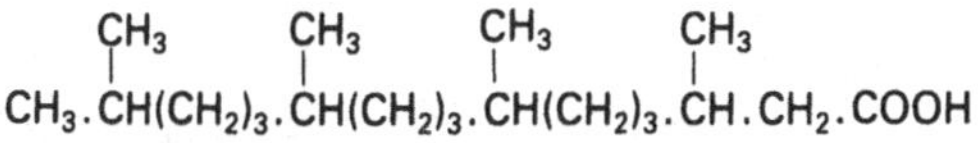

$$CH_3.CH(CH_2)_3.CH(CH_2)_3.CH(CH_2)_3.CH.CH_2.COOH$$

with CH₃ branches at positions shown

3,7,11,15-tetramethylhexadecanoic acid
(Phytanic acid)

Figure 8.12 Phytanic acid.

Table 8.8 Lipid storage diseases*

Lipid accumulating in tissue	Name of disease	Defective enzyme	Effects of disease
Sphingomyelin	Niemann–Pick disease (sphingomyelinosis)	Sphingomyelinase	Deposition of sphingomyelin in almost every tissue; loss of function, frequently fatal before third year
Gangliosides	Tay–Sachs disease	Terminal N-acetyl galactosamine cleavage enzyme	Accumulation of an abnormal ganglioside in tissue especially brain; impairment of mental and somatic function and vision; eventually demyelination
Glucosyl ceramide	Gaucher's disease	Glucocerebrosidase	Appearance of large lipid-laden cells in spleen, liver and bone marrow; pigmentation of skin
Trihexosyl ceramide	Fabry's disease	Ceramide trihexosidase	Skin rash; pains in the extremities, pyrexia; progressive renal failure
Sulphatide	Metachromatic leukodystrophy	Sulphatase	Impairment of motor function; ataxia; coarse tremor and progressive demyelination
Phytanic acid	Refsum's disease	α-Oxidation enzymes	Chronic polyneuropathy; night-blindness; narrowing of visual field; skeletal malformation

* See also Figure 7.15.

trihexosidase) for the hydrolysis of the substrate (galactosyl ceramide) that accumulates in the plasma of these patients (see also section 7.4.6). Maximum enzyme activity occurs 6 hours after infusion of the plasma and is detectable for seven days. The accumulated substrate decreases about 50% on the tenth day after infusion. Other methods involve the encapsulation of the appropriate enzymes in liposomes (section 8.5) which can be targeted to the appropriate tissue, or the future use of appropriate messenger RNA or specific DNA molecules to direct the synthesis of the correct enzyme.

8.15 TOXIC EFFECTS OF LIPIDS

8.15.1 Cyclopropene fatty acids

These compounds inhibit the desaturation of stearic to oleic acid (section 3.2.4). The effect is to alter membrane permeability as demonstrated by pink-white disease. If cyclopropene acids are present in the diet of the laying fowl, the permeability of the membrane is increased, allowing release of substrates, including pigments, into the yolk. Rats die within a few weeks of consuming diets containing 5% of dietary energy as sterculic acid (from *Sterculia foetida* oil, Table 3.4) and at the 2% level, the reproductive performance of females is completely inhibited. By far the most important edible oil containing cyclopropene fatty acids is cottonseed oil in which the concentration ranges from 0.6 to 1.2% although processing reduces the amount in the oil sold for human consumption to 0.1–0.5%. Man has been eating cottonseed oil for years in such products as margarines, cooking oils and salad dressings but the intake of cyclopropenes is very small. On this basis, it is presumed that low levels have no adverse effects but whether prolonged ingestion of larger amounts by man would be deleterious is unknown. Although the effects of cyclopropene fatty acids may be of little practical significance, they do illustrate the vital importance of the correct profile of lipids to membrane integrity.

8.15.2 Long chain monoenoic acids

Some natural edible oils contain, in addition to the widely occurring monounsaturated fatty acid, oleic acid, appreciable quantities of monounsaturated acids with 20 or 22 carbon atoms. Biochemically, these are usually formed by elongation of oleic acid, so that the double bonds are in positions 11 and 13 respectively. Examples are gadoleic acid (*cis*-11–20:1) in certain fish oils and erucic acid (*cis*-13–22:1, Table 3.2), characteristic of seed oils in the plant family *Cruciferae*. Herring oil also contains cetoleic acid (*cis*-11–22:1) and hardened fish oils contain a range of long chain monoenes (section 5.1.3). The most important commercial source of long chain monoenoic acids has been rapeseed (*Brassica napus*) containing up to 45% erucic acid. When young rats were fed diets containing more than 5% of dietary energy as rapeseed oil, their heart muscle became infiltrated with triacylglycerols. After about a week, the hearts contained three to four times as much fat as normal hearts and although, with continued feeding, the size of the fat deposits gradually decreased, other pathological changes were noticeable, such as the formation of fibrous tissue in the heart muscle. The biochemistry of the heart muscle was also affected. Mitochondrial oxidation of substrates such as glutamate was reduced and the rate of ATP biosynthesis impaired. The degradation of triacylglycerols containing erucic acid by lipases is slower than with fatty acids of normal chain length and this may have contributed to the accumulation of lipid deposits.

Although there are wide differences between species and the response is much weaker in older than in younger animals, the results have been taken very seriously from the point of view of human nutrition, as rapeseed is now a major edible oil crop on world markets. It is not known whether similar lesions occur in man, although in some countries rapeseed oil has been consumed for many years, albeit not in the high concentrations employed in animal feeding experiments. The toxicity is likely to be less in a mixed diet with adequate quantities of essential fatty acids and a good balance of the more usual fatty acids. Despite the lack of evidence for harmful effects in Man, it has been thought prudent to replace older varieties of rape, having a high erucic acid content, with new varieties of zero erucic rapes.

8.15.3 Isomeric fatty acids

Most unsaturated fatty acids found in nature contain double bonds in the *cis*-geometrical configuration and in rather specific positions in the chain, determined by the specificities of the desaturases. We use the term isomeric fatty acids here to denote fatty acids with double bonds in the *trans*-geometrical configuration and double bonds in unusual positions in the chain (Figure 5.1). Table 3.2 shows some naturally occurring examples and Table 5.3 some examples of isomeric fatty acids arising as a result of food processing. Their contribution to the diet was touched upon in section 5.1.3.

Worries about the possible toxicological effects of *trans* fatty acids began with the publication of results from an American laboratory that suggested that pigs fed hydrogenated vegetable fat for eight months had more extensive arterial disease than those on a control diet. More recent epidemiological studies have suggested that there is a correlation between mortality from arterial disease and the consumption of hydrogenated fats in the UK. There is no reliable evidence, however, that *trans* fatty acids have an hypercholesterolaemic effect nor that they enhance platelet aggregation. These are the most likely mechanisms that might implicate them in the pathology of vascular disease. It is likely that the experiments that demonstrated increased arterial lesions in pigs fed *trans* fatty acids could be explained by a relative dietary deficiency of linoleic acid since another experiment from the same laboratory, in which diets were more than adequate in their linoleic acid content, was not able to make a distinction between the pathological effects of *cis* and *trans* acids. Indeed, insofar as there are any adverse effects of *trans* fatty acids *in vivo*, they are most likely to be due to the general competition with essential fatty acids for desaturases (section 5.2.2(a)) thereby causing an increase in the requirements for essential fatty acids. Experiments with rats have shown that increasing the concentration of dietary *trans* fatty acids causes a decrease in the concentration of essential fatty acids in heart muscle and that this effect is accentuated when the diet is marginal in its essential fatty acid content.

The problems that arise in interpreting the results of experimental work on *trans* fatty acids are illustrative of the difficulties of translating the results of animal studies to practical human nutrition generally. How representative is one species of another? What concentration in the diet is appropriate in the test experiments? Scientifically, there may be merit in examining the effect of pure isomers included in a chemically well defined diet. Practically, human beings eat a poorly defined mixture of substances in the presence of a very complex diet. What should the control diet be against which comparisons are made?

8.15.4 Oxidized and polymerized fats

When fats containing unsaturated fatty acids are stored in the presence of oxygen at room temperature they gradually oxidize, first with the formation of hydroperoxides which may be subsequently converted into a variety of low molecular weight degradation products (section 3.3.4). After prolonged exposure to oxygen, polymerization occurs. This is the reaction that takes place in the hardening of varnish which contains a high proportion of linolenic acid. The ease of peroxidation rises disproportionately as the number of double bonds increases.

When lipid hydroperoxides are ingested, they are rapidly degraded in the mucosal cells of the small intestine to various oxyacids that are rapidly further oxidized to CO_2. There is no evidence for the absorption of unchanged hydroperoxides nor for their incorporation into tissue lipids. Although it is unlikely, therefore, that lipid hydroperoxides are toxic by virtue of their assimilation into the body, there is some evidence that they may damage the gut by potentiating the growth of tumours.

Although lipid hydroperoxides are not absorbed, some of their lower molecular weight breakdown products are absorbed and may be toxic. Thus, when experimental animals were given hydroperoxyalkenals, there were increases in the relative weight of the liver, increased concentrations of malondialdehyde, peroxides and other carbonyl compounds in tissues and decreases in tissue concentrations of α-tocopherol and linoleic acid.

The toxicity of oxidized cholesterols has also been demonstrated in several studies. When rabbits were given diets containing reagent grade cholesterol ($250\,mg\,kg^{-1}$) the frequency of dead aortic smooth muscle cells was increased and focal intimal oedema was induced. The active factors were found in a concentrate containing 25-hydroxycholesterol, 7-keto-cholesterol, 7α- and 7β-hydroxycholesterol, cholestane-3β,5α,6β-triol and the 5- and 7-hydroperoxides. The authors suggested that studies in which atherosclerosis had been induced in experimental animals by giving diets containing cholesterol that had been stored in air at room temperature, should be re-evaluated, since the cholesterol would almost certainly have contained significant quantities of oxidized sterols. These might have been responsible for inducing atherosclerosis rather than pure cholesterol. Since the publication of this work, more has been learned about the role of macrophages in scavenging lipoproteins that contain modified lipids or proteins and this could provide a functional explanation for the increased toxicity of these oxidized compounds (section 5.5.3). The subject is of more than academic concern since cholesterol is particularly susceptible to oxidation when present in dried foods that may be stored for extended periods at ambient temperatures, such as dried eggs and milk powder.

Heated fats do not contain high levels of peroxides since they are rapidly decomposed but they contain, instead, a range of polymerized compounds. The extent of this polymerization may be 10 to 20% under normal household or commercial cooking practice but up to 50% when the oil is severely abused. There is little evidence that the dimeric and polymeric materials produced by prolonged cooking of oils are absorbed or are toxic to experimental animals or Man.

8.16 SUMMARY

Under normal conditions, the membrane bilayer is in a fluid state. Both lipids and proteins are capable of diffusing within the plane of the membrane and lipids show rotational and vibrational movements. It is believed that membrane lipids can modulate protein activity (such as in transport) but the extent of these interactions is still an area of controversy. Some uncertainties about the above roles may be due to the subtlety of the interactions that may require particular lipid classes or even molecular species.

To keep their membranes functional, organisms have to maintain membrane fluidity. For those organisms dependent on environmental temperatures, this may require the alteration of membrane lipid composition. The most common change is an increase in unsaturation at lower temperatures but other alterations contribute to the overall change in fluidity. The importance of membrane fluidity is emphasized in the phenomenon of frost sensitivity or resistance in plants. The sensitivity of plants to freezing damage appears to be predictable on the basis of their thylakoid lipid composition.

Membranes, organelles and indeed, sometimes, cells are not static but are capable of changing shape and of fusing. For some membranes, fusion is a very common event. The process seems to be aided by the ability of lipids to undergo polymorphism. In particular, adoption of hexagonal H_{II} structures is needed.

The shape of the complex lipid molecules can give rise not only to polymorphism but also to their ability to pack well into curved regions on bilayers. In addition, cooperation of membrane lipids and proteins can be important in controlling membrane or cell shapes and properties.

The ability of membrane lipids to form bilayers has been exploited in the use of liposomes. Studies with liposomes have provided much important information about membrane properties and, in addition, they are used for drug delivery and in enzyme-replacement therapy. When liposomes can be specifically targeted by the use of antibodies to particular cells, then their use in medicine is likely to become even more widespread.

Membrane receptors are proteins (usually glycoproteins) embedded in the lipid bilayer of the membrane. They contain sites that interact with molecules external to the cell to initiate a number of different events, for example, transport of the molecule through the membrane or, if the molecule is a hormone or autocoid, the triggering of second messengers within the cell. The functioning of membrane receptors appears to be regulated in part by the microenvironment provided by the membrane lipids.

Intrinsic proteins may be anchored in the membrane by several different mechanisms. Attachments through interactions with lipids include fatty acid acylation of the aminoacyl terminus or the formation of thiolesters with cysteine residues. In addition, a number of proteins are covalently linked to phosphatidylinositol.

Inositol lipids play an important role in controlling cellular metabolism. They do this through the inositol cycle in which the key event is a phospholipase C-stimulated breakdown of phosphatidylinositol-4,5-bisphosphate. Agonist binding to receptors causes the stimulation of this substrate-specific phospholipase C. The breakdown of phosphatidylinositol-4,5-bisphosphate gives rise to the generation of diacylglycerol and inositol-1,4,5-triphosphate, both of which act as second messengers. Inositol-1,4,5-triphosphate causes calcium release from the internal cellular stores and, hence, activates the calmodulin system leading to an increase in the activity of various protein kinases. Diacylglycerol stimulates a different protein kinase, protein kinase C, and, in addition, can be catabolized by lipase to release arachidonate which contributes to eicosanoid formation. The continued breakdown and resynthesis of the phosphorylated inositol lipids causes them to have particularly high turnovers in cells.

Many types of immune reactions involve the rapid division of lymphocytes and other cells that form an integral part of the immune system. These processes naturally involve a high level of membrane turnover and renewal, which in turn requires enhanced lipid turnover. In addition, attention has focussed on more specific roles of lipids. Glycolipids, for example, are important constituents of the blood group antigen system and may also have a role in the aetiology of some autoimmune diseases. Moreover, polyunsaturated lipids act as precursors for eicosanoids, certain of which act as powerful stimulators or antagonists of cell-mediated immune reactions. Dietary lipids may influence immunity through their modification of the spectrum of eicosanoids produced by different tissues.

During the development of tumour cells, dramatic changes take place in the pattern of membrane glycosphingolipids, suggesting important roles for these lipids in the aetiology of cancer. Antibodies against glycosphingolipids have been used for therapeutic purposes and there is hope that it will be possible specifically to destroy malignant cells and so avoid the unwanted side effects experienced by patients undergoing conventional chemotherapy. The role of dietary lipids in the aetiology of cancer has been much discussed but mechanisms are still obscure. Influences on membrane function cannot be ruled out but there may be more specific effects on eicosanoid production, which may in turn influence the immune reaction against cancer cells. Oxidized lipids may cause damage to DNA and proteins, thus increasing the potential for tumorigenesis. The well known protective effect against cancer of high levels of vitamin A may derive from the antioxidant activity of this nutrient.

Oxidation of the polyunsaturated acyl groups of lipids by chemical or enzymic means can give rise to hydroperoxides. When these interact with transition metals, highly reactive free radicals are generated that can cause structural changes in proteins, including cross-linking and fragmentation. Organisms have a number of ways of reducing the accumulation of toxic lipid hydroperoxides, which may involve the use of antioxidants such as tocopherols, enzymes such as glutathione peroxidase, as well as turning over peroxidized fatty acids by accelerated phospholipase A_2 activity.

Lipids are key constituents of the surface layers of organisms. In mammalian skin triacylglycerols, wax esters and squalene are major components. Release of non-esterified fatty acids by bacterial action can give rise to acne vulgaris, a common complaint of puberty. Two other skin diseases may involve more specific aspects of lipid biochemistry. In psoriasis, there is evidence that changes in inositol lipid metabolism may contribute to this condition while a deficiency of the Δ6-desaturase which converts linoleate to γ-linolenate may underly the disease atopic eczema.

Efficient breathing requires a special lipid mixture known as pulmonary surfactant which contains exceptionally high proportions of dipalmitoyl-phosphatidylcholine. The secretion of pulmonary surfactant is carefully controlled and, when this secretion is impaired, as in premature babies, acute respiratory stress is found. Efforts are now being made to treat such respiratory distress either by hormones to hasten the development of the type II epithelial cells which make surfactant or by replacement therapy with artificial lipid mixtures.

When certain enzymes involved in the hydrolytic or oxidative breakdown of cellular lipids are absent from tissues, their lipid substrates tend to accumulate, resulting in a lipid storage disease or lipidosis. All are rare but the most important involve the accumulation of glycosphingolipids in the central nervous system. Enzyme replacement therapy is now becoming a practical possibility in some cases.

Finally, certain lipids may be toxic to cells. Cyclopropene fatty acids can act as desaturase inhibitors while oxidized lipids, as we have already seen, can damage proteins and DNA. Other toxic effects may occur, not by such specific mechanisms but because overconsumption in the diet alters the balance between essential and non-essential fatty acids resulting in an increased requirement for the essential fatty acids.

REFERENCES

See Chapter 6 also for references to lipids in membranes.

General functions of structural lipids

Cullis, P.R. and Hope, M.J. (1985) Physical properties and functional roles of lipids in membranes, in *Biochemistry of Lipids and Membranes* (eds D.E. Vance and J.E. Vance), Benjamin/Cummings, Menlo Park, CA, pp. 25–72.

Carruthers, A. and Melchior, D.L. (1986) How bilayer lipids affect membrane protein activity. *Trends in Biochemical Sciences*, **11**, 331–335.

Owen, J.S., McIntyre, N. and Gillett, M.P.T. (1984) Lipoproteins, cell membranes and cellular functions. *Trends in Biochemical Sciences*, **9**, 238–242.

Pugano, R.E. and Sleight, R.G. (1985) Emerging problems in the cell biology of lipids. *Trends in Biochemical Sciences*, **10**, 421–425.

Storch, J. and Kleinfeld, A.M. (1985) The lipid structure of biological membranes. *Trends in Biochemical Sciences*, **10**, 418–421.

Effects on membranes during temperature adaptation

Lynch, D.V. and Thompson, G.A. (1984) Retailored lipid molecular species: a tactical mechanism for modulating membrane properties. *Trends in Biochemical Sciences*, **9**, 442–445.

de Mendoza, D. and Cronan, J.E. (1983) Thermal regulation of membrane fluidity in bacteria. *Trends in Biochemical Sciences*, **8**, 49–52.

Murata, N., Ishizaki, O. and Nishida, I. (1988) Chilling sensitivity and phosphatidylglycerol biosynthesis, in *Plant Membranes: structure, assembly and function* (eds J.L. Harwood and T.J. Walton), The Biochemical Society, London, pp. 223–230.

Lipids and membrane fusion

Fisher, D. and Geisow, M.J. (1986) Molecular mechanisms of membrane fusion. *Biochemical Society Transactions*, **14**, 241–257.

Papahadjopoulos, D., Poste, G. and Vail, W.J. (1978) Studies on membrane fusion with natural and model membranes, in *Methods in Membrane Biology* (ed. E.D. Korn), *Vol. 10*, Plenum, New York, pp. 1–121.

Wilschut, J. and Hockstra, D. (1984) Membrane fusion: from liposomes to biological membranes. *Trends in Biochemical Sciences*, **9**, 479–483.

Lipids and proteins interact to determine membrane structure and shape

Eisenberg, D. (1984) Three-dimensional structure of membrane and surface proteins. *Annual Reviews of Biochemistry*, **53**, 595–623.

Op den Kamp, J.F.F., Roelofsen, B. and van Deenen, L.L.M. (1985) Structural and dynamic aspects of phosphatidylcholine in the human erythrocyte membrane. *Trends in Biochemical Sciences*, **10**, 320–323.

Liposomes and drug delivery systems

Gregoriadis, G. and Allison, A.C. (eds) (1980) *Liposomes in Biological Systems*. Wiley, New York.

Klein, R. and Schmitz, B. (eds) (1986) *Topics in Lipid Research* (various chapters on membrane structure and liposomes). Royal Society of Chemistry, London.

Ostro, M.J. (1987) Liposomes. *Scientific American*, **256**, 102–110.
Patel, H.M. and Russell, N.J. (eds) (1988) Liposomes: from membrane models to therapeutic applications. *Biochemical Society Transactions*, **16**, 909–922.

Membrane receptors

Anholt, R. (1981) Reconstitution of acetylcholine receptors in model membranes. *Trends in Biochemical Sciences*, **6**, 288–291.
Carpenter, G. (1987) Receptors for epidermal growth factor and other polypeptide mitogens. *Annual Review of Biochemistry*, **56**, 881–914.
Gilman, A.G. (1987) G Proteins: transducers of receptor generated signals. *Annual Review of Biochemistry*, **56**, 615–649.
Goldstein, J.L. and Brown, M.S. (1984) Progress in understanding the LDL receptor and HMG-CoA reductase, two membrane proteins that regulate plasma cholesterol. *Journal of Lipid Research*, **25**, 1450–1461.
Hjelmeland, L.M. and Chrambach, A. (1984) Solubilization of functional membrane-bound receptors, in *Membranes, Detergents and Receptor Solubilization* (eds C.J. Venter and L.C. Harrison), Alan R. Liss, New York, pp. 35–46.

Inositol lipids play specific roles in membrane protein anchoring

Ferguson, M.A.J. and Williams, A.F. (1988) Cell surface anchoring of proteins via glycosyl phosphatidylinositol structures. *Annual Review of Biochemistry*, **57**, 285–320.
Low, M.G., Ferguson, M.A.J., Futerman, A.H. and Silman, I. (1986) Covalently attached phosphatidylinositol as a hydrophobic anchor for membrane proteins. *Trends in Biochemical Sciences*, **11**, 212–215.
Schmidt, M.F.G. (1982) Acylation of proteins: a new type of modification of membrane glycoproteins. *Trends in Biochemical Sciences*, **7**, 322–324.

Inositol lipids and second messengers

Berridge, M.J. (1987) Inositol triphosphate and diacylglycerol: two interacting second messengers. *Annual Review of Biochemistry*, **56**, 159–193.
Berridge, M.J. (1984) Inositol triphosphate and diacylglycerol as second messengers. *Biochemical Journal*, **220**, 345–360.
Cockcroft, S. (1987) Polyphosphoinositide phosphodiesterase: regulation by a novel guanine nucleotide binding protein, *Trends in Biochemical Sciences*, **12**, 75–78.
Hokin, L.E. (1985) Receptors and phosphoinositide-generated second messengers. *Annual Review of Biochemistry*, **54**, 205–235.
Houslay, M.D., Wakelam, M.J.O. and Pyne, N.J. (1986). The mediator is the message: is it part of the answer to insulin's action? *Trends in Biochemical Sciences*, **11**, 393–394.

Majerus, P.W., Wilson, D.B., Connolly, T.M., Bross, T.E. and Neufeld, E.J. (1985) Phosphoinositide turnover provides a link in stimulus-response coupling. *Trends in Biochemical Sciences*, **9**, 453–456.

Role of lipids in immunity

Gurr, M.I. (1983) The role of lipids in the regulation of the immune system. *Progress in Lipid Research*, **22**, 257–287.

Johnson, P.V. (1985) Dietary lipids and immunity. *Advances in Lipid Research*, **21**, 103–141.

Curtiss, L.K. and Edgington, T.S. (1981) Biological activity of the immuno-regulatory lipoprotein, LDL-In, is independent of its free fatty acid content. *Journal of Immunology*, **126**, 1382–1386.

Lipids and cancer

Carroll, K.K. (1985) Dietary fat and breast cancer. *Nutrition Update*, **2**, 29–47.

Cohen, L.A. (1987) Diet and Cancer, *Scientific American*, **257**, 42–48.

Hakomori, S-I. (1984) Glycosphingolipids as differentiation-dependent, tumour-associated markers and as regulators of cell proliferation. *Trends in Biochemical Sciences*, **9**, 453–456.

Kritchevsky, D. and Klurfeld, D.M. (1986) Influence of caloric intake on experimental carcinogenesis: a review, in *Essential Nutrients in Carcinogenesis* (eds L.A. Poirier, P.M. Newberne and M.W. Pariza), Plenum, New York.

Wood, R. (1973) *Tumour Lipids*, American Oil Chemists' Society Press, Champaign, Illinois, USA.

Free radicals and cellular damage

Esterbauer, H. and Cheeseman, K.H. (eds) (1987) Lipid peroxidation: biochemical and biophysical aspects. *Chemistry and Physics of Lipids*, Vol. 44.

van Kuijk, F.J.G.M., Sevanian, A., Handelman, G.J. and Dratz, E.A. (1987) A new role for phospholipase A_2: protection of membranes from lipid peroxidation damage. *Trends in Biochemical Sciences*, **12**, 31–34.

McBrien, D.C.H. and Slater, T.H. (eds) (1982) *Free Radicals, Lipid Peroxidation and Cancer*, Academic Press, London.

Wolff, S.P., Garner, A. and Dean, R.T. (1986) Free radicals, lipids and protein degradation. *Trends in Biochemical Sciences*, **11**, 27–31.

Lipids and skin diseases

Faber, E.M. and Cox, A.J. (eds) (1982) *Psoriasis: Proceedings of the Third International Symposium*, Grune & Stratton, New York.

Harwood, J.L. (1986) Skin problems, in *The Lipid Handbook* (eds F.D. Gunstone, J.L. Harwood and F.B. Padley), Chapman and Hall, London, pp. 542–543.

Manku, M.S., Horrobin, D.F., Morse, N. Kyte, V. and Jenkins, K. (1982) Reduced levels of prostaglandin precursors in the blood of atopic patients: defective delta 6 desaturase as a biochemical basis for atopy. *Prostaglandins, Leukotrienes and Medicine*, **9**, 615–628.

Perisho, K. *et al.* (1988) Fatty acids of acylceramide from comedones and from the skin surface of acne patients and control subjects. *Journal of Investigative Dermatology*, **90**, 350–353.

Yardley, H.J. and Summerly, R. (1981) Lipid composition and metabolism in normal and diseased epidermis. *Pharmaceutical Therapy*, **13**, 357–383.

Pulmonary surfactant

Harwood, J.L. (1987) Lung surfactant. *Progress in Lipid Research*, **26**, 211–256.

Harwood, J.L. and Richards, R.J. (1985) Lung surfactant. *Molecular Aspects of Medicine*, **8**, 423–514.

Robertson, B., van Golde, L.M.G. and Bakenburg, J.J. (eds) (1984) *Pulmonary Surfactant*, Elsevier, Amsterdam.

Lipid storage diseases

Brady, R.O. (1978) Genetics and the sphingolipidoses. *Annual Review of Biochemistry*, **47**, 687.

Toxic effects of lipids

Barlow, S.M. and Duthie, I.F. (1985) Long-chain monoenes in the diet, in *The Role of Fats in Human Nutrition*, Ellis Horwood, Chichester, pp. 132–145.

Billek, G. (1985) Heated fats in the diet, in *The Role of Fats in Human Nutrition*, Ellis Horwood, Chichester, pp. 163–172.

Gurr, M.I. (1987) Isomeric fatty acids. *Biochemical Society Transactions*, **15**, 336–338.

Gurr, M.I. (1988) Lipids: products of industrial hydrogenation, oxidation and heating, in *Nutritional and Toxicological Aspects of Food Processing* (eds R. Walker and E. Quattrucci), Taylor & Francis, London, pp. 139–155.

Mattson, F.H. (1973) Potential toxicity of Food Lipids, in *Toxicants Occurring Naturally in Foods*, National Academy of Sciences, Washington, D.C., pp. 198.

Senti, F.R. (1985) *Health Aspects of Dietary* trans *Fatty Acids*, Federation of American Societies for Experimental Biology, Bethesda.

Taylor, C.B., Peng, S.K., Werthesen, N.T., Tham, P. and Lee, K.T. (1979) Spontaneously occurring angiotoxic derivatives of cholesterol. *American Journal of Clinical Nutrition*, **32**, 40–57.

Vles, R.O. (1975) Nutritional aspects of rapeseed oil, in *The Role of Fats in Human Nutrition* (ed. A.J. Vergroesen), Academic Press, London, pp. 433–477.

Index

Printed in the United States
by Baker & Taylor Publisher Services